WHAT PEOPLE ARE S

THE NEW PARADIGM

"At a time in human history when interest in organized religion declines while individual commitment to a 'spiritual life' continues to grow, Sharon Miller's book offers a fresh perspective on life's greatest mysteries: who God is and why we are here. Lovers of philosophy and those committed to a reflective lifestyle will find in this book a transforming vision of life fit for a more enlightened future. If you want to get serious about your purpose in this life and who God is (and isn't), consider this book as a good place to start and move forward."
Karol Jackowski, author of *The Silence We Keep, Ten Fun Things to do Before you Die,* and *Forever and Ever Amen.*

"This is a fascinating book because it manages to combine material from so many different sources: the natural and social sciences, philosophy (both ancient and modern), theology (e.g., a panentheism in which God is the "soul" of the world and the world the "body" of God), even archetypes derived from nature-astrology. Whether all this fits together perfectly is ultimately less important than that its originality and creativity will make readers of the book think furiously about what makes sense not only on a theoretical basis but more practically in terms of their personal spirituality."
—**Joseph A. Bracken, S.J.,** Emeritus Professor of Theology, Xavier University, Cincinnati, OH, author of *Christianity and Process Thought: Spirituality for a Changing World,* and *Does God Roll Dice.*

"To the skeptic, the integration of astrology and science seems to be the ultimate absurdity. Ironically, however, there is a delicate interplay between the ridiculous and the sublime. My own mentor, Arthur M. Young, spent two decades in the workshop where he invented the first helicopter to receive a commercial license — the Bell Model 47. Only then did he feel he had earned the right to explore the realms of cosmology that are the subject of Sharon Miller's book. The result of his profound thinking is published in his classic text, *The Geometry of Meaning.* Others have also paved the way for such exploration, including the great sociologist, Theodore Roszak, in his pamphlet *Why Astrology Endures.* Roszak pointed out that astrology, like Sanskrit, is a language of the soul. Sharon Miller's approach to understanding the deeper meaning of astrology builds upon both ancient philosophy and modern science. Her work deserves to be studied and remembered as a contribution of lasting value for those who choose to grapple with sublime paradoxes."

Jeffrey Mishlove, PhD, Program Dean in Transformational Psychology, University of Philosophical Research, author of *The Roots of Consciousness* and *Thinking Allowed.*

The New Paradigm

A Spiritual Scientific Cosmology
Utilizing Twelve Fundamental Archetypes
At Base of the laws of Physics

The New Paradigm

A Spiritual Scientific Cosmology
Utilizing Twelve Fundamental Archetypes
At Base of the laws of Physics

Sharon Miller

BOOKS

Winchester, UK
Washington, USA

First published by iff Books, 2013
iff Books is an imprint of John Hunt Publishing Ltd., Laurel House, Station Approach,
Alresford, Hants, SO24 9JH, UK
office1@jhpbooks.net
www.johnhuntpublishing.com
www.iff-books.com

For distributor details and how to order please visit the 'Ordering' section on our website.

ISBN: 978 1 78099 967 8

A CIP catalogue record for this book is available from the British Library.

Design: Stuart Davies

Printed in the USA by Edwards Brothers Malloy

We operate a distinctive and ethical publishing philosophy in all areas of our business, from our global network of authors to production and worldwide distribution.

CONTENTS

Preface

The ancient Greek philosophers like Plato and Aristotle tried to understand the precise meaning and implication of words like 'absolute' and 'infinite.' The purpose of doing this was, not just to understand what the words meant, but to figure out whether or not we experienced anything with these qualities. Plato was particularly concerned with knowledge—how we come to know things and make sense of what we are thinking about. Everything in the material world is relative and constantly in flux. Nothing stays the same from moment to moment. He decided that if we were to understand when something was, 'yellow' for instance, or if we were to understand what was the same about two yellow things, we had to have access to something that was a standard for this experience—and specifically an absolute standard that was the same for all yellow things so that we could reference it in our minds. Plato did not know how many absolute standards there were, but he felt very strongly that they had to exist. His thoughts began a movement that exists to this day, and one that has divided philosophy into two camps—that of the absolutists, who believe that we must have absolutes if we are to have knowledge, and that of the relativists who believe that 'yellow' is just an ideal abstraction that we make in our mind based upon all the yellow things that we encounter. Plato called his absolute standards 'Forms.' Later philosophers spoke of them as 'universals.' I like to speak of these as 'archetypes' and this book is based upon twelve fundamental universals or archetypes.

Although Plato believed the archetypes facilitated knowledge, he did not think they existed just in our mind. They gave qualities and form to things in the physical world, as well as helping to shape and form our thoughts.

Because he believed in the real being of these archetypes,

Plato concluded that they needed something to give them their being. And it was not just the archetypes that needed something to give them their existence. The material qualities produced by the archetypes were material *types* of being, which made him think that something called 'absolute Being' might be their ultimate source.

According to Plato, this principle of pure Being also gave being to us. Thus, not just our various qualities, but we ourselves as an individual depended upon this substance and power for our existence. He called the principle that gave being to the archetypes, to the actual material qualities of experience, and to us, the 'Form' of pure Being, which unfortunately did not help later readers understand what this substance and power might consist of. (Plato looked upon absolute Being as a power as well as a substance, since it had the power to give Being or existence to things.)

Plato believed that these Forms, including the Form of pure Being, had to be unchanging in order to provide for a stable identity. What did not change and yet existed was eternal in nature. In terms of qualities, the very purpose of the archetypes was so that something in the constantly changing universe could still have a definite identity. In terms of individuals, it was so that, not merely the qualities of a system, but the system itself could have definite being and identity. If we are just a cluster of constantly changing qualities, it may not be possible for us to exist or be, much less have a specific nature.

As with Plato, Aristotle saw whatever was at the base of everything as a substance, and specifically as a substance that gave a being its being. He called it 'ousia' which means 'substance.' But 'ousia' was not a substance as we use the term because obviously something that is pure, absolute and eternal Being is very different from things in a relative, changing, material world of flux. This description hardly fits the qualities we possess—and it hardly fits us as an individual either, since we are subject to death.

The early Greek philosophers believed that absolute and eternal Being existed. The question they wondered about was what it consisted of and, perhaps even more importantly, how it related to matter, and where such a thing might exist. Plato decided that the Forms, including and especially the principle of pure, absolute and eternal Being, had to exist in a world apart from the material world. He called this the realm of the Forms. But he also said it was the realm of the soul. That was because he concluded that we had something absolute and eternal about us.

So, do we have something called a soul? And if so, what is it and where does it exist? I say 'if so' because relativists have traditionally insisted we do not have to posit anything absolute in order to be or be what we are.

Not only did the Greeks try to figure out how absolute Being related to matter and where absolute Being existed; the very idea that this substance was pure Being brought up the question of whether or not anything less than absolute Being was 'pure,' such as the *types* of being that make up the material world. Furthermore, because 'being' relates to 'reality' and hence to that which is real, many of the early Greek philosophers (and later 'Neo-Platonic' philosophers) came to the conclusion that absolute Being was real, and conversely that the relative, changing material world was unreal, or at least possessed a 'measure' of unreality, the latter of which was Plato's stance.

The conclusions reached by Plato and the other Greek philosophers dramatically affected the subsequent development of Western civilization, and continue to do so to this day. The extent of the influence of these ideas on human history cannot be underestimated. That is because of what these thoughts of absolute and eternal being have to do with the subject of spiritual things. Plato's 'Form' of pure Being has been correlated to God. The Forms in general have been connected to the soul. Aristotle said 'ousia' was whatever gave being to a being, but also that ousia was God. These thoughts are linked to the theological question of whether or not God exists, if so how our soul is related to God and, most importantly of all, the question of how anything spiritual, including God, is related to our physical bodies. It is the question of how God relates to the material universe.

The early Christian apologists—those who decided how it was best for Christians to visualize or image God—looked to Plato, Aristotle and the later Neo-Platonic philosophers to help them with this task. (This is true of the Muslim and later Hebrew religious traditions as well.) It was decided that God was infinite, absolute, eternal and perfect. [He also was unified or one, since he was the absolute (undifferentiated) basis of everything.] This absolute and infinite Being was not subject to division and separation in the way the lesser, multiple beings on the earth were.] Because it was believed that these qualities ascribed to God could not exist in a relative, changing, imperfect world, the conclusion was that God had to be transcendent to the material world. This view of God as a transcendent deity is called 'theism.' Due to the development and power of the Jewish, Christian and Muslim religions, theism became the accepted view of God, and continues to be the predominant view today.

Prior to this image there was a view of God called 'pantheism.' (The twelve universals utilized in this book actually derive from the old pantheistic nature-astral religons.) Pantheism entailed the belief that God was totally immanent within the world. God ('theos') was often seen as the entire material universe ('pan' or 'all'). Because matter is finite, relative and far from perfect, this contrasted with the view that God was infinite, absolute and perfect. Due in large part to the influence of the Greek philosophers, theists later condemned pantheism as a false way to visualize God.

Pantheism also involved polytheism. Multiple beings within the created universe were seen as divine, including to some extent human beings. What is interesting is that the same Greek philosophers who lent their ideas to the monotheism that later arose were actually pantheists, including Plato himself. According to Plato the oldest god—or rather goddess—was the earth goddess Gaia (*Timaeus* 40c). She was not just goddess of the earth—she *was* the earth. All other beings, including human beings resided within her.

Gaia was the oldest of the gods in the visible heaven of our world, which included the twelve 'Olympians' (Jupiter, Venus, Gaia, Uranus, Mercury, Saturn, and others). At the same time there was a heaven of heavens, and Gaia was encompassed by an even older, larger god—the god of the universe. He was the first-begotten. All the gods, including Gaia, and the people who resided within her, also resided within him. According to Plato, this high-god was the soul of the universe.

Because there were different levels of gods, pantheism entailed a 'hierarchical' view of creation. The higher the god the more power they had to influence affairs. Within the temples dedicated to these gods, people often brought gifts to earn their favor or to placate them. [Plato on the other hand claimed the gods acted from utter goodness, and were too good to be influenced by gifts, sacrifices or anything of that sort (*Laws* 10:906-907). He speaks disparagingly of the way the Greek poets (e.g., Homer) made them out to be less than good, and less than divine.]

In addition to the hierarchical view of creation, another interesting aspect of the Greek cosmology was that human beings were not just parts within Gaia or the soul of the universe. In the very last book Plato wrote, the *Laws*, he and his companions within that dialogue have a discourse about the common belief of their time that everything that moved had a soul including people, and that all these ensouled systems were gods because of their soul. Per Plato, all things are 'full of gods' (*Laws* 10:899b). The soul was the source of self-movement and of creation (*Laws*

10:896). Because of this, each person was a self-moving mover and a self-creator. [In the *Laws* Plato speaks in a humorous manner about people making it easy for God as creator because they are responsible for their own creation (*Laws* 10:903-904).]

Although Plato was a pantheist, the reason the philosophy present in his ideas encouraged theism was because of his belief that the material body was not divine. Only the soul was. The body was something in opposition to the soul (*Phaedo* 80, 94). The soul was entombed in the body (*Phaedo* 81). The body was not pure like the soul and was a source of unreality. Although the soul was involved in moving and shaping the body, and was the principle of creation (*Laws* 10:899c, 10:904a), it was immortal, while the body was mortal (*Timaeus* 41d). Thus the soul was very different from the body.

Plato's way of imaging God and his relationship to mankind was actually a 'panpsychism' ('all is soul') rather than a true pantheism that included the physical universe; but it at least put God into the world, making of all creatures gods via their souls, and in fact putting a soul into everything that could move by its own volition. This included every living thing (*Laws* 10.896a, *Laws* 12.966e). Furthermore, the soul was constantly involved with the body. Its role was to mold it and make it what it was (*Epinomis* 981c). It was even a repository of memory (*Epinomis* 981d)—a book of life (*Philebus* 38e) which retained the character as well as molding and shaping it (*Laws* 10.90a4c). The retention made it possible for a person to reincarnate through multiple lifetimes (*Laws* 10:903d). Because the soul recorded the character and formed the body, Plato maintained that it took the body to the level and type of world that was in accordance with the character of the person (*Laws* 10.903-904). According to Greek mythology, even Hades was simply a lower level material world. Per Plato, the place that a person resided in depended upon their own (self-created) character, and the soul was the repository of that character.

According to Plato, the body could not exist from moment to moment without the soul, but the soul could exist without the body, and actually was better off without it. The goal in theism is heaven. Plato's ideas advanced these later theistic views by claiming the soul needed to be liberated from the body. Whether in Platonism or theism, matter was seen as inherently evil, and the source of pain and suffering.

Historically, pantheism was eradicated in lands that were later occupied and dominated by theists. This was accomplished often through killing pantheists. A more civilized way might have been to eliminate pantheism through faith in the truth. Truth has to do with real being, and the truth has the power to uphold itself. On the other hand, when something true is suppressed, it will find a way to rise again.

There is logic in both theism and pantheism. The monotheism inherent in theism, for instance, helps explain unity in the world. When there is only one God, it is possible to view him as, not merely the source of all being, but also as the source of ultimate unity. Pantheism on the other hand was accused of being a type of polytheism. And in a way it was. However, the concept of encompassment by the most-high god provided for an objective reality within which all things were included as parts within a whole. The unity involved would seem to water down objections that pantheism was not ultimately a monotheistic ideology.

There was actually something more behind the objections to pantheism than merely the plurality issue. This was the belief that it was wrong to speak of any being other than God as a god. When theism became the dominant way to image God, it became blasphemy to claim the existence of 'gods.' To do this was to usurp a quality that only God had. It was to insult God.

Apart from the political and emotional fanaticism that charges of blasphemy were later to be associated with, there actually was another logical argument involved in the theistic opposition to the existence of multiple gods. In theism God is presumed to have the characteristic of being infinite as well as absolute. The most-high god in pantheism would have a type of infinite nature. A universe has nothing outside itself. If the universe is boundless, then it is infinite. This is not a transcendent kind of infinity of the type that theists prefer to attach to the image of God. Nevertheless, it is a type of infinity. Lesser gods on the other hand would be less than the universe as a whole. Therefore it was believed that they could not be infinite, and hence could not be gods.

All of the above problems weakened the logic of pantheism. But of all the difficulties, the idea of the perfection of God was one of the greatest. If something is perfect, it does not have anything evil in it or about it. If God exists, it is important to see his nature as good. Plato said that the Form of the Good was the highest Form—higher even than the Form of Being. Plato's idea of pure absolute Being and pure absolute Goodness was incorporated into theistic views of God. Pantheism is particularly susceptible to 'the problem of evil,' since the material universe has things in it that are extremely bad, producing pain and suffering. Although pantheism has the greater theoretical burden in this respect, theism itself has not entirely resolved the problem of evil, even though theism places God outside the material universe. That is partially because theism sees God as the creator of the universe. If God created the world, and the world has evil in it, then it would seem as though God created evil as well as good. If God did not create evil, is there something that exists but is not his creation? The problem of evil poses serious theoretical problems relative to the absolute nature of God as well, since what is 'absolute' is absolute in power as well as in substance, i.e., 'omnipotent.' If God is opposed to evil, why not stop it? Why would a good God create evil or, in being omnipotent, not have the power to prevent it? The theistic answer to this has traditionally been a reference to human free will.

There certainly seems to be something sound about this

answer. People seem to be capable of creating evil—of doing wrong. But that suggests that 'creation' is going on right now. If people create bad things, or create anything in their life circumstances, good or bad, then they are creators. To see people as creators makes gods of people, and usurps God's power as the creator. (Theists have struggled with this problem for thousands of years, especially since thoughts of an unacceptable type could lead to charges of blasphemy.)

Both theism and pantheism have their strong points philosophically as well as their weak points. Theism can utilize the concepts of absolute, infinite, and eternal being—and a very strong case can be made for the existence of these things, not just in order to affirm the existence of God, but so as to account for thinking, truth, ethical standards, and so forth. For thousands of years relativists and absolutists in philosophy have been divided over the question of whether or not there is such a thing as absolute truth, devoid of relative context. Is there ever a real truth of the matter? Or is truth just a relative thing, so that what is true for me can be false for you? Or take ethics: Is that which is good an absolute, or is it relative, so that what is good for me can be bad for you? In terms of collectives, is it possible for total goodness to exist so that what is good for me will be good for everyone? Because the material world is relative rather than absolute, pantheism inclines towards the philosophical camp spoken of as relativism, while theism inclines towards the camp of absolutists. Because they are absolutists, the theists to some extent have an advantageous philosophical position (in both epistemology and ethics) via their acceptance of the reality of a standard of absolute Being and absolute Goodness. They just cannot connect these absolutes to the material world.

Morally speaking (in terms of *practical* ethics), both theism and pantheism have their good and their bad points. Theism devalues the material world. It is seen as something essentially evil and as something to escape from. It also devalues the present in favor of the future. Heaven is the goal, and heaven can never be achieved on earth. Paradise is a transcendent realm just as God is a transcendent deity. It is not the body that goes to heaven. It is the soul. How that soul is attached to the body is not explained by theism—not even how a particular soul which is eternal can be so and not be infinite. Only God is infinite. (There was logic to the objections against polytheism. In Chapter 3 on '**Spirit**' we will talk about how intertwined and interdependent the qualities of infinite and eternal being actually are. If a particular soul is going to be unending in time, it needs to be unending in space, i.e., to somehow be infinite.)

Morally speaking (again in terms of *practical* ethics) theism embodies a strange paradox. You would think practicing theists would not be materialistic. However, the three major theistic religions have not been faring very well in terms of convincing people that the material world is of little to no value to them. People value the material world, but in the sense that they want to possess as much of it as they possibly can. This is especially true of Christian countries. Christianity is the predominant religion in first world countries. The greed for material things is growing despite the fact that Jesus said it was more difficult for a rich man to get into heaven than for a camel to go through the eye of a needle. Is the desire for riches an atheistic development? How could it be, when many Christian churches teach that owning an abundance of material things and having a lot of money is a good thing—a reward on earth so to speak for being true believers. To make matters more confusing, many atheists have the opposite view in regards to the exploitation and consumption that being rich cannot seem to avoid.

One possible response to theism is atheism—the belief that God does not exist—or agnosticism—the belief that, if God does exist we cannot know anything about him since we ourselves are relative, finite creatures. Agnostics are often relativists. A type of theological agnosticism-relativism is at base of a movement known as secular Humanism, which arose in the twentieth century. As this relates to ethics, according to Humanists people are capable of living an ethical life without recourse to religious dogma or any type of belief in a supernatural being. Humanists do not devalue the material world, and in fact the secular humanists I know are very much involved with trying to produce a better world. In addition to actively promoting the good, humanists do not project value onto some unseen world in the future. On the other hand, when it comes to theory, the question that relativists are constantly confronted with is how there can be such a thing as something which is 'good,' or any type of 'value' if the Good does not exist—some standard by means of which we can ethically judge things in the world, to see when one thing is bad versus good—or even when one thing is better than another.

In terms of practical ethics, humanism may be faring better than theism, and pantheism might be as well—if there were any left of them, and theists had not virtually eliminated them from the face of the earth. True pantheism does not devalue the world; nor is value placed just in human beings as it is in the theistic religions. In pantheism, the world is spiritual or divine in essence, and each system within the world is visualized as a part of the divine. When every living thing is seen as spiritual in essence and intrinsically valuable, it is difficult (in good conscience) to plunder, tyrannize, subject and destroy pieces of the divine. But again, in regards to theory, there is the question of how pantheists can effectively deal with the absolute-relative, infinite-finite, perfect-imperfect problem—especially in regards to the problem of evil. If the world is finite, relative and imperfect, why not just call it nature. Why call it God?

So what is the answer—or does it even matter? Well, the answer more and more seems to be that it matters very much. The devaluation of the material world and the anthropocentrism of theism—the devaluation of non-human systems—has led to an ecological crisis that, within a century, may lead to an uninhabitable earth if excessive consumerism is not curtailed. [In addition to ecological problems, the devaluation of other human beings has led to economic and social injustices on an unprecedented scale, with 34,000 children a day world-wide dying of malnutrition and other preventable diseases, while in first world countries average caloric consumption and subsequent obesity has sky-rocketed.] There is a growing need to come up with a different way to image God, not so much for God's sake as for the sake of the earth and the human population that needs the earth in order to survive.

For the past two thousand years theism has been the dominant theological belief. There can be no doubt as to how religion has shaped our past history. The manner in which we image and worship God will shape our future as well—but which image? Will we go back to pantheism? Or turn to atheism? Or move on to something else?

When a major need arises, there is something in people that seems to have the capacity to respond in and of itself, almost instinctively, to fulfill that need—and especially in response to critical need. (Call it survival instinct if you will.) Right now there is such a strong movement going on all over the globe that it is almost as though billions of people are spontaneously and concurrently reaching inside themselves, trying to figure out who and what they are, including their relationship to God. Whatever is happening, it is so pervasive that it almost seems to be a response to an internal voice calling to people individually and collectively to wake up and discover their true nature—the more so because it is arising in so many areas, and does not have a single base or origin. People have always wondered about these things, but this is different. It does often include a search for a better image of God; but that is almost peripheral to the heart of the movement, which is centered upon trying to figure out if there is something spiritual or divine within people, and within other systems. This is an effort to try to understand matter—whether matter is inert or whether it is enlivened with spirit and soul.

Although the primary impetus creating this movement seems to stem from love of life and the fear that all life may be at risk right now, the new physics is also a major contributing factor. People love truth as well as life and there is an incredible amount of new data emerging relative to the nature of the physical universe. The scientific motivation for this movement began in biology as well as in physics when biologists became capable of exploring the subjects of 'life' and 'awareness;' but when it really took off was at the beginning of the twentieth century when physicists discovered that some sort of awareness was present at the quantum level of reality.

It is because this movement is grounded in science that it is an attempt to understand the physical universe. At the same time, it is because science is moving into metaphysical areas that it is adding fuel to the need for spiritual understanding.

Because of the apparent existence of metaphysical things within the physical world, such as life and consciousness, this movement could conceivably entail a reversion back to pantheism. The ecological crisis is encouraging a pantheistic paradigm as well. And yet, there are theoretical difficulties with pantheism.

But neither can the image of God called theism remain strong when evidence is mounting that there are indeed metaphysical realties within the physical world. It is possible to go back and forth with these pros and cons since, while pure theism as a way to image both God and the world appears to be failing, there are many concepts in theism that are logically strong.

When there are opposing views, if history is any indication, collective beliefs tend to move back and forth from one extreme to the other. (This is what a paradigm change involves.) Sometimes, however, a different type of change occurs. Instead of going back and forth between extremes, a way is found to see both as true. In this way the two poles can be united, thereby moving them both up to a new, higher level of understanding. That is what is going on right now, and the technical term for the dialectical unification taking place between theism and pantheism is pan*en*theism.

In the nineteenth century a philosopher named Karl Krause coined the term 'panentheism' to express the possibility of a reconciliation between theism and pantheism. In panentheism, something within the world is seen as a true part of God. Some panentheisms just see the spirit or soul of an individual as existing in God. (These often connect 'soul' to 'mind' and are called panpsychisms.) Others see the physical universe as also within God. Whichever type of panentheism is involved, the world and the individuals within it have something about God within them, and that is very different from theism. This is what links panentheism to pantheism. What makes panentheism different from pantheism and links it to theism is that, even when the entire world is visualized as in God or as God, God is still seen as something more. There is transcendence as well. This concept of transcendence concurs with theism.

The panentheistic paradigm is growing very fast. The details of this cosmology are still not clearly articulated, but they are coming together, and it is possible to begin to see the particulars within it. Because of the role of science in producing any cosmology, those panentheisms which are philosophically the strongest tend to include the physical universe. Within these, the most consistent elements revolve around four basic principles: One is an organic view of reality. Another entails a hierarchical perspective—seeing the world as composed of systems nested within other systems. Closely related to that concept is a microcosm-macrocosm perspective. And the last principle has to do with recognition of the freedom and creativity of people. And of course, grounding all four of these concepts is a spiritual view of the universe. In this new collectivizing cosmology, every system in the world is seen as divine, and as a 'piece' of the divine. It is the 'piece' aspect that gets into a different view of God; but again that is almost peripheral to what is going on.

In terms of the organic aspects of this new paradigm, when looking at the body of a human being it is hard not to be impressed by the incredible complexity and order. The order is centralized, and very dynamically so. Not only does a single, unified being possess this order, all the parts act as one. Biologists are trying to understand what it is that centralizes the parts. The very notion of 'centralization' presupposed a 'center' with the power to do this. But a true center is what physicists refer to as a 'singularity.' It is a zero point. Zero and true geometrical points are not physical at all. That is because physical things are quantified and extended. There are currently a number of biologists who believe that there is something nonphysical—a spark or 'spirit' of life—actively centralizing the parts of a body. These biologists are referred to as 'vitalists.'

In addition to the center of a system, the actual ordered parts of a system have historically presented an enigma. This is the question of whether parts self-order or whether the very personal (and very organic) order in living systems might involve something more than the sum total of the material parts—possibly a field of some sort. These thoughts are especially important to those biologists who refer to themselves as 'organicists.'

Because non-physical things can be related to spiritual concepts, if a biologist is a vitalist or organicist, and they are interested in these possible connections, they may refer to the vital principle as 'spirit,' and the organic unity behind the material parts (the 'something more') as 'soul.' (Thus it becomes possible to see where all this is headed.)

In addition to being centralized, the order of a living system is also hierarchical. A human being is composed of a number of systems, composed of organs, composed of parts, composed of cells, composed of molecules, composed of atoms—themselves composed of parts. The number of layers, and the diversity of the centralized order, is a measure of how high on the scale of evolution a living system exists. In looking at a human being it is difficult not to wonder if there are higher-level beings in higher dimensions with degrees of complexity we cannot even begin to imagine. It is extremely unlikely that we are the epitome of creation.

When thinking about how higher-level dimensions encompass lower-level dimensions, and about how we encompass our atoms, if the mind is allowed poetic license, it is possible to envision ourselves as atoms or cells in still higher level beings, including the earth and the universe as a whole. This strange mental wandering or 'poetic license' is one of the fastest growing ideas in modern times. An organic view of the earth, for instance, is a position now held by some of the most popular modern ecological advocates, such as environmentalist James Lovelock (*Gaia, A New Look at Life on Earth*).[1] Lovelock gives us a picture of the earth as a self-regulating system with the characteristics of a living organism.

The emergence of an organic perspective in geophysics is a part of the new paradigm, and is itself a re-emergence and an awakening, rather than an entirely new movement, which is why Lovelock included in the title of his book the name of the Greek goddess of the earth 'Gaia.' This is the old pantheistic nature religions coming back to haunt us. In addition to the old religions, there is a renaissance of old philosophical ideas as well. This organic view of huge systems that are not normally thought of as living or organic has existed in the field of serious thought for thousands of years. Plato, as noted, claimed that both Gaia and the universe as a whole were living organisms (*Timaeus* 30c, *Philebus* 30a). It is not just the earth that has interdependent, coherent parts. The universal laws of nature fit together and act together as a whole, and it is hard to believe that the universe is not a true system in every sense of the word.

In looking down towards the atomic level of being rather than upwards towards the earth or the universe as a whole, there is something about the quantum level world that also pulses with life, and even with awareness. An atom is a centralized system with very orderly interconnected parts. Today there are biophysicists who believe that the organic unity to be found in a living cell begins in and actually exists within the molecular level of being (e.g., Mae-Wan Ho, *The Rainbow and the Worm*).[2] The well-known quantum physicist David Bohm (*Wholeness and the Implicate Order*)[3] was interested in behaviors suggestive of organic unity present in plasmas, which are simply closely interacting groups of ions. Quantum physics has led to an increasingly popular holistic, even organic view of reality leading to a proliferation of popular books on this subject directed towards a general, non-scientific audience (e.g., Gary Zukav, *The Dancing Wu-Li Masters*;[4] Fritjof Capra, *The Tao of Physics*[5]). The upsurge of interest has been especially encouraged by the intriguing fact that some sort of awareness—a knowledge of being watched by us—appears to be present in atoms and molecules—witness the number of books now available discussing the link between quantum theory and consciousness (e.g., Amit Goswami, *The Self-Aware Universe*). [6]

As with Plato's belief that the cosmos was an organism, the organic view in quantum physics and chemistry is not a new one. The early Greek philosophers advocated what is called 'hylozoism'—the belief that everything in the material world was teeming with life. Aristotle claimed that the entire idea of 'substance' had to do with being a being. We normally use the term 'being' to refer to a living being, with 'life' and 'awareness' as attributes supposedly exclusive to this type of system. There are different grades of life and order, but the organic order we connect to 'life' may not be an emergent quality. Furthermore, in terms of awareness, there exists something called physical 'perception'—the fact that any true system seems to be aware of its physical parts. We may be aware of our thoughts and hence of the parts of our mental 'body' as well as of our physical body; but if an atom were not cognizant of external physical conditions there would be no chemical reactions. What is more, it responds as a whole, with some sort of sentience in regards to its internal physical parts. If it did not have this, it would not remain a true, internally coherent and centered system for very long. And it cannot be ordinary chemical or electromagnetic interactions at base of this sentience and its response, since electron levels, for instance, change instantaneously. No ordinary electromagnetic interactions are instantaneous.

When something appears to be moved by a non-physical (non-electromagnetic) cause or to consist of a non-physical substance, the cause or substance can be looked upon as 'metaphysical' [from 'meta' ('beyond') and 'physics' (the material world)]. What in philosophy is called metaphysics interfaces with biology, and possibly even chemistry, when life or the spirit of life is posited as a non-physical substance. It is the same with trying to understand what awareness is.

The so-called 'new sciences'—in particular quantum theory and relativity theory—have brought metaphysics back into a scientific community that was trying very hard to expunge such ideas. We now must take into account infinite things such as singularities, the possibility that the world was created from nothing and the fact that there appears to be some sort of primordial awareness at the quantum level of being.

Metaphysics interfaces with theology as well as science when people are trying to understand if a metaphysical reality like God exists, and what that substance or being might have to do with our own being—especially when the concept of life or awareness is an attribute of this substance. We think of God as being the source of life and as being aware. If it takes something metaphysical (and possibly spiritual) to unify a centered system, then any and all systems may need a spirit of life or a spark of divinity within them in order to even exist as material systems. Because these things may involve metaphysical realities, it is becoming more and more difficult to separate science from God—just as it has always been difficult to separate philosophy from God when discussing anything absolute (e.g., Plato's pure Being and Aristotle's 'ousia'). Scientific and philosophical concepts which posit the existence of metaphysical realities within the world lead very easily to panentheism. Panentheism has become a view held by some of the most popular writers in areas of science today (e.g., physicist Paul Davies, *God and the New Physics*).[7]

The new scientifically oriented panentheisms tend to center around life and awareness. It is via the life concept that they have an organic or organism motif. Everything from the cosmos to the atom can be viewed as a living system with organic properties. These properties appear to lead to the layering mentioned, and New Age scientists want to know why. There seems to be some sort of universal law of nature involved, since it is not merely internal parts that are hierarchically arranged. Even as independent wholes in and of themselves, systems occur in hierarchical layers nested inside one another. Thus our bodies, which are true systems, are composed of atoms, which also are true systems. Moving upward in the hierarchy of being, we are parts of the earth, and our earth exists within the solar system. Above us, our solar system along with millions of other stars exists within our galaxy. The galaxies themselves exist within meta-galaxies, which exist within meta-meta-galaxies, and on and on beyond the ability of our telescopes to follow.

It was only very recently that physicists discovered meta-meta galaxies, and the vast extent of the nested order above us. The basic concept, however, has been with humanity for millenniums in both the West and the East. What is going on is more aptly compared to the aforementioned awakening after a long period of sleep—and that may have to do with the pantheism attributes within panentheism. In Western philosophy the idea of hierarchy was a part of the metaphysical system of Plato and the later Neo-Platonic philosophers who visualized souls as occurring in a nested series of spheres.

Nor are these ideas exclusive to the West. His Holiness the Dalai Lama, in *The Universe in a Single Atom: The Convergence of Science and Spirituality*, prefaces his book with a quote from Buddhist scripture, *The Great Flower Ornament*, an ancient text which states, "In each atom of the realm of the universe, there exist vast oceans of world systems."[8] Astronomers with their meta-galaxies, and physicists, with their ideas of 'many-worlds' and 'many-dimensions,' are merely redefining and awakening the ancient wisdom.

It is the organic perspective (the centralized order in each system), along with the nested motif, which leads to the microcosm-macrocosm concept, and that too is very old. The saying, "As above, so below; and as below, so above," is credited to the Egyptian sage Hermes Trismegistus. If all systems, at whatever level, are organically organized along similar principles, we may be able to use one system as a model to understand other systems in other areas and at other levels of being. The organic paradigm may even be useful to understand interactions within groups. In the human sphere of experience the organic paradigm is gaining ground in political science and sociology, wherein a family or even a larger 'body-politic' like a nation can be visualized as having organic properties requiring respect for each part, and coherence between the parts if it is to live and flourish.

Given the link between politics and economics, and given our current global political-economic situation, the rebirth of an organic paradigm in these areas could be extremely valuable. The current movement towards globalization is helping us learn that we are all truly interdependent.

In terms of an area mentioned earlier, that of our current ecological crisis, panentheism is a way to see, not just the earth as an organic whole, but also every system on it as divine, and hence intrinsically valuable. Panentheism is emerging as the theoretical base behind a growing ecological movement consisting of any number of activists who are trying desperately to awaken a sleeping populace to a different way of looking at and treating the earth, including all the disappearing species that are an integral part of this ecosystem (e.g., Sallie McFague, *Models of God*[9] or Rosemary Ruether, *Gaia & God*[10]). And as with all the other areas mentioned above, an organic ecological paradigm is not new. The American Indians embraced this belief long ago.

In addition to the organic motif, the microcosm-macrocosm concept, and hierarchy in nature, the fourth precept mentioned is also extremely important to the future of the human race, and that is the freedom that being 'living' suggests. Living things, when in their role as a part within their external world, may indeed be subject to external forces, but in their wholeness and in their essence they have, not only a self-regulating capacity, they also have a self-creating capacity—by a degree equivalent to their degree of evolution. As highly evolved organisms, we have a very strong creative will power—a growing capacity to create ourselves and co-create our external world.

This creative aspect of the organic motif within panentheism has made it especially popular in the new psychology. At base of both an emerging new holistic medicine and a new psychology is the belief that people are powerful, self-creating agents. This counters the old theistic and the old scientific paradigms. In example, Bruce Lipton (*The Biology of Belief*) has worked with experiments that remove the DNA of living cells which, although devoid of DNA, still respond as living things without these instructional molecules to guide them.[11] DNA is apparently not the cause of who we are. We are. Lipton is working to awaken people and to free them from the lingering mechanistic materialistic paradigm which visualizes human beings as machines, with nothing within them to rule over the web of mechanical causes in their physical bodies. Both science and theism have been guilty in this respect—the latter because God is the creator. A conviction is growing that the old paradigm has separated people from the fullness of their real nature and potential power as creators.

[Many new thought physicians and therapists working in the fields of medicine and psychology emphasize the panpsychic aspect of their panentheism. The mind can control the body, and positive thinking laced in with a different belief system can work 'miracles,' which is not surprising if we ourselves are divine.]

The ideas behind panentheism actually encompass any discipline where people are visualized as self-creators, and as having something non-physical or spiritual within them by means of which to manifest this power. However, despite the fact that there is a burgeoning interest in areas that have little to do with religion, because metaphysical realities are involved, these New Age pioneers bridge the gap between science and theology. For that reason, they enhance the very different view of God that is emerging in religious circles.

In terms of traditional religions, Panentheism is enabling theologians to combine theism and pantheism, and thus bring God back into the world (e.g., Marcus J. Borge author of the national bestseller *Meeting Jesus Again for the First Time*[12] and *The God We Never Knew*[13]). It also provides a chance for them to combine theology with science (e.g., biochemist and priest Arthur Peacocke, *Theology for a Scientific Age)*.[14]

Philip Clayton and Arthur Peacocke as editors of *In Whom We Live and Move and Have Our Being: Panentheistic Reflections on God's Presence in a Scientific World* are typical of many books in print providing a literary forum for influential theologians and scientists eager to discuss their thoughts on the correlations they have seen between Christianity, science, and panentheistic spirituality.[15] Contributors in *In whom We Live and Move and Have Our Being* include Joseph Bracken, S.J. (*The One in the Many*),[16] Niels Henrick Gregersen, research professor in science and theology, and associate editor of *The Encyclopedia of Science and Religion*[17], Ruth Page (*God and the Web of Creation*)[18], and Robert Hermann, Damon Runyan fellow at MIT and himself a recent host-editor for *Expanding Humanity's Vision of God: New Thoughts on Science and Religion*.[19]

The 'new thought' churches are way ahead of the old ones. The new thought movement in religion began to emerge already at the end of the nineteenth century. Some of these have now organized into churches, such as the Unity Church, the Church of Divine Science, and the United Church of Religious Science. The latter group prefers to speak of their churches as Centers for Spiritual Living, as does the Agape International Spiritual Center founded by Michael Beckwith (*Spiritual Liberation*).[20] There is a realization that 'church' has become synonymous with jelled doctrines and stringent codices. (Organic, living things need to grow if they are to flourish.) The 'new thought' churches revolve around a spiritual philosophical movement based upon the core concept that God is immanent as well as transcendent—that each person is divine—and because of this that each person has the capacity to create their lives and their circumstances.

In addition to traditional churches and even new thought churches, belief in a divine within is now finding a way to express itself via Eastern spiritual practices as well. It was in Hinduism that yoga first arose as a way for an individual to connect to the divine within and hence to Brahman. (The 'atman' within each person is seen as pure absolute Being, and *is* Brahman.) Hinduism is a panentheism. It is when God is within that it is possible to make meaningful personal contact. (It is a philosophical truism that the known must be in the knower.) Here in the west there has been an explosion of interest in meditation, yoga and other practices aimed at producing a personal experience of the divine within. These often involve increasing the power to be found within creative centers in the body and mind. People believe that there is an inner path through which they can discover the essence of their being. These are sometimes referred to as New Age practices, but they are of increasing interest within traditional religious circles as well—and specifically as practices designed for the laity (e.g., 'centered' prayer). There was always a mystical tradition in Christianity, Judaism and Islam, but it was reserved for cloistered monks in sequestered orders. The movement going on today involves ordinary people living in the day-to-day world.

In addition to the introspective manner of finding God as the personal within, there has also been an increase in extroverted practices that include experiencing other people as divine, or experiencing 'Mother Earth' as divine. These include feelings, thoughts, words and actions that engage the divine and further an awareness of the mutual interdependence of all creation. It involves a universal compassion perspective, which in turn interfaces with sociological, ecological, and even feminist concerns via thoughts on the 'womb' of creation or the earth as our mother.

Whether in terms of practice or theory, the interest in all these subjects is expanding at an exponential rate. The increasing number of books written by people who want to share what they have found in panentheism—whether or not they articulate it as such—is a measure of the value people are finding in this spiritual view of themselves and of nature.

Because of the swell of interest, many authors as noted are serving as sources of forums, allowing other authors to come together in their books. Ervin Laszlo and Kingsley Dennis, editors of *The New Science and Spirituality Reader* [21] provide a literary forum for a range of philosophers, scientists and New Age practitioners, including New York Times bestselling author Gregg Braden (*The Divine Matrix*), [22] best-selling author Deepak

Chopra (*War of the Worldviews: Science vs. Spirituality*),[23] Kingsley Dennis (*New Consciousness for a New World*),[24] and Peter Russell (*From Science to God*).[25] In the introduction, Laszlo and Dennis discuss their belief that the mechanistic, materialistic paradigm simply can no longer support itself given the knowledge we have available about the nature of the world. In addition, there is the desire of people to find meaning, purpose and spirituality within the framework of their current, material lives as opposed to some future realm. In a foreword Deepak Chopra states that we need to consider the possibility that the universe is a conscious living organism representative of the organisms it encompasses, and that science, instead of ignoring or denigrating the world wisdom traditions in this area, may need to sincerely search for whatever may be valid within these ancient ideas.

And it is not just via books that these people are sharing their zeal for a new world view. Kingsley Dennis is cofounder of WorldShift International. Barbara Marx Hubbard (*Birth 2012 and Beyond: Humanity's Great Shift to the Age of Conscious Evolution*) is president for the Foundation of Conscious Evolution.[26] Astronaut Edgar Mitchell founded the Institute for Noetic Sciences. The list goes on and on: The International New Thought Alliance, The Association for Global New Thought, The Templeton Foundation, The Shift Network, The Institute for Religion in an Age of Science, etc., etc.

All the books, foundations and movements dedicated to these motifs in one form or another are evidence of a true phenomenon arising spontaneously within our time. All ages have their specific spirit-of-the-times. This particular renaissance is about spirit itself—its presence in the world and within us.

In addition to all these burgeoning branches of thought moving and growing into so many different areas, there is a convergence going on as well. In *Consilience: The Unity of Knowledge*,[27] Edward O. Wilson expresses so well the urgent need in modern society for all these areas of knowledge to converge, from physics to the social sciences, to ethics and religion. Panentheism is making this possible. The commonality between biophysics, ecological and sociological organicism, religious thought, and even the practices of mysticism is sensed but is only beginning to be articulated. This global phenomenon is calling for and actually demanding a really strong metaphysical system of thought capable of, not merely demonstrating the unity at base of all these ideas, but also the philosophical foundation—the logic of the ideas.

There are a number of modern-day philosophers trying to formulate a logical system of thought that will do this. The English philosopher Alfred North Whitehead may be looked upon as the father of one of the most popular modern models of panentheism. Whitehead wanted to correlate modern science and theology. Quantum physics is the way we now describe our world. Quantum theory speaks of the ultimate 'atoms' underlying the world experience ('quanta') as tiny bits of process—hence the title of Whitehead's book, *Process and Reality*. [28] According to Whitehead, ultimate reality resides in the moving, changing material world—and that reality includes God. Process theology entails a true panentheism, since there is the claim, not just that God is within a special part of the world (e.g., as a life spirit), but that the entire world is God, including and especially the physical universe that science studies. Whitehead's philosophy was further articulated by a graduate assistant of his, Charles Hartshorne (*A Natural Theology for Our Time*).[29]

Process theologies demonstrate a real concern with the manner in which theism has, not only taken God out of the world, but also with how it has taken the focus from the here and now, and placed it in an afterlife. Like humanists they tend to believe that we need to concentrate on this world with all of its problems. Unlike humanists, who are inclined towards agnosticism and atheism, panentheists are interested in bringing God into the picture (e.g., *Beyond Humanism* by Charles Hartshorne).[30]

In bringing God down to earth, process panentheists have given up the belief that the body of God is perfect, unchanging and absolute. If God is in the world, and especially if he is the world, the belief is that he has a body that moves, changes and is far from perfect. What the new panentheisms do is to create a picture of God as relative, changing, and contingent—i.e., not omniscient and not omnipotent (e.g., Charles Hartshorne, *Omnipotence and other Theological Mistakes)*.[31] Within these process panentheisms, God is very much involved with things that are bad, such as emotional and physical pain and suffering. As with theism, process panentheisms do not visualize God as causing bad things to happen. Rather they describe the systems within him as true creators themselves. They believe that God allows the systems within him to create as they will. It is this that causes evil. However, unlike theism, which then removes God from the world, in panentheism the world is God's body. Thus he too suffers. A painful, sick body is a painful, sick body and, in a process panentheism that truly looks upon the material world as the body of God (or upon systems like ourselves as 'cells' in the body of God), then God has a body that is partially 'bad.' Philosophically, it would seem as though there is no other answer to the problem of evil. Process panentheists tend to accept the evil in the world as necessary for contrast, or even necessary in order to appreciate the good things that do exist in the world. Whitehead argued that, what seems to be bad or evil, may have some greater or ultimate good that we just cannot see.

Process panentheisms are currently the most popular type of true panentheisms. They do posit absolute realities, such as archetypes. (Whitehead calls these 'eternal objects.') But the body of God is seen as the physical universe. Hence, they entail a theology that is very relativistic in nature, since physical matter is a process of dynamic (relative) interrelationships. Philosophers with a process theology see no way that matter can be infinite, absolute and perfect.

It is because matter is relative that an absolutist type of panentheism does not work very well. In *Panentheism: The Other God of the Philosophers,* John Cooper classifies Plato as a panentheist. But in Plato, the god of the universe is the *soul* of the universe. I tend to agree with the process theologies that a panentheism is not a strong panentheism if it does not include the physical universe in the body of God. [That is what 'pan' ('all') means.]

The question that arises is whether or not process panentheisms are the answer. If they are, then there is no need for a new or different model. If the philosophy behind process panentheisms is sufficiently strong, then the new paradigm which the world needs already exists in Whitehead's process panentheism. Philosophers who have followed Whitehead's line of reasoning essentially just enlarge upon his basic system of thought.

I believe that process panentheisms are not adequate, either philosophically or in terms of the outcome of the view they advocate. Relative to the latter, I worry what will happen to the human race (and its home, the earth) if it elects to incorporate this particular view of panentheism into its collective theological ideology. Process theologies bring God down to the level of the material world with the idea of truly advocating for matter and for this world. The intent is very idealistic, but I think in the end it will hurt matter and the material world just as theism did. That is because this perspective also gives up on matter. I fear that, if we accept that the world is naturally evil, we will become inured to the pain and suffering that goes with evil.

There is nothing good about children dying or about illness. It is only when a person believes that illness is a necessary part of existence that philosophical excuses must be made for it like, "Well look at the strength this person achieved by overcoming this or that." Or, "Well, it is a good thing for sick things to die; then (as per Darwin's thesis of 'survival of the fittest') a greater good will come from their elimination."

And again, this is not so much about God as it is about us.

Matter' is not something external to us. There is no matter that is not the part of some system. It is demoralizing to believe that we have a naturally dark and bad side which must exist if there is to be contrast. If we are to be true advocates for matter (and for all the systems in the world, including ourselves) the worst thing we can do is to say these are *naturally* evil—or even that they *naturally* possess a partially evil nature.

Some modern panentheists try to escape from the question of evil by reverting to a panpsychism. But this in essence is what Plato did. Because most panpsychisms tend to blur the distinctions between soul and mind, they usually entail cosmologies wherein 'all is mind.' The reasoning is as follows: If, rather than believing the brain to be the mind, mind is looked upon as its own reality, and as non-physical in nature, it may be possible to come up with a different way to deal with both the relative-absolute problem and the problem of evil. If the mind is non-physical, perhaps it is not finite or relative. And if mind is soul, then soul has been linked, not only to a non-physical substance, but also to goodness and spirituality.

The purpose of a panpsychism is just as idealistic as the purpose of a process panentheism. It is people who have minds, and one of the purposes of panentheism is to connect the divine to people. In the end, though, panpsychism does the same thing as theism. To say only mind is real or good is also to desert physical matter, and to forget that there is mental illness, and illogical and even degenerate patterns of thought that are just as incoherent right now as physical matter is managing to be. Even when a proper distinction has been made between mind and soul, to say only soul is good (or real) is also to desert matter instead of redeeming it. Buddhism did this to Hinduism, and theism did it to the Christian, Jewish and Muslim religions.

I believe that a different course is, not only possible, but desperately needed. And what is needed is not to bring God down to the world, but rather to bring matter up to God. Via a different type of panentheism, which includes the latest data arising in quantum theory and relativity theory, matter can be shown to be infinite and absolute, and that sounds like a contradiction in terms, but it is not. In addition to being absolute, matter can also be shown to be good, and with the (natural) potential to be wholly good. And that sounds like a denial of reality, but it is not. Most process panentheists believe that 'absolute' good is an ideal that can never be achieved in a relative world. I disagree. The material expression of goodness in the world is wholeness. It is possible for our bodies to be whole. In fact either they are whole or they are not. (Either a finger is missing on a hand or it is not.) There is nothing relative about that. Wholeness is a material absolute. (It is especially easy to visualize real wholeness in terms of a living organism—and the strength of a good panentheism resides within its organism concept.)

In addition to our own body being whole, it is also possible for our collectives to be whole. There are many, many things about the world which indicate that it is created to be whole. For instance, the laws of nature are integrated integrals and are naturally coherent. Coherence is what 'wholeness' is all about. In an organic, microcosm-macrocosm paradigm, where matter is centered upon individual systems, the laws are centered upon us. These processes and powers—these laws—are just what is meant by matter. We are naturally coherent. It is incoherence that is unnatural.

It is possible to have a different kind of panentheism, one which elevates matter to its rightful place, and that explains how and why matter itself is infinite, absolute and eternal. This is the way to find our true divinity. We will not find that divinity, or our goodness, by believing that wholeness in ourselves and our collectives is a will-o-the-wisp that can never be truly achieved. Wholeness is a material ideal, and one that can be achieved—is in fact naturally achieved by a healthy body. A reachable ideal has the potential to both ground hope while concurrently elevating it. The 'ideal' aspect of a reachable goal is important because people follow ideals with passion, and we need that passion if our material bodies (and their home, the earth) are to be saved.

People all over the earth are sensing how important that it is to save matter. *The New Paradigm* is an effort to grasp the essence of this spirit-of-the-times, and help give it life and definition. Defining occurs, not only through articulation, but also by giving the parts of a concept place—by showing where everything fits together. In philosophy (and in science) when this process involves a world view, it is called creating a cosmology. When something called spirit is an integral part of the cosmology, it is referred to as a metaphysical system. The purpose of a metaphysical system is to serve as a model or picture that people can use to better explain and conceptualize a world view that includes spirit. If it is a good, sound metaphysical system it serves as a 'proof' of the validity of the view it articulates. This can help people when they are trying to find the truth about themselves, about their fellow creatures, and about their relationship to God.

The New Paradigm set forth in this book is a philosophical metaphysical system, and it is a panentheism. It is not a book about God, even though Chapter 8 is on '**God**.' It is a cosmology with the purpose of explaining who and what the systems called 'people' are, how they are related to other systems, and (incidentally) how they are related to their creator.

If there is to be a model of creation that justifies the efforts of so many people currently working for and searching for a new world view, it needs to be a really strong metaphysical system. *The New Paradigm* is able to reconcile the absolute-versus-relative dichotomy that has plagued philosophy for millenniums. In the process it is able to deal with the major philosophical issues in theism versus pantheism, and in panpsychism versus process panentheism. It is able to do this with a simplicity and strength that is satisfying to both logic and common sense.

The method used entails a set of three tools that include first of all the ideas of the new sciences, secondly a set of tightly interwoven premises and conclusions that make up a strong philosophical system of thought, and last but not least an important tool that is as old as mankind. *The New Paradigm* utilizes twelve primordial ideas or archetypes to construct a very new and very old cosmology. These ideas are the same archetypes that were the basis of the old, pantheistic nature-astral religions. These are products of the wisdom traditions that Chopra makes reference to—the wisdom that science now needs to rediscover. What we refer to as 'myths' were once organized religions that were just as interwoven with philosophy as our own religions are. The thoughts in Plato's books are replete with references to the pantheistic religion of his day. They helped him formulate his thoughts about infinite and eternal realities—the same thoughts as were used by later religious apologists to justify many current beliefs.

Unfortunately for the millenniums to follow, even Plato could not handle the problem of infinite-finite and absolute-relative connectivity, or the problem of evil; and in the end he and his followers contributed to the theistic rejection of the material world—both of its ultimate reality and of its goodness. We are that world. Furthermore, we are good and we are real—our bodies as well as our souls. It is possible to construct a new and better paradigm that does not follow down this path of negation. And it can be done, paradoxically, with much the same base as Plato worked from, including the ancient wisdom that threaded its way through his philosophy.

Thanks to political and religious persecutions of the old pantheistic religions there is not a great deal of information left in regards to these cosmologies. The data is fragmented, and what is available is often veiled and esoteric rather than clearly and openly set forth.

There is, however, something available today that can provide a very systematic and very complete cosmology, although it is not thought of in cosmological or religious terms. The old religions were not merely nature or pantheistic religions. They were nature-*astral* religions. The same archetypes at base of the old religions can be found in modern-day astrology. Thus Plato spoke often of the god Hermes (Mercury) and the star sacred to him, as well as of Zeus (Jupiter), Aphrodite (Venus), Cronus (Saturn), Poseidon (Neptune), Uranus, and of course Apollo (the sun). In the old religions these gods were regents to the World-God (when perceived as 'parts' of the world) and regents to the Sun-God (when in their role as astral deities and 'parts' of the solar system). The old religions did not survive, but astrology somehow did, and the archetypes at base of its 'rulership' system provide a way to formulate a philosophical metaphysical system. *The New Paradigm* utilizes these archetypes to form a picture of how the world is made. In Chapter 2 the twelve archetypes at base of astrology's rulership system are discussed in terms of their being a philosophical category system. A category system is a list of the most fundamental qualities at base of the world experience, such as mass and spatial extent.

The model of the universe which can be created from a look at these principles is not a reversion to an archaic, unscientific view of the universe since, in the process of creating this paradigm, each archetype is linked (in Chapter 2) to a specific law of physics. The difference is that the laws of nature can be shown to be spiritual laws—and spirit itself becomes something that is in the world, making of each system—from an atom to the cosmos—a spiritual being in the truest sense of the word.

Nor is the model of the universe which can be created from a look at these principles a pantheism. We do not have enough extant data on the old religions to know if they entailed pantheism or panentheism. On the other hand, when the astrological principles are used to create a cosmology, they very definitely lead to a panentheism.

The cosmology that emerges from the use of these archetypes winds up having the four concepts mentioned previously—those that make a panentheism philosophically strong. They include an organic motif, hierarchy in nature, the microcosm-macrocosm thesis, and personal freedom. Because the cosmology that develops includes the physical universe, it is stronger than the panpsychic panentheisms from the past. Because, via these same archetypes, the relative-absolute paradox is reconciled, the cosmology is also stronger than the new process panentheisms. When the concepts all come together, as they do, the ideas in this system of thought conform to all the known and trusted criteria for finding and knowing the truth—fourteen of which are discussed in Chapter 1, including scientific empiricism, comprehensiveness, simplicity and logical coherence.

I did not invent this cosmology. I found it in the archetypes. What I am responsible for, however, is one very important thing, and that is (hopefully) a persistent state of clarity and readability. When reading Whitehead's *Process and Reality* it takes a while to figure out what he means by an 'eternal object' or a 'prehension.' (Whitehead does not clearly define and differentiate between concepts like awareness, mind, soul, emotions, feelings, physical responses, etc. These are all 'prehensions.') *Process and Reality* is not a straight-forward and clear system of thought, and for that reason different ideas Whitehead articulated can be interpreted in different ways. Common language is called 'common' because it is how we relate to one another and share ideas. Words are how we articulate and define concepts. The emerging new paradigm deserves clarity, along with the utilization of common words that can be used for a common vision. In philosophy in particular (clearly defined) common terms are to be preferred over professional terms or manufactured terms. Furthermore, because simplicity is a criterion for truth, when something is true it can be stated simply and clearly. In *Consilience* Wilson says that Western philosophy is not helping with the vital need for a new world view, partly due to its "involuted exercises" (philosophers writing for other philosophers), not to mention "professional timidity."[32] *The New Paradigm* is a little academic at times, in Chapter 5 for instance on '**Soul**,' where the concept of 'wholeness' is approached by beginning with simple premises that lead to others based on a logical progression of thoughts; but for the most part it is a persistent attempt to be practical and readable. Nor is *The New Paradigm* 'timid.' There is no professional timidity about offering to the public a serious philosophical metaphysical system based even just partly on astrology, something that is currently thought to be nonsense in academic circles.

Our world needs help. The spiritual nature of every 'bit' of matter needs to be redeemed so that the earth can live and be our mother for generations to come.

Individuals need help. They want to know who and what they are. The American writer and philosopher Henry David Thoreau (1817-1862) said that people were living lives of "quiet desperation." The philosopher known as Philo of Alexandria (20 BC - 50 AD) recommended kindness because as he put it everyone was fighting a terrible battle. They were right. Life is difficult—and short. People want to know if they have a spirit and, if so, what it is. They want to know if they have a personal soul and, when they die, if they will still be themselves. They want to know why there is sickness, death, and other bad things.

If philosophers have some thoughts available that can help change our view of the earth, and help people know who they really are, they need to be put forth both courageously and clearly—in a way that everybody can access.

—Sharon Miller, www.newparadigmforlife.com
info@newparadigmforlife.com

Part I

Constructing a Cosmology

The purpose of *The New Paradigm* is to assist with the development of a new, spiritual cosmology to replace the old materialistic one. In the current paradigm it is claimed by scientists that the laws of nature make the world what it is. In *The New Paradigm* the laws of physics will also be seen as foundation principles by means of which the world is created and characterized. The difference is that these will be shown to be spiritual laws—produced by a substance called 'spirit' and continuing to express the nature of that substance. Chapters one through eight are devoted to the philosophy at base of this new way of looking at the universe.

In this metaphysical system there are twelve basic principles that are behind the laws of nature. I refer to them as archetypes. An archetype is a primal type of pattern which gives character to material things. From a particular type of archetype all material things of the same species take their nature. This is not the first time that these particular twelve archetypes have been seen as foundation principles. They were revered thousands of years ago in the ancient nature-astral religions that we now call the myths. They were 'revered' because it was understood that these principles were spiritual in character.

These were the religions of people who worshiped a high god whom they connected to the sun. The sun god was worshipped literally around the world from Japan to the Americas. In these old religions there were invariably twelve deities that were ministers to the high god. Thus there were twelve Adityas in India, twelve Asir in Scandinavia, twelve Tangaro in Polynesia and Melanesia, twelve Olympians in Rome, twelve Titans in Greece, etc. These twelve gods were involved with twelve fundamental types of being by means of which the world was created. The old religions were overrun and virtually exterminated by the Jewish, Christian and Muslim incursions. What sacred books were available were generally destroyed so that the ideas might be expunged from all records wherever and whenever possible. What data still exists of these old religions is very fragmented and of course very antiquated.

There is, however, something that is connected to the old solar myths, and that is still with us today in the modern world, and that is astrology. Our planets were named after Roman gods—the 'Olympians.' In astrology each planet is linked to a particular astrological sign. The Roman gods and the twelve asterisms of astrology are connected to the same twelve regent deities that were worshiped the world over by similar or etymologically linked names. And it is not just their names that are linked. These gods had rulership over the same departments of life—the same categories as exist in astrology, where everything is said to be 'ruled' by one of twelve signs (and its corresponding planet or luminary).

Unlike the old myths, astrology is alive and flourishing today. By looking at astrology's rulership system it is possible to get a good idea of the fundamental archetypes that existed in the old religions. These can be used to create a cosmology, and that is what **Part I** consists of.

Each of these twelve archetypes is also correlated to a specific law of nature. Thus despite the antiquity of the myths and the questionable nature of astrology in academic circles, modern science will not be left behind.

[The cosmology that can be constructed from the twelve archetypes is referred to in this book as the astrological cosmology (abbreviated as the 'AC') in honor of astrology and its ability to survive.]

Chapter 1

A New Paradigm

Do you want to know who and what you are—what your relationship is to the universe, and to the source of the universe? Would you like to know whether or not 'God' exists? Would you like to know if you have a 'soul' as you may have been told? Are you interested in finding out if there is any purpose to life—in general as well as to your own? Do you want to know what the origin is of that which we call 'good' versus that which is 'bad,' and whether or not the bad things in the world have to exist along with the good? This book then is for you and for anyone who is seeking answers to these ultimate spiritual questions.

You may have been told that it is not possible to know the truth about these things. If you believe that, you are defeated before you begin. That is simply not true. You have been told this by people who adhere to the current materialistic model of the universe wherein spiritual things are seen as totally separate from the material world—unknowable and unreachable. It is possible to have a new and very different scientific paradigm wherein matter is seen as composed of the very essence of spirit. Once the paradigm changes, then matter itself can be perceived as having answers to these questions.

Astrology can assist in the establishment of a new paradigm. It has a great deal to say about the laws that are shaping our universe. That is because the twelve signs that are at base of astrology are themselves expressions of what scientists refer to as laws of nature—and actually are the same laws that physicists are studying—plus some they have not even discovered yet. From these twelve principles or laws it is possible to construct a new theoretical model of the universe. This model is what philosophers refer to as a metaphysical system. It is a 'system' of thought because the concepts underlying the paradigm are all interconnected logically, giving the various parts of the model rational coherence. It is a 'metaphysical' system because it claims that the material world arises from something non-physical. I speak of this model as the astrological cosmology or AC.

Writers of so-called 'New Age' material have been claiming for some time now that we either have entered or are about to enter a new age. They speak of the new age as being related to a paradigm shift, in other words that an entirely new way of looking at the world is about to arise. What these people believe will happen with the change varies from person to person. There is a great deal of talk about people becoming more spiritual, although not in the traditional sense of an increase of interest in church, synagogue or mosque attendance. Many speak of a rise in the level of consciousness by humankind in general. But again, what this will consist of is vague. Relative to the role of science in this change, some new age advocates are obviously frustrated with the seeming inability of the scientific establishment to change its views from a materialistic focus to a spiritual one. And yet, it is science that creates models of the universe, and paradigm shifts are generally the result of new scientific discoveries. If a new model of the universe does arise, I cannot imagine that people would go against science in order to adopt this model, nor would we want them to. Science is the way we collectively look for the truth about the nature of the universe. If the new paradigm is to be a spiritual one, then presumably science will be a major part of it.

The astrological cosmology affirms the existence of spirit. It further claims that matter is produced from the substance of spirit. It does not, however, oppose science.

There is evidence arising right now in the sciences, and even in mathematics, that will corroborate the existence of spirit and thereby greatly assist in the establishment of a spiritual paradigm. In science it is quantum theory and relativity theory that are moving us towards spiritualism. For instance, a particle of light (a photon) is known to have infinite energy at its center. Furthermore, the field potential of a photon in wave-form is everywhere at once. The photon has an infinite number of possible placements. Being infinite is hardly a material quality and yet light, the basis of all material creation, seems to have these characteristics.

In mathematics it is the new mathematics of infinity that is giving us our first glimpse of the new spiritualism. This area of math is also sometimes referred to as the mathematics of the continuum. The two names are integrally related because mathematicians have discovered that we cannot talk about the material quality of spatial wholeness (a 'continuum') without hypothesizing the infinite as the basis of this characteristic in matter.

Although there is currently an explosion of new ideas arising in science, it is not possible for scientists to establish a strong spiritual paradigm right now or a logically strong cosmology of any sort. That is because a cosmology is based upon how the world works, it works in accordance with the laws of physics, and science does not yet have a complete set of the laws of nature. The twelve basic symbols that astrology uses called signs are expressions of twelve potentials that exist within light for qualitative differentiation in the universe. The signs are not just one particular manifestation of some of the laws of nature; they are a complete set of the laws underlying our world. Through astrology there is an opportunity for science for the first time to have the complete set as well. In addition, because astrology explains each law so well—including their interrelationships to one another—this is also an opportunity for physicists to understand each individual law in a way they never have before.

The principles underlying the signs produce the most common and basic things we experience, such as the ability to move, to have mass, and to possess a spatial extent. More importantly, their character shows them, and the things they produce (such as mass), to be composed of a spiritual substance. They reveal that matter is produced from the infinite. Studying the signs makes it possible to understand the role of spirit in producing the world. One of the signs (Leo) even shows that, in addition to its role in the composition of matter, spirit also has a pure, unchanged, non-material presence in the center of systems.

There are no words to express how important astrology can be to every area of human endeavor. The human interests we call science and math are emphasized in this book. However, astrology also illuminates the areas in philosophy referred to as ontology, epistemology, ethics, and even religious philosophy. There is no way to include in one book all the areas of life that can be enriched through astrology. Because it expresses in such a simple and complete manner the first principles at base of our world, it is the science of sciences and the philosophy of philosophies.

If you have never taken a serious look at astrology you may have an entirely different opinion of it. Webster's defines astrology as "The pseudo-science which treats of the influence of the stars upon human affairs, and of the foretelling of terrestrial events by their positions and aspects."[1]

First of all, it is true that looking at the stars and foretelling events on earth is a part of astrology. This is possible because the solar system is a group probability field—a ray of light in waveform—and one that is linked to the probability fields of atoms and even to that of people. The emergence of quantum theory in the twentieth century has given us our first opportunity to learn about probability fields. Astrology is our chance to really understand these fields, including how they record information via a natural 'language' that exists in light.

But information about how probability fields record information is not the most important thing that can be learned through astrology. What is important is that any ray of light has this language in it because it is the language of the laws of nature; and what the laws of nature indicate, when they are explicated through astrology, is that there is such a thing as 'spirit,' and that it is active in producing the world.

Webster chose the word 'pseudoscience' to describe astrology because that is the prevailing belief. The word pseudo comes from the Greek pseudēs ('false') and pseudein ('to deceive'). A pseudoscience is something that pretends to be a science but is not. Astrology is not a pseudoscience. It is the oldest science; and the ancient wisdom underlying astrology can be the basis for a new science. If you will put aside any preconceived notions you may have about astrology, I promise to give you via astrology some philosophical food for thought, and possibly some answers to ultimate and spiritual questions you may have been searching for all of your life.

I believe that, with three basic tools, a new paradigm can be constructed—one that will be strong enough to intellectually support itself. The first tool is the twelve principles underlying astrology. These comprise what in philosophy is referred to as a category system. A category system is a list of the most fundamental aspects of experience—the simplest qualities that all systems manifest. By looking at these twelve principles as a coordinated set it is possible to construct a theoretical model of the universe—and one that is a spiritual paradigm. The astrological category system is given in the next chapter, Chapter 2.

The second tool is science. According to physicists the characteristics of the world are due to natural law. The laws of physics are the category system that scientists have used to construct the prevailing materialistic paradigm. Spirit and spiritual matters have no part in this cosmology. Chapter 2 compares the astrological category system with that of modern science. There are some close correspondences in that each of the basic laws of nature in physics correlates to one of the astrological signs (inertia to Taurus, the second law of thermodynamics to Pisces, the third law of motion to Libra, and so forth). What is different is that the astrological category system is able to show how and why each of these is a spiritual law. Furthermore, it includes several laws that are not yet known to exist. Some of these involve systematic processes and states of affairs in the world that are observable as regularities—and that a number of scientists have suggested might involve laws of nature, but which have not yet been officially recognized as such by mainstream scientists. One of these is the law that in astrology is represented by the aforementioned Leo principle. This law produces processes of centralization in systems. The centralized parts in a true system, be it an atom or a human being, can be visualized in a spiritual paradigm as due to spirit in a very pure form acting from the center of the system to unify the parts.

In addition to astrology and science, the third tool that will be utilized in this quest for a new paradigm is philosophy. The role of philosophy is to gather whatever data is available and put it together into a coherent whole. Science is a high calling, but science is specialized and hence compartmentalized, while philosophy has the freedom to include ideas from all the sciences, even to incorporating non-scientific data if it is intelligent and coherent. Philosophers are free to go where scientists would not dream of setting foot, and that includes the possibility of utilizing ancient sources of wisdom.

The Story of Creation

The twelve laws of nature being utilized to construct the astrological cosmology give a very explicit description of how the world is made. What these archetypes reveal is that something exists which I will refer to in this book as the 'substrate.' I call it the substrate because it stands under the material world. Everything is composed of this substance. It is also the potential for anything and everything that can ever exist in the material universe. It is a 'potential,' however, only from the material side of creation. In and of itself this substance is real, not a potential.

This substrate is what 'spirit' consists of. [For those who are familiar with Buddhist philosophy, spirit in this form corresponds very closely to a Buddhist description of spirit as pure undifferentiated Being and pure undifferentiated Awareness. In terms of what physicists would call this substrate, the 'quantum potential' is a good name. The well-known physicist David Bohm used that term frequently to refer to the field potential at base of the universe. This field is metaphysical but real.]

In addition to the archetypes indicating that spirit as the substrate is real, they also explain how material creation arises from out of this field. According to the principles, spirit is by nature creative. It creates by creating individuals. During a process of creation, a certain set of potentials is singled out from the infinite set. This is what a 'soul' consists of. It continues to be a material potential, just like spirit, but is itself a real substance—again just like spirit.

The material body of a particular soul entails the actualization of that soul's potentials. Thus body and soul are very closely connected. And yet the soul is a 'piece' of the substrate, and so is very spiritual in nature as well.

Most religions claim that the soul exists, and that it is spiritual in nature. But while spirit has been connected to a universal substance, the soul has traditionally been connected to individuality, since each person is said to have an individual soul. Given that individuality is connected to being material, you would think the soul would have been more clearly connected to the body, but in most organized religions there has been a disinclination to see soul as connected to matter. That is partly because matter is problematical (presumably the source of sickness, death, etc.), whereas soul is perceived as spiritual in nature. But even more directly it is because, if the soul is truly spiritual in nature, it must somehow be infinite and eternal like spirit. If soul were too closely linked to the material body, it would (presumably) not have these characteristics.

When Webster's attempts to differentiate between spirit and soul, it says "soul is the preferred term when the connection with the body is in mind … 'spirit' when an *opposition* to that which is material is in mind (italics mine)."[2] While Webster recognizes that there is a difference between spirit and soul, spirit at one point is called soul.[3] These things tell the story of where we stand today. Not only have people seen spirit as in opposition to matter, they have seen soul in a fuzzy manner as something individual and connected to the body (in some way), but also as metaphysical in nature and connected to spirit (in some way). Webster's role has always been to document our common ways of looking at and defining things. People in general are unclear about these two metaphysical concepts—both individually and in their relationship to one another—and it is easy to see why. If we do not clarify these terms in language and thought, we cannot expect more of a dictionary.

In terms of how the soul is related to the body, the archetypes indicate that the soul of a person is the 'wholeness' of the material parts. And the wholeness is not just the sum of the parts. It is a true substance of its own—a field that it may be possible to also come to understand as the soul of the subject. When a cluster of potentials is selected out of the substrate, and these have actualized materially into a centered system, the cluster then becomes a 'whole' with 'parts.' And the wholeness is not only then differentiated from the infinite substrate that it arises from; it is different from the parts as well. Being the wholeness of a set of material parts, the soul intermediates between spirit and matter. It links the absolute to that which is relative, and it provides a means for that which is universal to individuate.

Just as spirit as the substrate may be visualized as a field (the quantum potential of quantum field theory), in a similar manner the soul may be looked upon as a field as well, but as a limited set of potentials. From a scientific perspective it may be compared to a specific 'probability' field connected to a particular individual particle or system.

But that is not the end of the story. What this cosmology indicates is that, when an individual is created, their material parts surround a piece of the substrate. That point or piece of spirit continues to exist in the center of the system in a pure form. In other words, spirit as the substrate is the potential for material actualization, and the soul is a specific set of those potentials, but each created system also continues to have a piece of pure spirit within them as well—a piece that is not the soul and is not actualized as matter. This is their individual spirit. Both an individual spirit and an individual soul are a 'piece' of spirit as the infinite set. Spirit and soul are both a means to individuation, but the one remains totally non-material and hence is a spark of pure spirit at the center of the system, while the other (soul) is the bridge between spirit and matter.

Because the spark of pure spirit has absolutely no material qualities, not even that of a spatial extent, it is infinitely 'small.' And since the only thing with no extent at all is a (truly non-extended) point, it can be visualized as residing at the precise center of an individual. In a highly evolved system we speak of it as the life of the body. [The archetypes indicate that this piece of pure spirit is, not only the life of the physical body, but also the spark of individual awareness that centralizes the processes of the mind as well. Hence, even more than the soul, this spark of the divine is the source of a (real) individual ego.]

Chapter 3 is dedicated to trying to understand exactly what spirit is, while Chapter 5 involves a look at the nature of soul. The chapter in between these two, Chapter 4 ('**The Role of Science in the Astrological Cosmology**') is dedicated to the second major tool being used to construct *The New Paradigm*. It entails examining the latest data arising in physics, and incorporating this new knowledge into that of the ancient wisdom. Of particular interest is relativity theory and quantum theory.

Relativity theory posits the existence of 'singularities.' A singularity is a quantity of energy that is either infinitely small or infinitely large. In relativity theory, a particle begins as a dimensionless point or power center which eventually becomes an energy field with a specific extent. While the ideal particle is infinitely small (as a true 'center' naturally is), the ideal field is infinitely large. (The quantum potential is infinite in potential.) Spirit, because it is both infinite in potential and individualized in the center of a system, expresses both types of infinities—the infinitely large and the infinitely small. Via relativity theory it is possible to link the concept of spirit to particles and their fields, and in this way to matter.

In a similar vein, quantum theory can help explain what the soul is. Quantum theory is an effort to understand light. Although soul like spirit is metaphysical, it is much closer to being material than spirit in that it is, not only linked to a particular material body, it is also apparently composed of light. This gives an added material characteristic to soul. Matter is composed of electromagnetic vibrations, which are generically referred to as light. So is the soul. At the same time, the soul would not be different from matter if it were not different from the light making up the material parts of a body. At the beginning of the twentieth century, physicists discovered that light was dual in nature. Via the famous double slit experiment, physicists came to see light as both a spread out (metaphysical) wave and a precise confined (physical) particle. Light in particle form (as a photon) is energy, which in turn is the basis of matter. Unlike light in particle form, the real nature of light in waveform is not yet clear. Physicists are in process of trying to understand what it is. Because it is not physical energy-matter, it needs to be recognized as metaphysical, and that is what is currently causing so much confusion relative to its ontological status. Light in wave-form is spoken of as a probability field. As such it appears to be capable of directing the materialization of particles. (Not merely photons, but also atoms and molecules have a probability field as well.) As best physicists can determine, a probability field is something about a particle that affects where it will manifest in the external world, and what its internal structure will consist of. Due to the ability of the wave-form to direct the parts of a whole, a carbon molecule with an incredibly complex internal structure (consisting even of sixty or more atoms) will manage to have its own particular internal character. Thanks to its wave-form, a carbon molecule will not be a water molecule. Because of the ability of light in wave-form to direct the placement of the parts of a system, it bears a resemblance to the wholeness of a whole.

The above is only a synopsis of the story of creation as depicted through the archetypes. As can be seen, both relativity theory and quantum theory have much to contribute to this new paradigm. We will return to these powerful new scientific theories in Chapter 4 and to a more extensive look at the old archetypal view of creation in Chapter 2.

Truth and the Reconciliation of Paradoxes

As the new paradigm is developed, one of the things that will become evident is the tremendous power it has to reconcile paradoxes. There are two sides to every story, and this is true of a cosmology as well. The opposing sides often present as the two horns of a dilemma. Formulating a cosmology can be like engaging in a bull fight. If one or the other horn is grasped as the truth of the matter, and the other denounced as false, the one not grasped may come back to haunt the matador, even to gore him.

A paradox is a little different from a dilemma in that, what on the surface of things presents as two opposing sides, actually has the potential to be unified, much like the horn of a unicorn which is dual at its base but a unified point at its tip. Matter is dual in nature, but it is spiritual in essence. The unicorn is a symbol for the reconciliation between matter and spirit.

In the preface it was mentioned that philosophers generally tend to align with one or the other of two camps—that of the relativists versus that of the absolutists. The absolutists affirm that infinite and absolute being exists. But they also tend to believe that this state of being is the ultimate or only (real) reality. The relativists, on the other hand believe that the relative world of change and flux—the material universe—is the only reality.

The AC is able to show that both are real, and that matter is an expression of that which is infinite, absolute and eternal. The manner in which the AC does this will become clear in the chapter on '**Soul**.' It is precisely because soul mediates between spirit and matter that it has the type of nature which can solve this absolute-relative dilemma. New information and knowledge can raise the elements of a paradox up to a higher level of understanding—a place where the seeming opposition is reconciled.

Because of this characteristic of the truth, contradictions and inconsistencies in a paradigm are a signal that it is time for a change. In addition to those presented by the absolute-relative controversy in philosophy, the new physics is currently dealing with some major contradictions of its own. In relativity theory the ideal particle is a dimensionless point. Tiny particles manifest 'singularities' at their center. Quantum theory currently opposes relativity theory by claiming nothing can be non-extended or infinitely small in the way that a true center as a singularity can be. It insists that particles begin with a certain very small extent. The search for a unified field theory is the search for a means to reconcile relativity theory and quantum field theory. What the problem boils down to is basically trying to explain how a very small extent (a quantum) might derive from something that is infinitely small and basically non-extended (a singularity). We will return to that problem in Chapter 4 on **The Role of Science in the New Paradigm**. Science could greatly benefit from a cosmology which has the potential to reconcile this paradox.

In addition to the absolute-relative and infinite-finite problem, another paradox the AC has the potential to reconcile is that of individuality versus universality. When New Age advocates speak of a new, spiritual paradigm emerging, they connect the concept of 'spirit' and being 'spiritual' to a substance and state that unifies everyone. They correlate the idea of a new paradigm to a chance for people to finally see that they are one.

Buddhism has encouraged the idea that individuality opposes universality, and hence true spirituality. The idea is that any spatial division or qualitative differentiation separates things. Matter is differentiated, and individuals are separate from one another. Spirit on the other hand is seen as an undifferentiated, undivided substance. The goal of Buddhism is to renounce the individual ego and the material world in general. The very term 'ego' has been connected to selfishness. The ego and the material world are seen, not only as in opposition to spirituality, they are said to be inherently unreal. That is because spirit is visualized, not just as an undifferentiated substratum, but also as ultimate and absolute Being, so that anything individual or differentiated then is seen as partaking of non-being.

Quantum field theory has reinforced the view that individuality does not exist. There is actually no such thing as a true individual system in the current scientific paradigm, not even an atom. When, via the electron microscope, scientists were finally able to look into the composition of an atom, what they saw were atomic particles within the atom—protons, neutrons and electrons. What the 'atom' then came to be seen as was a cluster of interactions between these particles. After that initial look into the composition of the atom, later it was possible to look into the composition of the particles within the atom, and they presented as composed of more primary particles involved in interactions between those particles. In going farther and farther down, scientists finally came to visualize the interactions as the reality instead of the so-called 'particles.' In other words, instead of seeing the world as composed of systems such as atoms (people, etc.) *possessing* qualities now there is no 'atom' at all as a system or subject in possession of qualities. Rather the world is seen as composed of a vast field of energy interactions with some more or less intense clusters that we give the name 'atom' to or 'atomic particle' or 'person.' These energy interactions are the laws of nature and they are the qualities the world possesses. They are what we speak of as matter. But only the world as a whole possesses these. There is no true subject.

One of the most important concepts that arises when looking at the archetypes—and at Leo in particular—is that spirit, not only unifies the parts in an individual, it is also responsible for the existence of individuals in the first place. When comparing the astrological cosmology to that of the current scientific cosmology, the primary difference is not just that one is spirit-based and the other matter-based, but also the difference between there being individuals—systems or subjects *in possession of* qualities—versus a universal, disembodied electromagnetic field.

The problem of whether or not there are subjects in the world is actually not a different problem from trying to arrive at a 'spiritual' versus a 'materialistic' model of the universe. That is because having a 'spirit' is the means for a cluster of qualities within an electromagnetic field to be a subject. The astrological cosmology indicates that there is something (a force of unity) at the center of an atom that centralizes its parts—something that it may be possible to (rationally and legitimately) define as an individual spark of spirit. (The astronomical body connected to Leo is the sun, and from antiquity the sun has been said to rule the spirit in a person. Leo has also been said to rule centers. In esoteric tradition the center of the physical body is said to be the 'solar' plexus.)

As a law of physics, this law is not 'spirit.' All natural laws are material laws. They have to do with processes in matter. Nevertheless, the center-seeking processes produced in a system by this law make it appropriate to speak of it (and hence of the sign Leo) as an 'immediate mode' of spirit. Of all the twelve basic laws that govern our world this one best explains what spirit is in terms of how we ourselves may experience it. It is this law that helps define spirit as a means to individuation. An understanding of this law can bring back the concept of true individuals. Atoms and people can become real again and no longer merely intensified clusters of activity within a unified electromagnetic field.

What the astrological cosmology indicates is that spirit is an undifferentiated, non-individuated substance—but that it is also a creative power. To 'create' is to create a material universe. 'Matter' is always qualitatively differentiated. Furthermore, there is no matter that is not the part of some system—be it an atom or a human being. Even the photon (which is energy rather than matter) is a system. It is a ray of light with parts (electric and magnetic parts, themselves involving positive and negative poles).

In the process of material creation, the paradox of individuality-versus-universality is reconciled. This occurs in part because an individual spirit both is and is not a 'piece' of spirit as a universal substance. It is matter that allows a system to experience spirit in a partial manner. Hence it is the body that produces what is, for that body, an individual spirit. In essence, spirit at the center of an individual is not apart from the spirit in every other individual or from spirit as the substrate. Thus everyone and everything is united at heart. Spirit is an infinitely 'small' point in the center of each person. It is the same metaphysical substance that is in the center of every system. And all these individual spirits are the same infinitely 'large' potential at base of all creation. Thus all systems are not only united spiritually with one another, they are united with the substrate as well, precisely because the two 'poles' of spirit (universal and particular) are one.

The paradox of individuality versus universality is reconciled by soul as well as by spirit. The unity of material creation takes place via part-whole duality and hierarchy in nature. Each system is not only a whole *with* parts. It is also a part within higher level wholes, and ultimately a part withhn the universe as a whole.

But there is more to the universal nature of each soul than just that. By linking soul to wholeness, the archetypes explain what 'wholeness' is in a way that has not been explicated for thousands of years. They show how a personal soul, which is only a particular, limited set of potentials can, at the same time, be infinite. And I will not spoil the surprise element of the paradox by explaining prematurely how soul reconciles this seeming contradiction. The answer is in Chapter 5, which is specifically dedicated to the subject of '**Soul**.' In accomplishing this, and precisely because soul is the bridge between spirit and matter, a better understanding of the soul reveals that matter itself is infinite and eternal, which sounds like a contradiction in terms, but is just another paradox that can be reconciled. What I will say at this point is that the astrological principles never cease to amaze me relative to the depth and breadth of what they can explain—and how they do it. The cosmology that can be extrapolated from the ancient wisdom at base of the twelve signs is a more rational, coherent and surprising cosmology than any presented by any philosopher, ancient or modern—and that includes those two incredibly wise men, Plato and Aristotle. In addition to being a philosophical delight, the elements of this cosmology in no way contradict the latest ideas arising in physics, and in fact help reconcile contradictions that do exist.

Searching for Truth

We have just been talking about a major criterion for truth—the fact that it has the ability to reconcile paradoxes. Contradictions can be a sign that something is wrong. When there are two opposing sides, one may be true and the other false. On the other hand, truth just as often lies in the middle of extremes and overcomes seeming oppositions—and not as a watered down blend of the two, but as a true reconciliation. When contradictions are looked at as polarities instead of oppositions, they can sometimes be united into a higher level of understanding.

A criterion for truth is a means to judge it. Philosophy in its sub-discipline called 'epistemology' actually studies the nature of truth—the signs to look for as we pursue the truth about anything. If you really are searching for truth, you must become a demanding reader, and you begin by consciously understanding every known criterion for that which is true—before you read a book—even and especially before reading this book. When there are two competing paradigms, people have to choose, and that includes the reader as well as philosophers and scientists. The best way to make a good choice is to understand as much about truth as is humanly possible. We all have a duty to know what truth is and how it can be found—and to use this knowledge in a *conscious* manner. Some of the attributes of truth are listed below. As for my role in actually respecting and using these attributes or criteria, this book proceeds on two levels. One entails building a new, spiritual paradigm based upon a comprehensive look at the available data. The other involves explaining why it is very likely that the constructs and conclusions are true based upon what is known about truth. That is a very different process from many books, and so it becomes important to understand the process as well as the content. That type of process and grounding mechanism is especially important in this particular book because *The New Paradigm* links science and philosophy to questionable ideas like ancient archetypes and astrology.

Relative to the reader's role, if at the end of this book you think that a spiritual paradigm may fulfill these criteria for truth better than the current materialistic paradigm, consider the possibility of allowing a new, spiritual paradigm access into your own body of beliefs.

Truth is not an amorphous thing. It has features, and its characteristics are known. They have been for millenniums.

There are, for instance, *mathematical proofs*. The renowned mathematician Georg Cantor proved that it takes an infinite number of points to make a continuum continuous. The reason that mathematical proofs are a means to truth about the world is because matter itself is rational. Its qualities and parts are coherently interactive, and the interrelationships that numbers possess relative to one another reflect the nature of the material universe.

There are also *empirical* tests for truth. We observe what is going on in our world. Sometimes we set up experiments to see what will happen under certain circumstances. The famous 'double slit' experiment is the basis of quantum theory. This experiment observes light as it goes through two slits. It proves that light is both a discrete particle and a spread out wave. *Empiricism* is primary to the scientific method of proof, but science utilizes all the other means as well. That is because observations have to be interpreted, and different scientists can interpret the same data in different ways. The other means to truth assist with correct interpretation so that the explanations can be truly strong.

What is true is *consistent* and *enduring*. If we do the double slit experiment sixty years from now it will still work just as it did sixty years ago. If we do the same experiment a hundred times, light will still act like both a particle and a wave. (Endurance is one of the reasons extremely ancient ideas should never be dismissed out of hand. There is such a thing as the 'perennial wisdom.')

What is true is *coherent* with other known things about the universe. That is because we live in an orderly universe. The attribute of coherence as a mental tool is closely related to 'being logical'—from the term logos, which refers to the rational principle in both mind and matter. As logic is to the mind, so law is to matter. Each part of the world, each law of nature, is an integrated integral. The laws interact and fit together like the parts of a whole—because that is what they are. Because matter is just what is meant by 'law,' the natural world is naturally lawful and coherent when created right and naturally coherent intellectually when seen correctly. When two bits of data about the world do not cohere with one another (assuming they cannot be reconciled in some way), one is 'false'—and the one that is false is the one that does not fit with the others. No piece of information should conflict with our overall understanding of the world. [Since the material world is lawful and rational, some have supposed that something 'spiritual' might be inherently irrational, but (in a down-to-earth spiritual paradigm) spirit is the source of matter, and matter is an expression of spirit. From this perspective, knowledge of spiritual matters will not conflict with facts about the material world, but rather augment and enrich that data.] If there is to be a coherent model of the universe, then everything we know about it must be capable of orderly integration.

Truth is also *sound*. Soundness is related to coherence but with the added quality of attention to detail. In a structure that is sound, each element is secure. When working out the attributes of a new paradigm, even if the picture being presented is true, it is not sound if each new brick in the structure is not carefully worked and mortared into place. That requires thoroughness, sometimes even redundancy, since each new element introduced then must be again related to all the others that have been put in place thus far. In this book we will go over some data again when the whole is looked at from a new direction. (We may even go over these criteria for truth again, since they are so important.)

Truth is *simple*. Given two explanations for something, the simplest one is generally the true one. The reason that simplicity is a criterion for truth is because, first of all, there are only a few simple laws of nature that have produced the world. Secondly, these are unified with one another because unity is at base of their creation. The search for truth is a search for the highest level of simplicity and unity. To give an example: Material causes and interactions have traditionally been seen as contact initiated. One thing touches another when it 'causes' it to move or change. Causes do not occur at a distance. Yet physicists have recently discovered that two things can sometimes be mutually affected (instantaneously) that are on opposite sides of the world and not in material contact with one another. (They call this non-local causation.) If we find that there are causes or interactions that occur when two things are not in material contact, the simplest explanation may be that there is a third, non-material cause or substance that is between them and makes the connection possible.

Truth has *patterns or regularities* to it, and *regularities that repeat* consistently. For instance, relative to fields of electromagnetic interactions between systems, we know that there is a microscopic electromagnetic-like field within the nucleus of the atom, an electromagnetic field between atoms, and a gravitational field at the macrocosmic level of being (which itself may be electromagnetic-like). This is a pattern of ever higher repetitions, and the levels are hierarchically arranged. ('Hierarchy' refers to a series of repetitions of some state of affairs where each level is nested within the one above.) It is the same with the hierarchy we observe relative to systems themselves. These occur in a nested hierarchy of nucleons within atoms within molecules, within mega-molecules, within cells, within organs, within systems ... In the heavens above us there are solar systems existing within galaxies, and the galaxies are within meta-galaxies, which in turn are within meta-meta-galaxies, and on and on, beyond the ability of science to follow their progression upward. Patterns that repeat themselves are caused by something consistent and important, and the pattern is a signal to look for a cause.

Truth is *beautiful.* This sounds more like an aesthetic perception than a guide for truth, but it has been remarked upon so often by renowned scientists and mathematicians that it bears mentioning here. Truth and beauty both involve simplicity, patterns and coherence. Patterns in nature are beautiful, and the beauty is reflected even in the mathematics of a theory in physics. If the math is cumbersome and ugly, the theory is suspect. The special theory of relativity is a beautiful and straightforward four dimensional equivalent to the Pythagorean Theorem, one of the simplest, most elegant, and yet far-reaching mathematical formulas in history.

Truth is *comprehensive.* No model of the universe will be complete. We are constantly growing in knowledge. The point is to encompass everything that we do currently know, and in a simple and coherent fashion. The more things the new paradigm can explain (as opposed to the old paradigm) the more likely it is that it is true.

Truth *embodies reality,* which is to say true being. It is *the knowledge of that which is real.* Because of this, truth is *powerful.* The truth upholds itself with the strength of its structure and nature. Because it embodies true being, that which is false cannot stand against it.

Truth is *fruitful.* Reason builds upon reason, and truth upon truth. Like a well-rooted tree branching ever upward and outward, a firmly planted truth leads to still more ideas and revelations because of its connections to the rest of reality. Things which are true embody pure Being—the spirit of Truth. When knowledge sprouts from this seed, there is an organic quality to the truth, helping it to live, to endure, and to grow.

Because it is fruitful, truth *furthers the good* and has the power to transform the world.

These then are the characteristics of that which is true. The pages ahead will continually discuss them, respect them, and adhere to them. If you find that this new spiritual paradigm fulfills these criteria better than the old materialistic, mechanistic paradigm, consider the possibility that it is true.

Before moving on to Chapter 2 there is a particular procedural element that I feel the need to mention. When something is written in philosophy that is purely hypothetical, any such statements are usually prefaced with, "It could be … ," "It may be … ," "It might be … ," and so forth. I try to do this in the beginning, but after a while I cannot keep it up. It is not only tedious to preface everything with "It may be that," it feels dishonest. The things I will be sharing with you I have come to really believe to be true. As each chapter unfolds I promise to continue always to point out the concurrent criteria for truth that each added concept fulfills, but as the comprehensiveness of the overall case builds you will see me fail to use "It may be … " more and more. Do not let this discomfort you. The most important thing about truth is that it entails a personal journey that you must work for and come to know in *your* heart. The spirit and ground of truth in the center of you must ring true. Listen to your own heart. I am merely speaking mine.

Although each person must find their own way to the truth, there is a truth of the matter, as well as a common way to find it. Taoists call this the Tao, which means 'Way.' They see it as something that governs people in right ways of thinking as well as right ways of living. We are each a part of the Tao. We each have our own path to follow. But precisely because each part *is* a part, our unique path to the truth is still a collective journey. Because of our ultimate union, we can help one another. There are other people thinking about how the world is made and the meaning of life. As we move along in this book I am going to reference those people who have helped me along the way. Even when I do not reference them, I am including their books in the bibliography. I do not recommend any of them unreservedly.

Each person only has part of the truth, including myself. But that is where comprehensiveness and the reader's own ability to see any underlying coherence in different parts of the truth are important.

Chapter 2

The Astrological Category System

The purpose of *The New Paradigm* is to assist with the development of a new, spiritual cosmology to replace the old materialistic one. The signs of the zodiac are the means to do this. For that reason, this chapter is the cornerstone of this book. These twelve fundamental principles are examined at the end of the chapter, with each one discussed individually, including the sign Leo that was mentioned in Chapter 1 as having to do with the spirit in a person.

As important as it is to explore the spiritual nature of these principles, it is equally important to understand from the beginning what they are—the very idea for instance of what it means to say there are twelve fundamental types of being creating our world. Why only twelve? What are they? Where are they located? Are they something up in the sky (as astrology would indicate)?

The questions that came to mind when I first started studying the signs, not just as an astrological phenomenon, but as themselves representations of ultimate principles behind the creation of a world, included the following: (**a**) In what way is astrology involved? (**b**) How are these principles connected to the laws of physics that scientists see as responsible for the characteristics of the world, and to light, the substance from which everything is supposedly composed? (**c**) Is it possible that there are just twelve simple and fundamental types of being? (**d**) If this is true, what are the criteria for figuring out what the simplest types of being consist of? For instance, are these simple in that they are not combined with other things? (**e**) If these exist, and if they are simple and fundamental (and only twelve altogether) how does complexity arise in the world? Is it, for instance, through combining simple things? (**f**) Can really simple (uncombined) things exist in the relative, material world? (**g**) If these can produce a world that is actually spiritual in nature, what makes them capable of doing this? How are they related to concepts like spirit and soul?

Relative to the first question (**a**), these principles are not merely up in the sky as an astrological phenomenon. As will be discussed more in depth in Chapter 9, the astrological phenomenon involves a ray of light that is a unifying factor within our solar system. Everything is composed of light, so these fundamental principles are present in every ray, whether it entails those involved in the nucleus of our atoms, or those above us, even beyond us in higher dimensions.

Relative to the second question (**b**), the twelve signs of the zodiac do appear themselves to be an expression of twelve fundamental types of being by means of which the world is created. The simplest and most basic things we possess and experience are manifestations of these fundamental principles, such as our ability to move, to cause other systems to move, to possess a spatial extent, and to connect spatially with other systems. These twelve types of being are the laws of nature that create form in matter giving it the qualities it possesses. And it is not that these are external forces having an effect on matter. The laws of nature are no longer viewed in that way. They do not just cause things with qualities to move and change in certain ways; rather they cause systems to possess the qualities they do because they are the qualities expressed in a dynamic fashion; hence the 'laws' exist internally as well as externally. The totality of these laws is just what is meant by 'matter.'

These laws of nature are inherent in light. Physicists now view the universe as composed of a vast field of electromagnetic energy vibrations that are generically referred to as light.

We are composed of these energy vibrations. Hence the twelve types of being are, not only inherent in light, they are the parts of our bodies.

Although the twelve principles do correspond to laws of nature, it is more useful to look upon them as basic types of being rather than as laws of nature. For instance, one principle makes it possible for motion to exist; but we do not have a 'law' of motion. We have laws that indicate that things move in this and that type of way. Each type of being is a power as well as a substance (precisely because light is a power as well as a substance); but it is too limiting to look upon them as types of movements (or types of forces), when motion itself has to be accounted for (and 'force'). The idea of their being types of being widens the breadth of what they are. At the same time, they do account for every known law of nature.

Examining the character of the laws of physics through and by means of the signs of the zodiac presents a somewhat different view than that of a normal scientific one. First of all, the entire set of these twelve types of being show clearly that they derive from a substance that can be looked upon as pure Being, rather than as a type of being—and as a substance that is metaphysical or spiritual in nature. Secondly, the type of being called Leo indicates that, in addition to matter being composed of the substance of spirit, spirit in a pure form also exists in an unchanged state within material systems at the center of their nexus of parts. Although Leo is itself merely a material law of nature—one involved with processes of centralization in matter—by studying this sign it is possible to know that spirit exists, what it is like, and where it exists within our own bodies. In addition to affirming, via the sign Leo, that we have a spirit, there is a principle called Aquarius that makes it possible to know that we have a soul. This sign explains how the soul is different from spirit, what the soul does, and where it is situated, including its relationship to the body. This thought, however, gets down to the last question and, however important a spiritual view of the laws of nature may be, it is important to ask the simplest questions first.

The first of these is (**c**), "Is it possible that there are just twelve simple and fundamental types of being?" The answer resides in the fact that (despite the complexity in the world) there are only a few basic laws of nature. Because there are not that many laws of nature, and because these are just what is meant by matter and its qualities, this is one piece of evidence that material bodies may be composed of just a few simple types of being.

Long before modern science began to refer to 'laws of nature,' and to see these processes as just what is meant by energy-matter, philosophers had been asking questions like the above, and actually had constructed sets of the most fundamental principles they could come up with. They refer to them as category systems, since the idea is, not only to devise a set of the most fundamental qualities of experience, but also to categorize these types of being relative to one another, and (ideally) to explain how the complex day-to-day world is produced through the set involved.

As for the fourth question, (**d**) what the criteria might be for figuring out the simplest and most basic types of being, to decide this it is necessary to think about what it means to be 'simple' and 'basic.' 'Simple' implies uncombined with other things, and 'basic' has a similar denotation with the added idea that this is something from which other things are derived. One way to come up with a list of fundamentals is to consider the simplest things that everything possesses, like a spatial extent and the ability to move.

When the actual astrological category system is discussed, it becomes evident that there are two 'domains' or meta-categories that I call the 'kinematical' and 'quantitative' domains respectively. The former has to do with our capacity to move, and to move other things. The latter has to do with measurements and 'how much.' (Our spatial extent and how far we extend out into space is one of the most basic types of 'how much' we possess.) These two domains are, not only very simple and fundamental;

they are so basic that physicists currently explain every quality in our world in terms of these two. Thus the quality of being 'yellow' is described as an electromagnetic ray with a certain frequency. ('Frequency' correlates the notion of a specific color to a wave of energy or process with a specific extent or size to its wave-like undulations.)

In the astrological category system there are three meta-categories. These are kinematics, quantity, and quality. As noted, I speak of them as 'domains.' The twelve basic types of being all fit into one of these three higher-level sets. Each domain has four types of being that operate within that domain. For instance, the four in the kinematical domain are: action, change, reaction-exchange, and non-change (inertia, rest). In the quantitative domain the four basic types of being present as attributes of space, since space is the simplest and most fundamental 'how much' that exists. The four principles in this domain are spatial extent, spatial content, spatial connection and spatial separation.

I prefer to call these domains rather than meta-categories because, although there are categories into which the basic types of being can be grouped, the three domains are not specific, material qualities such as being able to move. When constructing a category system, it is important not to mix categories. When classifying things, there can be categories and meta-categories. The Greek philosopher Aristotle constructed a ten-fold category system that included quantity, quality, relation, place, time, position, condition (state), action, passivity, and Being per se, the latter of which he called ousia (*Categories* 1b25). Action as the ability to move and move other things seems very basic. So also does inertia or passivity. But these complement one another, and might be better understood as belonging to a meta-category of kinematics, which can incorporate things like motion, change, inertia, and the reactivity and exchange of energy involved with causation. 'Quantity' and 'quality' also present as meta-categories. For instance, if you think of quantity in terms of spatial extent, it can include things like having a spatial extent, the separation of different spatial extents, and the ability to make connections between separated extents. As for ousia or 'Being,' this seems like a category even above kinematics, quantity and quality, since anything in any of the categories would possess being.

The fifth question that comes up is (**e**), if these types of being are simple and basic (and only twelve in number altogether) how does complexity arise in the world? If there are only a few basic laws of nature, and the complexity in the world arises from these, this needs to be explained. The astrological category system is extremely strong and fruitful in this respect in that it shows how material complexity arises from out of an underlying simplicity. To briefly summarize, complexity occurs by means of combinations, layering, and even differing vehicles. To begin with, although the twelve types of being are simples in and of themselves, it is possible to combine them. For instance, motion can be combined with the quantitative domain to produce things like 'speed' or 'velocity,' i.e., the *amount* of motion in a context, or how fast something moves relative to something else. This act of combination can be exercised in many different ways, as occurs for instance with circular motion. Moving in a straight line involves the categories of motion and non-change combined with the quality of direction. Moving in a circle, on the other hand, involves the categories of motion, change, direction, and constancy of direction of change. 'Constancy' is an expression of non-change. While the direction of motion in a circular motion is changing, the *degree* (quantity) of change of direction is not changing.

These simples and their combinations also manifest at different levels, and by levels is meant such things as complexity and evolutionary levels. There is a vast hierarchy of nested systems within our world, and two astronomically different levels include that of atoms versus that of people. In addition to atoms moving, people do as well—and they do so as a whole. Motion is a fundamental type of being at both levels. In addition, the more complex a system becomes the more expressions there are of a particular type of being. Consider for instance inertia. A human habit involves a lack of change, which is a human-level expression of the atomic-level law of inertia.

The basic concepts can even be used in a metaphorical sense. We can talk about people 'going around in circles' and being stuck on 'wheels' of self-undoing relative to drug abuse, etc.

In addition to combining and occurring at different levels, the basic principles are also utilized in different kinds of vehicles, such as the physical body and the 'body' of our ideas. The twelve basic types of being are indispensable to our formation of mental concepts, and that idea will be elaborated upon in the pages to follow.

Mental matter can be combined just as is the case with physical states of affairs. Once the mind has a certain set of basic ideas then it can use these in innovative ways. The mind is very creative and can even use the basic types of being in ways that are not a reality in the physical world.

Because the mind is so mobile and innovative, mental constructs can and often do precede physical ones. However physical states of affairs also precede and affect mental ones, as occurs when a certain state of the physical body affects the mind. These basic types of being are not purely mental concepts or physical facts, but rather the basis of both.

There is a school of thought in philosophy called 'mentalism' or 'immaterialism' which asserts that all reality is mental, and that the external world is a projection of mind. 'Physicalism' is the opposite view wherein mind is seen as simply a brain state. Most Western scientists adhere to physicalism. Many Eastern scientists believe a mentalist cosmology may be more appropriate. Truth generally reconciles extremes, and in the cosmology being formulated via the astrological category system there is both a physical realm and (in highly evolved systems) a mental realm. Astrology indicates that the law of inertia (which has to do with mass) and all the rest of the laws of physics exist in both the physical and mental realms. Thus it is possible to speak of mental matter as well as physical matter.

Different combinations of simple concepts, layering and differing vehicles are just a few ways in which simplicity expands into complexity. The above manner in which the AC explains complexity will be returned to in Chapter 5.

The very fact that the basic types of being can be expressed in so many different ways, and in particular that they can occur in either the mental or physical realm, makes it seem indeed as though both mind and body might be following the same design, perhaps in terms of certain fundamental archetypes. What makes this seem even more possible is the fact that there are many expressions of one particular type of being even at the same level of being and in the same realm. Take motion for instance. There are many different systems manifesting particular expressions of motion, and in many different ways. In fact, in the material world there are no two alike. Is our experience and idea of motion per se just an abstraction from these, or is there an archetype of motion that makes any particular kind of motion understandable to us? A particular archetype producing motion could be even more basic than motion itself. When looking at motion in an even more fundamental manner, it can be viewed as a 'going forth.' When using this archetype to conceptualize directions in space, the concept of 'forth-ness' can be used to understand anything 'in front' or 'to the fore' as opposed to 'in back of.' Relative to time, the same idea of forth-ness can be used to understand what a beginning is. [In astrology all things that go forth (like motion) or are to the fore (like beginnings) are said to be 'ruled' by the Aries principle.]

There are several reasons for postulating the existence of archetypes. The most important of these is the simplicity mentioned in the previous paragraphs. Closely connected to simplicity is the characteristic of having a definite identity. First principles need to have a distinctive identity. Of all of our concepts these must be the clearest. They need to be self-evident, self-explanatory and self-contained. Each needs to be only one thing, each has to have a strong identity factor, and each needs to be definite in nature. When manifesting in the material world, they are relationships and are relative to one another, thus forth-ness is related to back-ness and a beginning is related to an end. (They also, as noted, combine with one another to produce the complexity of the world.) Unless each of the basic types of being has a non-relative archetype behind it to give it a stable base, it may not be possible for materially actualized types of being to have a definite, understandable nature. The problem of identity includes the question of how the millions of actualized types of being of the same type can be related to one another, and how we can know mentally that they are related.

The Greek philosopher Plato thought about the need for the material types of being to have a non-relative (absolute), unchanging base somewhere. Plato was influenced by several Greek philosophers who had preceded him, including Zeno and Parmenides. These had addressed the issue of change as it relates to identity, and Plato expanded upon their thoughts, developing a hypothesis called the Platonic 'theory of Forms.' This theory has challenged philosophers for several thousand years now. It has divided even modern philosophy into two camps—that of the absolutists versus that of the relativists—with the absolutists agreeing with Plato that we need to postulate absolutes in order to make sense of reality.

Plato's hypothesis is a philosophical classic relative to the subject of identity, and what identity has to do with simplicity, unity, non-change, and absolute being. As identity relates to a simple kind of unity, Plato believed that for something to have an identity it had to be specific, and that something cannot be something specific if it is a multiplicity of things. An archetype in particular would seem to need to be a clear, simple, uncombined unity.

Plato also thought about the role of non-change. For something to have an identity it needs to be still, and be what it is. The material world involves constant change. Change makes things become other than what they were even just a second ago. Everything in the material world moves through space and time, and changes from moment to moment. Only something eternal (eternally the same) would not do this. Identity requires a persistent and enduring qualitative self-sameness. (Sameness is a qualitative type of unity.) According to the logical principle of identity a = a. What is, is. To be what it is something must not be in process of changing into something else. Identity requires stability.

As for the requirement of being absolute, physicists now view matter as a set of dynamic interrelationships. When material parts and qualities are relationships, they derive their nature from the things they relate to. According to the theory of relativity, any particular material thing derives its nature from its relationship to *everything* in the universe. For identity purposes, an archetype may need to be absolute rather than relative.

However, as soon as something is thought of as absolute, it presents as having a non-relative, non-physical nature. And that brings up the sixth question (**f**) can really basic, simple (uncombined) things exist in the material world? What is non-relative and non-physical is metaphysical. The concepts of spirit and soul are connected to metaphysical substances, and that brings up the last question, which is (**g**) what do these twelve founding principles in our world have to do with spirit and soul?

When trying to understand these things, Plato came to the decision that things in the material world had the character they did based upon absolute, eternal Forms that 'enformed' material things of that type. Plato believed that it was philosophically necessary to posit the existence of these Forms if things in the physical world were to have the nature they did and, mentally speaking, if we were to understand things in the world and form clear concepts.

Later philosophers began to call the Forms 'universals.' These were seen as absolutes, and the philosophers who agreed with Plato that they must exist may be looked upon as 'absolutists.' 'Relativists' on the other hand tend to believe in the existence only of the relative, material world. They believe that things can have an identity even though they are constantly changing. They hold that something can be known and be what it is even though its character is not only changing, but also is relative and depends upon everything else in the universe to be what it is. Relativists tend to believe that motion per se for instance is an abstraction we make in our mind based upon our experience of all the different kinds and different expressions of motion we encounter in the material world.

Absolutist philosophers, who believe the Forms really exist somewhere, are themselves divided into two different camps—the realists and the absolutist conceptualists. When philosophers use these two terms they are trying to decide *where* the Forms or universals exist—whether they exist in the physical world, in the mind, or somewhere else. Plato was a realist and believed that the universals or archetypes existed in a realm of Forms apart from both the physical and mental worlds. Absolutist conceptualists believe that the Forms exist in the mind. They are 'absolutist' conceptualists because they believe the Forms, which are absolutes, exist. They just differ in that they place them in the mind.

Because they were conceived of as absolutes, it was very difficult for Plato to say where the Forms existed in a material world, or in relationship to people, and he finally, as noted, just placed them in a realm of their own. Because mind is not physical, absolutist conceptualists on the other hand thought it would be better logically to place the Forms in the mind. However, some of these conceptualists apparently realized that thoughts entail motion, change, arrangements of complex ideas, and all the things that the physical world entails. They reasoned that there could be nothing eternal and absolute in a relative, changing world of thought. This led those absolutist conceptualists—especially the Latin medieval scholars referred to as the 'scholastics' (e.g., St. Augustine)—to speak of the Forms as existing in the mind of God since God was conceived of as absolute and eternal in nature.

While his thesis of the Forms had a strong logical basis, one thing that landed Plato into difficulty philosophically was his idea that everything had to have a Form if it was both to have a specific identity and (mentally speaking) to be identified as whatever it was. Thus circular motion would have to have a Form of circular motion, a table a Form of table-ness, etc. This led later philosophers to say that he was merely repeating everything in a world of the Forms just so as to explain how material things could have a clear, stable identity. This would require millions of Forms.

For this reason, the astrological category system, with its twelve simple Forms and their capacity (when material) to combine, is much stronger philosophically, and stronger also relative to where the Forms exist. In the astrological cosmology people have an absolute aspect to their nature.The archetypes exist in each person's spirit and soul. Both spirit and soul are material potentials in the sense that they are the substance from which actualized material types of being emerge. Spirit is unlimited in the number of potential types of being it serves as the basis for, while soul entails a specific set of archetypes. These are a set of potentials being actualized in a particular world. (In our three-dimensional world there are twelve in number.)

Because he could conceive of nothing absolute within the material world (or within people when they were in the material world), Plato separated his Forms from both. This especially bothered Aristotle who, although he was a student of Plato, realized that the Forms had to do with the qualities that things have. Aristotle reasoned that, in the real world of experience, there is no quality that is not possessed by some system.

In addition to Plato making the Forms disembodied entities, what also bothered Aristotle was Plato's treatment of Being per se as a quality. Plato actually postulated a hierarchy of forms. Ordinary Forms he said gave both character and being to material things. But the Forms themselves had to get their being from somewhere. Because ordinary Forms were involved with *types* of being, he believed that Being per se had to be a higher Form, so that the lower level Forms might exist. Aristotle objected to Plato making Being a Form and hence an archetype for a quality. Aristotle not only thought that qualities (and their archetypes) needed to be attached to subjects; he believed that Being per se was just what is meant by a subject. [In the astrological cosmology spirit and soul are the basis of being a subject, and these are both types of absolute being. The subject's spirit and soul are responsible for the Being (and the Unity) of their material parts, the types of being.]

Both philosophers thought about Being per se and realized it was something above and beyond the types of being. Plato conceived of it as a higher category—the Form of pure, absolute Being. And in one way it makes sense to see Being as a higher category, since in the astrological category system the twelve basic types of being are all types of *being*. On the other hand, there is the problem that Aristotle pointed out with seeing Being as a quality (or even an archetype for a quality).

Aristotle himself got into numerous difficulties when trying to figure this one out. Aristotle called pure Being 'ousia' which is translated as 'substance' (*Metaphysics* 1034a). Aristotle conceived of ousia as that which gives form to the body, and as existing in the *domain* of the soul, but he also wanted to relate it to subjects as well, and to the material world. If you are an absolutist philosopher, this is the question of whether Being per se (and the archetypes) exists merely in the domain of the soul or actually in each person's soul. It is also the question of what the soul consists of, where it exists, and what it has to do with *this* world and *this* body.

As with Plato, Aristotle realized that, of all the universals, pure Being needed to be non-relative (absolute) and non-changing (eternal). He began by visualizing absolute Being as a substratum. Plato had claimed that pure Being was the source of the being of the types of being. Aristotle likewise said that the substrate was the starting point of everything (*Metaphysics* 1034a). The substrate was the source of the qualities in our world—of the matter—but itself had no qualities and did not change. Rather, it persisted through change (*Physics* I,5-7) It had no relativity, opposites or contradictions connected to it (*Categories* 4a10).

One of the problems with the idea of an amorphous 'substrate' is that it leads the mind away from the idea of ousia somehow being the basis of individuals—of beings. It then does not seem to be personal—more like some sort of a prime matter.

When trying to figure out whether this substance was material or non-material, Aristotle played around with the possibility of ousia being a potentiality (*Metaphysics* Z,7-9). That way matter could possess ousia without ousia actually being matter. The reason that idea may have been rejected is because in the end Aristotle was persistent in his belief that it had to exist. How could Being not exist? Existence is behind the very meaning of the words 'to be.' Parmenides had said that Being is. What is, is. It is not a potential. [In the AC the soul is a set of *material* potentials, but it is a real substance in and of itself. Both spirit and soul are precursors to the creation of energy-matter. The soul is a particular field potential. In the language of physics, when referring to light, this soul potential is the probability field of a particular system. Furthermore, the soul is light in wave-form rather than in particle form. Some physicists believe that the wave-form of light does exist, others that it is just a potential that exists in matter (cf. Chapter 4).]

As noted, Aristotle felt ousia had to be in a being, and had to be the basis of being a 'subject.' Subjects exist. But, if absolute Being is in a subject or the basis of being a subject, the question keeps arising as to how anything absolute can exist within a person in the material world. In his theology Aristotle came to connect pure Being with God, whom he looked upon as the ultimate substance or ultimate ousia. (He called God the Unmoved Mover.) As noted, this works well philosophically since God is perceived as an absolute, infinite, eternal reality. (Aristotle also decided that God, if he 'existed,' would have to be pure actuality and not a potential in any way.)

In addition to using God as an explanation, these philosophical dilemmas are reconciled even better when individuals are perceived as having a spirit and a soul. Although Aristotle and Plato could not explain how something absolute and infinite could be in a subject, today we have a new mathematics called the mathematics of infinity and a new physics (e.g., relativity theory), both of which are in agreement that the infinite is real, and that it exists in the material world, within systems. We will be discussing these ideas in Chapter 4.

These problems that the ancient Greek philosophers grappled with are with us today, and they have to do with things that are of major importance to us. When considering the problem of identity, all these thoughts involve, not just the question of the identity of the qualities of experience, but also the question of our own identity and self-sameness (which relates to our being a 'subject.') If we are merely a cluster of constantly changing interrelated electromagnetic vibrations, it is not possible for us to have a precise, individual, personal identity. Mentally speaking, if we are just a cluster of ever-changing thoughts and perceptions, again our personal identity cannot exist. In the astrological cosmology 'spirit' and 'soul' become the basis of personal identity as well as the stabilization factors behind the twelve fundamental types of being.

In the next couple of chapters on '**Spirit**' and '**Soul**' we will be discussing some more ways in which spirit and soul may be helpful to us—even beyond establishing our personal identity. We may not be able, for instance, to understand the types of being and their interrelatedness to one another without something absolute to assist in this matter.

These efforts to understand how many fundamental types of being exist, whether these need archetypes, the special nature of pure absolute Being, etc., have occupied the minds of philosophers for thousands of years. As a specific branch of philosophy this is called 'ontology' (from the Greek 'ontos,' for 'being').

As its own area of thought ontology can be distinguished from for instance epistemology, the study of how we know things, or ethics, the question of right and wrong, good and bad. And although these various subjects can in a way be separated from one another, in another way they cannot, since the other subjects require philosophers to take a stance as to whether we need to postulate the existence of absolutes somewhere within a relative world if we are to have real knowledge (a 'truth' of the matter), and real and true standards for values. That is why philosophers have historically been divided into the two camps mentioned. If they are a relativist in one area they need to be in the other areas also—that is if their thought patterns are to be logical and consistent. This is why ontology is such an all-pervasive and important subject for both philosophers and scientists. The astrological category system is an invaluable (and empirically testable) philosophical tool that can be used to settle this debate. The category system shows clearly that absolute Being exists. It furthermore shows that absolute Being is the basis of what it means to be a subject, as well as being the source of the types of being. Absolute Being is the spirit in a subject. That spirit stabilizes and produces the subject's matter. It does so through the soul which is the wholeness of the individual, and is a specific set of potentials singled out from the infinite potentials in spirit as the substrate. In a similar manner as with spirit, soul has an absolute and eternal nature, but one which is a little different from spirit. From this category system a cosmology can be built that includes absolute, infinite and eternal realities. Because it is not only rationally defensible, but also empirically verifiable, I am confident that this powerful system of thought can (and will) eventually serve as the basis for a spiritual paradigm that has the capacity to overturn the old relativistic, materialistic scientific paradigm that exists today.

One of the most important aspects of this category system's value in regards to the study of Being is the relationship it indicates exists between Being and Unity. There is no word in philosophy similar to ontology to cover the subject of the study of Unity, but there should be, and so I am going to suggest 'unitology.' Next to ontology this is one of the most important subjects there is. There has not been a sufficient study of the relationship between Unity and Being. Furthermore Unity itself has not been studied in depth, especially the difference between the unity of a point (which is totally non-extended) and the unity of wholeness (which relates to a specific personal extent).

In the astrological cosmology, it is not just pure Being that is the highest principle. So also is pure Unity—and these two are seen as the same thing. In the next section of this chapter, the twelve fundamental types of being are examined individually. At the end of that section, the twelve types of being are shown to concurrently be twelve types of unity. If the types of being are types of unity, this is strong evidence that pure Being is pure Unity.

Being and Unity are not only different ways of speaking about the same thing; they are the substance of what it means to be a subject. The identity between Being and Unity and their nature as a subject can be seen in terms of spirit or in terms of soul. Soul provides for the Being of an individual and for that being's unity. The unity of soul is called wholeness. Spirit gives being too. It is the life of the body. It centralizes the parts of an individual. The nature of the Unity of spirit is different from wholeness. Pure Being of the spiritual type has no qualities, not even of extent, so obviously it cannot be divided. It is unity as a point when experienced as the center of a system.

Aristotle compares Being to Unity in his book called the *Metaphysics*. He considered Unity to be the highest universal along with Being (*Metaphysics 1053b*). He believed that unity and being had the same number of meanings and that these were complete correlatives. But he never clearly said they were the same thing.

Plato compares Being to Unity in his book called the *Parmenides*. This is the most confusing of all Plato's writings and it is a dialogue that never reaches any conclusions because, first of all, he does not see that Unity is not a quality. (The twelve basic *types* of unity are qualities.) Things get further complicated in the *Parmenides* because Plato also does not differentiate between wholeness (which allows differentiaton to exist within it as parts) and a unity that has no extent or differentiation. It is a very enlightening book to read, however, because it considers all the characteristics of unity, and presents all the difficulties that can arise if a philosophy does not define terms carefully and make clear distinctions. The chapters to follow will discuss how spirit, soul and matter are the same and how they are different. These are three types of being, and three types of unity.

Because of the relationship that exists between unity and identity, and because identity was so important in his philosophy, Plato decided that Unity had to be above even the Form of pure Being (which itself had to be simple and hence one if it was to have a precise identity). Thus he did not connect the two Forms.

(The idea of being 'above' and of levels has to do with priorities and dependencies.) The problem with Plato's decision is that, although Being cannot be simple and hence one unless Unity is prior to Being, the other side of the case is also true in that Unity, if it is higher than Being, cannot exist or 'Be.' And that is precisely why Being and Unity must be perceived as one thing. Furthermore, to satisfy the (legitimate) objections Aristotle had to Plato's theory of the Forms, Unity-Being needs to be treated not as a quality but as the essence of what it means to be a subject. These two logical requirements are precisely what the astrological category system fulfills.

In addition to being influenced in his thoughts by Zeno and Parmenides, Plato was also heavily influenced by the Greek philosopher Pythagoras (570-500 BC) who did connect Unity to Being. According to Pythagoras the number one represented the point, and was the ultimate constituent of all things. Although in the astrological cosmology the point is viewed as the ultimate constituent of all things, the number one is seen to symbolize wholeness or soul, and zero (the point) to symbolize spirit. This mathematical symbolic misunderstanding on the part of Pythagoras epitomizes the lack of clear distinction between spirit and soul that existed in Greek philosophy and actually still exists to this day.

And as if the lack of distinction between zero and one were not enough, there is also no distinction made between no-thing and nothing. Something can be no-thing (no material thing) and still exist. The Western philosopher/mathematician Gottfried Leibniz came to the conclusion that the world consisted of 'monads' that were dimensionless points, and yet (inconsistently) he looked upon zero as nothing, as the antithesis of matter, and as the principle of evil in the universe. It has been especially difficult for Western philosophers to see the difference between no-thing and nothing. This has contributed to a materialistic paradigm.

Although Plato and Aristotle were unclear about the relationship between Being and Unity, they did realize that these were both something above and beyond the types of being. The scholastics later spoke of 'Being,' 'Unity,' and the 'Good' as 'transcendental' categories. (Plato identified Unity with the Good, but as noted put the Form of Being slightly below Unity-Goodness.) For the scholastics, because the transcendental Forms were above and beyond the ordinary Forms by means of which we conceptualize, they were Forms that they believed had to be intuited. (Of course even to 'intuit' something a piece of it must be within us.)

It was because of the role that the Forms played in mental conceptualization and reason that led Plato and Aristotle to see such a close correspondence between Being and Unity. According to Plato the contemplation of unity guides the mind to true being (*Republic* 7.524e). He continually reiterated that understanding entailed comparison—seeing how things are different and how they are the same. Reason is seeing how it is that many different things are related to one another. (The things we speak of as 'qualitative 'sameness' and qualitative 'relationships' have to do with qualitative unity.) We are not just trying to find reality or real Being when we seek understanding. Knowledge is also a search for the Unity in things. To see the unity the mind needs to reach up. Conceptualization always entails finding unity, and the highest Form of unity is just pure undifferentiated Unity.

As a category system, astrology is unique in its ability to show that Unity is Being and vice versa. That is because it shows a descending series of the expression of Being down from spirit, to soul and then to matter (producing the material *types* of being). More importantly, it takes Unity down through these three layers as well. It is not just spirit that is (pure) unity, and not just soul that is unity (as wholeness). Each one of the twelve basic type of being is shown in the astrological category system to be a type of material unity as well as a type of material being.

[To take the time to show that Being is Unity is not for the purpose of making some sort of abstract philosophical point. The astrological category system indicates that pure Unity is the essence of what it means to be a subject. We are Unity-Being. Because of this, what this system of thought then also reveals is that, when our material bodies do not express the Unity which is our Being, they cannot remain in being. And that has very practical ramifications for all of us. Being whole and being alive are correlatives.]

The purpose and goal of philosophy is to investigate the nature of reality. Twenty five hundred years ago Plato and Aristotle brought to light these issues that to this day still divide absolutists and relativists, and that have not as yet been reconciled.

When it becomes evident that there is a category system, from which a cosmology can be constructed—one that solves these philosophical problems, and one where the category system itself (via astrology) can be proven to be true—this is an important find. When furthermore these solutions add logical weight to the thesis that the universe is a spiritual one, this makes the find even more valuable.

The rest of this chapter is dedicated to the actual category system. The explanation of the manner in which the types of being relate to spirit and soul is towards the end. That is because the twelve types of being are separated into the 'domains,' and the first two are the domains of kinematics and of space or quantification. The eight types of being in those two meta-categories are very basic in terms of describing the material universe. The materialistic paradigm is in agreement with that at least. Scientists currently describe the world in terms of these two domains, which together comprise the current concepts of energy and amounts of energy.

Although it is possible to describe the material world in terms of these two domains, what is being missed by proponents of the materialistic paradigm is the third domain of experience, which has to do with the fact that these two aspects of energy arise from absolute Being and are unified by absolute Being-Unity—and specifically by the absolute as it exists in individual systems. As Aristotle plainly saw, there are no material qualities that are not the qualities of some system. It is in terms of this third domain that it is possible to construct a spiritual cosmology capable of replacing the old materialistic one.

The astrological *cosmology* is a metaphysical system built up or deduced from the astrological *category system*. The actual cosmology is developed in the three chapters on '**Spirit**,' '**Soul**,' and '**Matter**.' Although the discussion in this chapter is merely a blue print of the metaphysical system, this introductory presentation is important because, in giving all twelve of the archetypes together, it is possible to show the interrelatedness between basic sets of principles. The four principles in each domain are closely interconnected. The qualitative domain is the last to be examined. It has principles in it that have traditionally been said to rule spirit and soul. By looking at the interrelatedness between these four it is possible to see how spirit expresses itself in two ways, how these two together create a soul and, via soul, how matter comes into being. To see these interrelationships is to understand how the material world is created from spirit, and that is what this book is all about.

The Astrological Category System: The Language of Light

An archetype cannot be expressed in words. Words are just that—expressions—and as such are material in nature. We think and verbalize in terms of actualized material types of being, and any material type of being is always merely an expression of one of the archetypes. Nevertheless, it is possible to get close to the essence of an archetype by examining types of being that are themselves extremely simple and fundamental. The place to begin is with simple systems. Human beings are complex. So are atoms. The simplest system is a photon. To examine the types of being in terms of a photon is to get very close to the nature of the archetypes. (To try to visualize them in terms of a photon's waveform is to get even closer, since the archetypes manifest via probability fields.)

Light is energy rather than matter; but it too is a fundamental system, perhaps the most fundamental system of all. It has parts (e.g. positive and negative elements, plus electric and magnetic components). And what has parts is a (centered) whole. The wholeness is a primordial form of soul, and the centering of its parts is an expression of the presence of spirit.

The 'language' of light is the qualitative component that it carries within it. It is a universal language by means of which the entire world is created—the language of the laws of nature. Law is not an emergent quality of matter. The laws of physics are to be found in a simple ray of light rather than in complex material particles like electrons, nucleons and atoms—although these too obviously follow and express the laws of nature, as do higher level systems like human beings.

The way in which the laws of nature arise in our world can be seen by comparing the parts of a ray of light to certain groupings that the signs belong to in astrology. Thus although astrology is a tool utilized to understand human beings, it is also useful to understand any and all systems, precisely because everything is composed of light. This way of explaining how the laws of nature arise in a simple ray of light is merely a hypothesis, but one I believe to be a valid one.

First of all, a ray of light has positive (+) and negative (-) components. These are primordial poles. Positive and negative poles are like two sides of one phenomenon, and yet they do entail a type of division. In their material form, the laws of nature are relationships, and in order for there to be a relationship, there must be some type of division in the world. At the same time the world and all the systems within it are one. Positive and negative polarity permits division *within* unity. These two poles are the Adam and Eve of the material universe. Their essential unity is evidence that spirit is pure Unity as well as pure Being, and that light, the means to creation, is a phenomenal expression of that essential unity. Together these poles are a fundamental 'whole.' (Adam and Eve are not just twin souls; they are one soul.)

In addition to positive and negative polarity, a ray of light has two basic attributes—electric and magnetic components. Each of these has positive and negative aspects.

Consider first, the electric component. This can be thought of as composed of positive and negative poles of light. We know that the potential for dimensions exist within light. Hence, in terms of spatial dimensions, the two poles can be thought of as creating, together, a one-dimensional system. If dimensions evolve or have any sort of sequential development, when a first dimensional system came into being, it was probably already beginning to utilize a higher second dimension-in-the-making by manifesting, not in terms of a static straight line (although a line is one dimensional), but rather (because of the poles and the animated nature of a dimension) in terms of a moving, curving *'S'-shaped* line that alternated between complementary poles, positive (+) to (-) and then back again. (The electric portion of an electromagnetic ray is visualized by physicists as an 'S' as it undulates through space-time.)

The electric component of an electromagnetic ray is prior to the magnetic component. The poles of electricity can be represented as + and - . Magnetism is looked upon as a derivative of electricity, and from what the astrological category system reveals, its poles are +/- and -/+. All matter is composed of relationships, and when there are two of two things, there are four ways for them to interrelate. As electromagnetic waves cycle through space and time they apparently manifest cyclic changes in an alternating process (+/+, -/+, +/-, -/-). These four become in a ray of light fundamental 'elements,' a set of internal interactions or relationships that are the beginning of what is meant by a 'part.' An electromagnetic ray presents as a two dimensional system with its electric and magnetic components arranged like two S's at right angles to one another (which is how these waves are described in electromagnetic theory).

The ancients did not understand electromagnetism, but they somehow knew that there were four primal elements involved with the creation of the world. They called them fire (+), earth (-/+), air (+/-) and water (-). The actual things on earth that we experience as fire, earth, air and water are not these primal elements. What the idea of elements depicts, when looking at how the world is created, is four types of differentiation in light by means of which the laws of nature come into being.

In terms of these four elements, an electromagnetic ray has four basic relationship potentials. As it interacts with (and produces) a three-dimensional system or world it creates twelve basic types of being. What the category system indicates is that we ourselves are in a state of evolution, as all systems and their worlds are. We have the potential to become a fourth dimensional being, a fifth dimensional being, and so on. However, systems in our world are currently limited to three dimensions. The four basic relationships mentioned above manifesting in three dimensions literally give three 'dimensions' to the four elemental interactions (4 x 3), producing twelve basic types of being. In the soul of a three-dimensional individual there are twelve basic relationship potentials.

If these are looked at in terms of cyclic interactions *with* matter (and as *producing* matter) they can be diagramed in the following way:

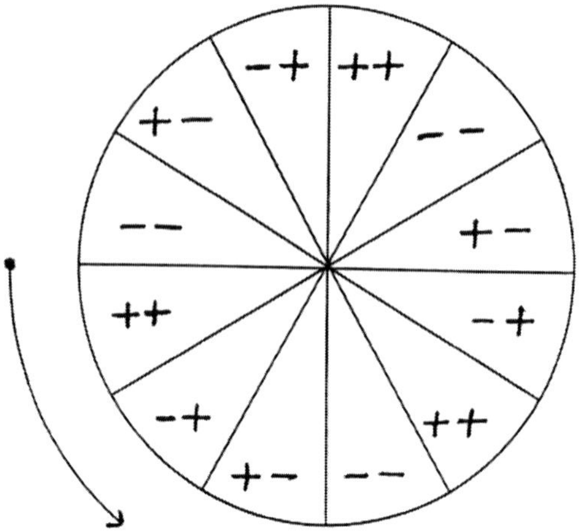

In three-dimensional matter the interaction is cyclical, and a circle can be used to represent the stages.

Of the twelve types of being produced, there are three that are purely positive, three that are purely negative, three that are *primarily* positive (with concurrent negative attributes), and three that are *primarily* negative (with concurrent positive attributes). These are shown below where the types of being are linked to the names they carry in astrology.

(+/+) ♈ (Aries)	(+/+) ♐ (Sagittarius)	(+/+) ♌ (Leo)
(-/-) ♏ (Scorpio)	(-/-) ♋ (Cancer)	(-/-) ♓ (Pisces)
(+/-) ♎ (Libra)	(+/-) ♊ (Gemini)	(+/-) ♒ (Aquarius)
(-/+) ♉ (Taurus)	(-/+) ♑ (Capricorn)	(-/+) ♍ (Virgo)

Astrologers call the three purely positive (+/+) signs 'fire' signs. These are Aries, Sagittarius and Leo. They refer to the three purely negative (-/-) signs (Scorpio, Cancer, and Pisces) as 'water' signs. Libra, Gemini and Aquarius they speak of as 'air' signs. Taurus, Capricorn and Virgo they call 'earth' signs. These signs are expressions of the twelve types of being that qualify the world we live in. Everything that we experience is produced by the twelve archetypes underlying the signs.

One of the simplest expressions of the archetypes is in terms of the spatial dimensions. Each higher spatial dimension gives more spatial relationship potentials—the things we call 'directions.' There are four relationship potentials for each dimension. The more dimensions a system is expressing, the more directions in space that being is able to experience and understand. The directions are relative and are relationships. They apparently arise in this way:

Places in Space
(Spatial Dimensions)

UNITY

Primal Side-ness
Negative-ness (-) & Positive-ness (+)
left (-) right (+)

HORIZONTAL	**VOLUME**	**VERTICAL**
to-the-fore (++)	out (++)	up (++)
back (- -)	in (- -)	down (- -)
back and forth (+ -)	in and out (+ -)	up and down (+ -)
neither back nor forth (- +)	neither in nor out (- +)	neither up nor down (- +)

The first type of direction is in terms of left and right. This is the most elemental form of differentiation we experience and is connected to the poles in electromagnetism. As dimensions develop it becomes possible to know what in-front-of (to-the-fore) versus in-back-of consists of, what 'up' versus 'down' is, and what 'in' versus 'out' is. The three purely positive (+/+) types of being called in astrology Aries, Sagittarius and Leo, produce our experience of in-front-of, out, and up—which are positive directions in space. The three purely negative (-/-) water signs called in astrology Scorpio, Cancer and Pisces, produce our experience of behind, in, and down. These are negative directions in space. (By 'negative' is never meant 'bad' or 'inferior' in the pejorative sense.)

Later we will talk about how the air (+/-) and earth (-/+) signs unite their respective parent signs to produce, for instance, our ability to relate or connect internal (Cancer) things with external (Sagittarian) things via the 'air' sign Gemini which makes 'in-

and-out' ('connections') possible. The earth sign in this domain (Capricorn) makes it possible for there to be a dividing line between things that are 'internal' and 'external'—producing places that are 'neither-in-nor-out' (boundaries and division).

HORIZONTAL	**VOLUME**	**VERTICAL**
♈ – in front of (to-the-fore)	♐ – out	♌ – up
♏ – behind (in back of)	♋ – in	♓ – down
♎ – back and forth	♊ – in and out	♒ – up and down
♉ – neither back nor forth	♑ – neither in nor out	♍ – neither up nor down

Not merely the spatial directions, but rather the characteristics of everything encountered in our world are produced by the archetypes. The character of each of the types of being can be expressed in terms of a word or two. These can be looked upon as 'key words.' I first heard the term 'key word' used by an astrologer by the name of David Davidson.[1] I do not know if he coined the term or not, but it is a good one. Key words give a general sense of the qualitative meaning reference—the actual archetypes—at base of the types of being. As soon as something is expressed as a type of being it begins to have a 'relative' 'relationship' quality to it, whereas the archetypes are absolutes. Notice that the four types of being in each dimension already bear a relationship to one another, and actually in terms of two descending levels. The vertical direction, for instance, is not only related to the horizontal direction (in terms of being a different, but related, direction in space), within the vertical dimension itself 'up' and 'down' are different (but related) principles.

Although these key words refer to simple material qualities, because they are so simple they work well as key words. Just how well they manage to express the nature of the archetypes becomes evident in Chapter 10, which discusses this category system as it relates to the complex, human-level world of experience. There it can be seen that these key words carry over extremely well even into that complex realm, making it possible to see that the words developed here are very close to the essence of the archetypes. The descriptive terms I like to use are these:

(+/+) ♈ (Aries) – forth-ness, going forth
(-/-) ♏ (Scorpio) – behind-ness, leaving-behind
(+/-) ♎ (Libra) – back and forth (equilibrium, reciprocity)
(-/+) ♉ (Taurus) – neither back nor forth (stasis, inertia)

(+/+) ♐ (Sagittarius) – externality (out), extent, extending
(-/-) ♋ (Cancer) – in-ness, inclusion
(+/-) ♊ (Gemini) – in and out (connectivity, connections)
(-/+) ♑ (Capricorn – neither in nor out (bounds, limits)

(+/+) ♌ (Leo) – superiority (up, over), superiorization
(-/-) ♓ (Pisces) – inferiority (down, below), inferiorization*
(+/-) ♒ (Aquarius – both superior and inferior
(-/+) ♍ (Virgo) – neither superior nor inferior

*The term 'inferior' means 'below' rather than inferior in the sense of less good.

As basic as the simple directions in space are, they are just that—directions in *space*. The fact that space exists precedes the fact that there are *types* of spatial directions. Space is a very fundamental existent but it is only one of three basic types of 'domains' that we experience in our world. These three include the domain of kinematics (including motion), the domain of space (and quantification), and the domain of qualification. The three domains and the astrological signs linked to them are listed below:

Kinematical Domain	Quantitative Domain	Qualitative Domain
Action (♈) kinetic energy	Spatial Extent (♐) spatial expansion	Law of Evolution* (♌) qualitative more-ness (up)
Force (♏) change	Spatial Content (♋) inclusion	2nd Law of Thermodynamics (♓) qualitative less-ness (down)
3rd Law of Motion (♎) reaction/exchange	Spatial Junctures (♊) connections	Law of Hierarchy (up & down) (♒) order
1st Law of Motion (♉) non-change/inertia	Spatial Division (♑) boundaries	Law of Qualitative Divergence (♍) qualitative differences**

*This is also a law of gravitation.
**(Also hierarchical *level* differentiation – neither 'up' nor 'down.')

The three domains are related to the simple places in space given previously, and in fact the relationship of each to a specific spatial direction helps show the simplicity of the fundamental archetypes, as well as how key words can express, for instance, both spatial directions and the more basic domains. Aries rules kinetic energy because motion is a going *forth*. Scorpio rules change, and in processes of change something is left *behind*. Libra rules the law of nature that states "for every action there is an equal and opposite reaction." This is Newton's third law of motion and entails a *back* and *forth* type of reciprocal relationship. Taurus rules the first law of motion, which is referred to as the law of inertia. Inertia entails non-change of motion. It involves a neither going *forth* (into newness) nor a leaving *behind* (of what was).

Although the second domain rules space per se, when it is compared to one of the three types of spatial dimensions (the basic directions), it most closely corresponds to that which we call 'out' and 'in.' The spatial extent of a system entails how far it extends outward into space. In terms of personal space there is a personal extent or 'outward-ness' as well as a personal 'inwardness' (content). When used to express that which is personal as it relates to an external world, a person has their own 'within' in comparison to what does not belong to them, in other words, there is the inwardness of the person compared to what is external or 'out' (the external world). Content and extent are polarity partners because personal content produces personal extent. In terms of the personal versus the impersonal, again there is no within without a without.

Spatial boundaries entail a reference to both of the 'parent' principles just mentioned—in-ness and out-ness. Boundaries involve places that are neither external to the object in question nor a part of its internal nature (neither in nor out). Spatial connections also entail a reference to both 'parent' forms of differentiation. They involve places that are both in and out. They connect the two. [We will talk more about these 'parent-child' relationships, (x + y) and (neither x nor y), later on.]

The qualitative domain can also be compared to spatial directions. It most closely relates to the places in space we call 'up' and 'down.' The qualitative domain has to do with all the parts of a system as they are qualities of (parts of) a subject. The whole is 'over' its parts. The qualitative domain also has to do, in a value sense, with that which is above something else qualitatively versus that which is beneath it. [The part-whole relationship is a primordial type of value in nature.] As Leo rules the high in place or space, so the superior in quality has a link to spatial highness. In terms of laws of nature, Leo is a principle of evolution, which makes systems move 'upward' qualitatively. Pisces rules the second law of thermodynamics which, in a disabled world, degrades or lowers things qualitatively; and to be 'lowered' qualitatively relates to the low in space. (The lowest point is a state of undifferentiated 'sameness.') It is because there is a high and a low in space and, more directly, because wholes are over parts, that individuals wind up being hierarchically arranged. Hierarchy in nature, as made possible through the Aquarian principle, involves the linking of wholes and parts—

the above to the below (both x and y). Qualitative separation, as ruled by Virgo, involves the fact that parts are apart from one another qualitatively speaking ('different'). The qualities of a system are its parts, and the parts are neither the wholeness (Leo) nor a state of undifferentiated sameness (Pisces). They express a condition of 'neither x nor y.'

The three domains are prior to the three kinds of simple spatial directions. Nevertheless, by keeping in mind the simple places in space as these relate to each of the archetypes, it makes it easy to visualize each in terms of one or two key words. The key words in turn help show the interrelatedness between opposite (complementary) principles within a particular domain, and that will become important later on when a cosmology is constructed from these archetypes.

As each domain is explored more in depth, it will become evident that some of the types of being correspond to known laws of nature. This is particularly true of the kinematical domain, since Newton's laws of motion have to do with motion. What scientists call a 'law' is currently conjoined with the idea of process and causation.

It is true that a law makes something move, and in a particular kind of way. It is the *kind* of way that brings in the qualitative aspect of law. A better way of looking at how the world is qualified is by seeing it as composed of twelve basic types of being. Because the domains interpenetrate, it is nevertheless true that there is a dynamic aspect to each type of being. For instance, the Sagittarian form of differentiation not only produces the extent we call 'space' and 'spatial extent,' there is also a dynamic *expansion* of space going on in our world. However, an over-emphasis on the kinematical domain takes away from the equally fundamental value of the other domains. In the spatial domain for instance, a particular type of motion does not cause spatial extent to come into being any more than it causes space to expand. Rather, spatial extent is a fundamental fact of our material experience. It is because these domains interpenetrate that there are processes in the spatial domain, such as the growth of our individual spatial extent and the expansion of the space of our external world.

Despite the fact that there is merit in speaking of the twelve archetypes as types of being rather than laws of nature, one of the strengths of this category system is its ability to bring science and philosophy a little closer. Philosophers have looked at experience and seen that the world possesses some very primary characteristics like spatial extent, spatial connections, motion, rest, etc., and have constructed category systems in terms of what they believe to be the most basic and fundamental types of being. But they have not correlated these to laws of nature. Physicists on the other hand have studied the laws of nature and have noted that these produce the characteristics of the world. But they have not connected the laws to philosophical category systems. As a result of this, physicists have missed out on some laws of nature that do exist, and that become evident via a category system. Scientists also do not have an adequate understanding of the interrelatedness between the laws of nature—between for instance the first and third laws of motion—something that a category system does make possible, as will become evident when the actual types of being are discussed individually below.

Although it is better to address the domains as equal and as interpenetrating rather than to see forces as primary, there is a sense in which one is prior to the other two, but it is the qualitative domain rather than the domain of motion and process. That domain has to do with the production of individuals, and it is individuals that possess the two attributes of motion and spatial extent.

Although the astrological signs are only one expression of the fundamental archetypes at base of our world experience, from this point forward I will refer to the archetypes by the names they possess in astrology. This simplifies the manner of writing. Later on in Chapter 9, the role that the archetypes actually play in producing signs, planets, etc., is discussed.

The Kinematical Domain

♈ (Aries) – MOTION (going *forth* in space and time; *forth*-ness in space or time, being in front)
♏ (Scorpio) – CHANGE (leaving *behind*), FORCE, POWER (the potency *behind* motion as a going forth; *behind-ness*)
♎ (Libra) – 3rd LAW OF MOTION (action exchanges and reciprocity; back & forth)
♉ (Taurus) – 1st LAW OF MOTION (inertia, stasis, neither back nor forth)

In the domain of kinematics the first type of being, Aries makes it possible for there to be motion in the world. Motion involves a relationship of 'going forth' relative to something else. In addition to motion, as this principle pertains to places in space, it produces the relationship of being to-the-fore in space or time.

[The purely positive signs in each domain are primary. In this case, Aries has to do with the priority of causes over effects, and the fact that we move forward in time (not backwards). This is a natural type of value in nature.]

Aries is a purely positive principle. Scorpio, the purely negative polarity partner of Aries, makes change possible. Change is the back side of motion, and involves the relationship of leaving-behind. Acting together, these produce what is called a 'force.' A force entails a *change* in the degree or direction of movement. In a world where (as per relativity theory) everything is interrelated, when one thing changes, everything else must adjust to accommodate the change and bring things back to unity as relative equilibrium. That is why a force is closely related to change as well as to energy as motion.

[Due to different levels of systems in the world, and different levels of parts within systems, there are different levels of kinetic energy. Kinetic energy at a lower level of being is said to be 'stored' or hidden energy, and this is referred to as power or potential energy. This is made possible through the Scorpio principle, as that which is 'behind' energy (behind 'going-forth'). This entire domain expresses that aspect of spirit which is its character as a power.]

The third type of being in this domain, Libra, is a recognized law of nature. The effects it produces in the world are referred to as Newton's third law of motion. This law is a result of the Scorpio and Aries principles combined, uniting back-ness and forth-ness, and thereby producing reciprocity (and equilibrium) between actions. The first part of this third law states that for every action there is an equal and opposite reaction (back and forth). The second part states that forces tend to occur in pairs. The Libra archetype is responsible for the pairing of all the laws and forces in the universe, as well as for the most primal pairing and interaction of all—that between the positive and negative forces in electromagnetic phenomena.

As with Aries and Scorpio, Libra is involved in the notion of a 'force.' When there is a change in a unified field, all other members of the field then must adjust themselves to the change. Libra rules this adjustment process.

Libra governs interactions and adjustments to change both between the parts within a system and between separated systems. As Libra relates to interactions between separated systems, it rules the concept of exchange wherein one system affects another causing it to move and change. There is a *reaction* and a subsequent *exchange* of motion-change (energy). It is this that we experience as mechanical causation.

The fourth type of being in this domain, Taurus, produces relative stasis in the world. This is referred to as inertia. As a law of nature it is referred to as Newton's first law of motion, which states that a body at rest tends to stay at rest, and a body in motion tends to stay in motion (at the same speed and in the same direction). At one time it was believed that rest or inertia was the natural state of things. Now it is known that process is prior. Unless some force causes a change, a system will continue moving in the same direction and at the same speed forever. Taurus is not a primary type of being like process and change. Like the Libra law, Taurus is a result of a combination of the Aries (motion) and Scorpio (change) principles. This archetype produces neither going forth (motion) nor leaving behind (change), in other words, a non-change of energy manifestation.

Taurus makes it possible for bodies to have mass. The primary attribute of mass is inertia. Mass is one of the most 'material' qualities that matter possesses. Mass is not the weight an object has, or some sort of spatial 'stuff.' Mass is a form of energy, energy is associated with motion, and mass involves the amount of inertia a system has. Thus it entails a resistance to change of motion. Currently physicists cannot explain what mass is. They simply know that it corresponds to inertia, and they know there is a close correlation between energy and mass via Einstein's famous equation $E=mc^2$ (energy is equivalent to mass times the speed of light squared).

It is possible to make a guess as to what mass is based on a knowledge of the character of each principle as gained through astrology. Taurus is the principle of non-change and, in order to have mass, there must be a constant. Because of the interpenetration of the domains, the first law states that inertia involves non-change of direction in space or of degree of motion. In a world of motion and change, the only way to have a constant is to have constant motion and constant change. Photons are energy. When two or more photons interrelate coherently, they create a center between them. The center is a force for Unity as well as being a substance. It keeps them together and, as moving particles, they cycle and circle together around this center. I suspect that they produce a massive particle when they fulfill the requirement of constancy relative to both motion and change, and that they must fulfill it in regards to the other two domains as well, i.e., in regards to quantity and to direction as well as to motion. By moving at a constant speed they do not change the rate of motion. In other words, they are moving, but at a constant speed. Likewise, by circling around a center they are changing directions, but the degree (quantity) of the change of direction (quality) is constant. Only a circle fulfills that requirement. As for the meaning of the equation $E=mc^2$, as with motion and extent, if mass comes about in this way, it involves a link between two different dimensions. When two or more two-dimensional photons join to produce a whole, a new level is created, the wholeness. What the astrological cosmology indicates is that levels correspond to dimensions and, when this happens, a new dimension is scaled. In this way a three-dimensional, massive particle is produced. [The speed of light (c) is a limiting factor between dimensions. Squares in mathematics can indicate a dimensional ascent (e.g., c^2).]

These and other thoughts will be developed more in the chapters to follow, which deal with the cosmology that can be deduced from this category system. I am not a physicist and these thoughts relative to mass are just a guess, but scientists would do well to study this category system. By combining their specialized knowledge with astrology, scientists could understand things that are currently not understood. This system is not just a guide for philosophers, but also for physicists.

The four archetypes that in astrology are called Aries, Scorpio, Libra and Taurus make it possible for the material world to have action (+), change (-), reaction (+/-) and non-change (-/+). Aries produces action. Scorpio produces change. Taurus unites these two in a primarily negative fashion, so that the change aspect (Scorpio) is primary, which is why inertia presents as non-*change*. At the same time the motion aspect (Aries) is also present as a non-change *of motion*. In Taurus the negative is primary (-/+).

In Libra the positive (Aries) aspect is primary (+/-). Libra makes re*action* possible. But the Scorpio principle is present as well in terms of a reaction *to change*).

[All twelve of the laws of nature show clearly that there is direction in the universe, as will be discussed more in depth in the chapter on first-final causation (Chapter 6). This is an important subject, since directed movements are a strong indication that there is something non-material making matter move towards itself as an 'end' cause. Absolutist philosophers are open to the existence of non-material realities, and hence are willing to admit that teleological or end causes might exist. Relativists discount any evidence of value in nature. The character of the third law of motion has been cited by relativist philosophers as a proof that there are no directed movements (or any such inequalities) in the universe. This law is referenced because it states that for every action there is an equal and opposite reaction. Through this category system it is possible to see that the kinematical domain does have an 'arrow' pointing in a particular direction. We move forward in time. Causes precede effects. The Libra principle is simply a child of its directed parents, allowing union between the two for the specific purpose of allowing reaction/exchange.]

The Spatial-Quantitative Domain

♐ (Sagittarius) – SPATIAL EXTENT; LAW OF SPATIAL EXPANSION (out-ness, that which is 'external,' extent, extending-ness)

♋ (Cancer) – SPATIAL INCLUSION (in-ness, that which is 'within,' inclusion)

♊ (Gemini) – SPATIAL CONNECTIONS (in and out-ness)

♑ (Capricorn) – SPATIAL BOUNDS & LIMITS (neither in nor out; exclusion)

The four archetypes in this group make it possible for the

material world to experience space and spatial extent. This domain may be referred to as either the spatial domain or the quantitative domain. Space is a geometrical equivalent to mathematical quantification, and is the most fundamental form of quantification.

Sagittarius is the purely positive principle in this domain and makes it possible for there to be 'outward-ness' and for a system to have this thing we speak of as an 'extent.' It also makes it possible for the external world to have an extent as well. This is spoken of as outer space because it is external to systems.

Just as the domain of kinematics can be expressed by the key words 'forth-ness' and 'back-ness,' (as these relate to simple places in space), in a like manner spatial extent is most closely correlated to the places in space we refer to as 'in' and 'out.'

Cancer is the purely negative principle in this domain and rules content. Extent and content—inward-ness and outward-ness—are two sides of the same phenomenon. In terms of a single system, the degree of its internal nature will determine the degree of its extent—how far out it extends in space. Relative to a system as it relates to its external world, there is a personal 'within.' 'Out' then is outer space, the external world.

That which is inner space versus that which is outer space can change over time. In the astrological cosmology the external world is perceived as a number of systems at a comparable level of development that have the potential to move upward, joining together into hierarchical concrescences, and thereby forming a higher level of being (and a higher dimensional world). What was outer space then becomes inner space. We are wholes, but we are also parts in a budding higher-dimensional form of wholeness. This evolution into ever higher levels, since it does not destroy the levels below, produces hierarchy in nature, a condition wherein there are systems nested within systems.

This will be discussed more in depth in the next section on the qualitative domain, which governs wholes, parts, and their relationship. It is the part-whole relationship and hierarchy in nature that produces a valid within and without in the spatial domain. That is why the qualitative domain may be seen as prior to the spatial domain.

When there is a valid within and without, as there is in a hierarchically constructed universe, there can also be boundaries between systems. Wholes are independent, and the existence of true boundaries is dependent upon wholeness.

Wholeness makes it possible for parts to be distinct as well as the whole itself. In a hierarchically constructed universe, the independence and completeness of wholeness at one level produces boundaries so that, when the whole is seen as a part, that part is set off from other parts precisely because it is a whole in another capacity.

It is Capricorn, the earth (-/+) sign in the spatial domain, which rules surfaces, spatial bounds, and limits in general. Spatial boundaries are places that are not outward or external to a system; but they are not really internal either. They are however more closely related to a situation's internal nature (-) than to its extern world (+), since they are the limits of that particular system. Thus this earth sign, like its comparable earth sign, Taurus, in the kinematical domain, is a primarily negative principle (-/+). Capricorn is a 'child' of its two 'parent' principles, the within (Cancer) and the without (Sagittarius). Whether something is 'within' or 'without' for a system depends upon the level of being it is involved in. This relationship of Capricorn to Cancer and Sagittarius shows clearly that what we call a 'limit' is a relationship (which in turn is a type of unity). The 'limitation' aspect of spatial boundaries is produced by the fact of the existence of two different levels of being.

[Next to mass, there is no more 'material' characteristic than the finitude it is believed the 'limits' aspect of matter creates. However, if the existence of limits (and hence of that which is finite) simply involves a relationship between two different levels of wholeness (and two different dimensions of light), then being finite and having limits is not as it seems. That which is finite becomes explainable through something that is infinite, and is itself not divided or limited in extent at all. (This concept of boundaries and limits being created by different levels of wholeness is returned to in Chapter 7 on '**Matter**.')]

The air sign in this domain is Gemini. This archetype makes it possible for there to be spatial connections. A spatial connection links something that is external ('out') with an internal domain and vice versa. Thus a link is a place that is 'both-in-and-out.' Like Capricorn (neither-in-nor-out), the Gemini relationship is a result of the combination of the two primary principles, Sagittarian and Cancer, but in a slightly different way. There is linkage instead of division.

Because of the interpenetration of the domains, as was noted, all of the principles have an 'active' component to them. Thus they are dynamic laws of nature as well as types of being. The Capricorn and Gemini relationships in the spatial domain are not currently recognized as laws of nature, although in chemistry there are dynamics at the surfaces of liquids and solids that appear to manifest these principles. The Sagittarian principle not only produces a spatial extent, it is a law of nature that impels a spatial extent to expand. The actual expansion is due to spirit creating energy and matter at the center of a system. Due to this increase in content, internal and external worlds grow, wholes and wholes-in-the-making grow, actualized worlds above us grow, and new individuals are constantly being created. Scientists tell us that the universe is expanding. But they explain the observable expansion of the universe in terms of a 'Big Bang' thesis. According to the Big Bang theory of creation, there is nothing new in the world. The exact amount of energy-matter in the universe always existed as it is. All this energy-matter was once packed into a tight nucleus of almost infinite density. Then supposedly it exploded. What we are observing, per this thesis, is the result of that outward momentum preserved via the law of inertia. What creates problems in this thesis is that, if it were true, everything would be moving away from the point of the explosion. Instead, the growth is taking place from every point in the universe. A 'point' is a center, and in terms of a spiritual new paradigm (which entails a 'constant creation' cosmology), every particle and atom in the universe is naturally growing from within as new photons are created at the center of systems. (New systems are also constantly being created.).

[A constant creation cosmology does not preclude a 'big bang' occurring when systems, via evolution, ascend together to a new, higher (unified) level and consequent higher dimension of being, at which time quite a bit of new matter is produced, since a new individual (and subsequent new external world) is thereby created. (Truth is not only simple and in line with what is empirically observable; it also reconciles extremes, including opposing scientific theories.)]

The Qualitative Domain

♌ (Leo) – LAW OF CENTRALIZATION & SUPERIORIZATION (individual wholeness, gravity, evolution)

♓ (Pisces) – 2nd LAW OF THERMODYNAMICS (universal undifferentiated unity, qualitative inferiority,* inferiorization)

♒ (Aquarius – LAW OF HIERARCHY (links between levels and types of being); LAW OF COHERENCE & ORDER

♍ (Virgo) – LAW OF QUALITATIVE DIVERGENCE (differentiation into levels and types of being)

*In the sense of a lower qualitative level, not in the sense of qualitatively less good or 'bad.'

This can be referred to as the qualitative domain for several reasons. First of all it has to do with the production of subjects, and subjects have part-whole duality. The parts of a system are the 'qualities' it possesses. The other reason for referring to this as the qualitative domain is due to the second meaning of 'quality' as a determinant of value. This domain has within it immediate modes of spirit and soul. Spirit and soul in turn entail absolute Being and Unity. There is a logical reason as to why Plato and Aristotle equated Unity, not just with Being, but also with the Good. The parts of systems possess within them an impetus to be and to be whole—also to move towards ever greater being and wholeness. Because of this, matter has inherent value. 'Value' means non-equality of state, direction, etc. Relative to states, it is good to possess a state of wholeness, and just to be (rather than to not exist or to not be whole). In terms of directed processes in the world, it is a good thing when processes promote greater being and unity. This domain is appropriately spoken of as the qualitative domain for both these reasons.

The first sign in this domain is Leo. It is a purely positive (fire) sign. The polarity partner to Leo is Pisces, which is a purely negative (water) sign. These two principles in the qualitative domain make it possible to understand what spirit is and how it creates the material world.

Astrologers have debated as to whether Leo or Pisces rules spirit. Both sides are right and both are wrong. They are both wrong in that no type of being is spirit or pure Being, precisely because pure Being is not a type of being. Both Leo and Pisces express qualities and processes in matter that are immediate modes of the effects of spirit. Both sides are right, on the other hand, in that these two signs express the two poles of spirit. They reveal that spirit manifests in two ways. First of all, spirit is a universal undifferentiated substance. It is the potential for everything that can ever be, but is itself not differentiated into this or that type of being. Aristotle's view of Being as the eternal, unchanging substrate was an appropriate visualization. Pisces governs this aspect of spirit, and rules processes produced in matter by the substrate. These are referred to by physicists as the second law of thermodynamics, and involve processes that move systems towards a state of less qualitative differentiation, in other words, towards universal unity expressed as undifferentiated at-one-ment.

The second type of being—Leo—is related to spirit as a principle of individuation, and to processes that promote differentiation and individuality as opposed to undifferentiated universality. When spirit creates an individual, a set of potentials is singled out from the substrate. The specific set is what a soul consists of, and the actualized types of being that are linked to a particular soul are the material parts of that whole. Within the set of actualized material parts, a spot of pure spirit shines through at the very center of the nexus of parts. It is a piece of the substrate that does not change into matter. This is the life spirit in a system. It is an individualized expression of spirit, and is ruled by Leo. [Although the spirit of life at base of a particular system resides at the center of that system, the paradox inherent in this state of affairs is that spirit actually has no material place that it resides 'in.' Place is a material experience, and spirit is absolute Being rather than a type of being. (We will talk more about centers being 'no-place' in the next chapter, which is specifically dedicated to the subject of '**Spirit**.')]

It is precisely because of this 'no-place' attribute of spirit that the seeming opposition of individuality versus universality is reconciled by spirit. An individual spirit is not divided off from spirit as the substrate (or from the spirit of any other person or system in the world). Extent is a material quality. What is non-extended cannot be divided. Spirit is experienced by a system as *their* spirit when a certain set of qualities surrounds it. That which surrounds a 'piece' of the substrate is the matter of an individual. It is the matter that makes for the individuality. In other words, these two different manifestations of spirit (Leo and Pisces) are only *relative to* the matter. Nevertheless, matter is real and, when creation occurs, a set of potentials is activated from out of the substrate producing an individualized expression of spirit. Thus spirit is both universal and particular.

Spirit is a power as well as a substance. Leo rules processes of centralization in systems that are due to spirit pulling the parts towards itself, which is to say inward towards the center of the system. Leo governs processes in matter that produce both individual Being and individual Unity. When pure Being-Unity unifies the material parts of systems making them whole it is an omega or end cause, pulling the parts to their origin which exists at the center of the system. It is then perceived as a principle and power of Unity. When pure Being-Unity gives being or life to a system it is an alpha cause—not only producing (via wholeness) a new being, but also converting spirit into energy at the center of the system. Spirit is then perceived as a principle and power of Being. Being and Unity are the same thing, but these are experienced as either Unity or Being from the material side of life, depending upon whether energy is materializing from the center of a system, or parts are moving in towards their center.

[As is the case with Leo, Pisces also governs processes in matter that give both Being and Unity (but of a non-individuated, undifferentiated sort). The Pisces principle in its role as the second law of thermodynamics,pulls the separated, differentiated qualities in systems towards a less differentiated state. As noted, the substrate is the 'end' of this process—a totally undifferentiated unity. (While the substrate pulls matter towards itself and hence towards unity, it is also the source of the types of being that are selected from out of this field and emerge during a creative process, adding being to the world.)]

The Leo law, in addition to being an attractive force that secures the wholeness of an individual, is also active in causing two or more separated systems to seek unity or wholeness relative to one another. Physicists refer to this as the law of gravity.

The law of gravity is also a law of evolution. Due to centralization, there is a constant pull within systems towards wholeness. The Leo law presses systems to add more and better material parts and functions, and hence to improve qualitatively. When a measure of wholeness has been achieved, the system looks towards higher levels of wholeness. It does so via external interactions. At any one point in time all systems are not only wholes, they are also parts within external wholes-in-the-making. Thus an atom is a whole but it unites with other atoms within the wholes-in-the-making we refer to as 'molecules.' In a like manner we are 'atoms' and we interact with other people in groups. This is our human-level 'external' world. These external wholes-in-the-making are also pulled towards (true independent) wholeness via the temporary spirit involved with any nexus of coherent interrelationships. In its form as a gravitational attraction between systems, the Leo law leads a system to join with other systems and in this way to eventually rise to a higher level of wholeness. [Thus one-celled organisms become many-celled organisms. We ourselves are composed of various bodily systems (like the endocrine system), each of which is composed of organs, which are composed of cells, which are composed of molecules, which are composed of atoms.] Evolution is due to spirit, as a force of Unity, drawing separated systems to unite and achieve ever greater and higher levels of wholeness. This produces hierarchy in nature with systems nested within systems.

The Pisces type of being is a principle of qualitative sameness and, as noted, is responsible for what physicists refer to as the second law of thermodynamics. This principle can, in a disabled world, produce a movement towards less and less qualitative differentiation. Scientists believe that the second law of thermodynamics will eventually lead to a state of undifferentiated chaos. It is believed that one day there will be no people, no atoms, not even massive particles. According to the astrological cosmology, however, the Piscean principle is not only opposed by a law of qualitative betterment, it actually complements the Leo principle, and is the back side of the evolutionary process. (In all three of these domains, the role of the purely negative principles is to assist the character and directional value made manifest by the purely positive principle in that domain.) The role of Pisces in uplifting matter becomes more obvious when the world is seen as hierarchically arranged, in other words, in terms of sets of these principles. (There is a Pisces principle in higher level sets as well as in lower level sets.) When systems in an external world act together coherently and seek to be 'parts' within a higher level of wholeness, they must find a common point of unity even though they are different. In order to do so (precisely because they are different at the level at which they exist), they must find a *higher* point of sameness, such as an overriding common purpose or function. When many different things are united for the same purpose the principle of sameness makes this possible. The Pisces principle of sameness is as essential for evolution as is the Leo principle.

When looking upon Leo as a law of evolution, it is also helpful to think of it as a principle of superiorization. This term corresponds to the high in space, and each of the domains can be correlated to simple places in space. In the qualitative domain Leo correlates to the high in place and Pisces to the low. This 'inferior' and 'superior' designation has nothing to do with good versus bad. There is nothing wrong with a lower level of being. These two together just make layers possible, with a subsequent hierarchical arrangement of systems.

It is the two other signs in the qualitative domain that make it possible to understand what soul is—both that it exists and is real, as well as what it is. And since 'soul' is an intermediary between spirit and matter, they also make it possible to understand how the material word is created from spirit. These two signs are Aquarius (+/-) and Virgo (-/+). Each combines its two parent principles [Leo (+) and Pisces (-)], but in a slightly different fashion.

It is spirit individuating that produces a body and soul, and both poles of spirit are used to do this. That is why the Aquarian and Virgo principles may be spoken of as 'children' of their two parent principles Leo and Pisces.

Aquarius is closer to the nature of soul, and hence may be spoken of as an immediate mode of soul. Virgo, on the surface of things, is an immediate mode of matter but, as will become evident, it also cannot be what it is, nor do what it does, without soul.

The soul is the wholeness of a system. That wholeness presides over the parts, giving them their specific, individual place within the 'form' of the body. Aquarius rules the effects of a whole on its parts (formal causation). For this reason Aquarius, as noted, can be spoken of as an immediate mode of soul. Of the two signs (Leo and Pisces) that are its parents, Aquarius is closest in nature to Leo. That is because soul (as with spirit when it manifests through the Leo principle) is a means to individuation. Nevertheless, wholeness involves a set of potentials taken out of the substrate and hence Aquarius is only primarily positive (+/-). It also has elements of Pisces in it.

Aquarius makes it possible for there to be a combination or union of the high and the low, and produces the actual connectedness between parts and wholes. It rules the part-whole duality that all true systems possess. Each system in a hierarchically constructed universe is both a part and a whole, and which it is depends upon the level of observation and experience. Hence, as an active principle or law of nature, Aquarius produces the state of affairs referred to as hierarchy in nature. This law of physics has not been recognized by science to date, despite evidence of the existence of hierarchy in the universe virtually everywhere.

The parts of any whole are produced by the twelve fundamental types of being. They are what is meant by a 'part.' Each is a way of looking at and expressing pure Being. Each type of being is a type of unity as well—a perspective on pure absolute Unity. The fact of the essential oneness at base of the twelve different types of being makes it possible for there to be coherence and unity between different things. This produces law and order in the world. Because Aquarius rules part-to-whole relationships it governs order and coherence per se—and governs it in both the mental and physical realms. Coherence is the 'cup' of life that produces a proper material container for spirit. [In mythology Aquarius is called the Precious Urn, and the god Uranus is said to be in charge of the urn. (In astrology Uranus is the planet connected to Aquarius.) Uranus is pictured holding a sacred vessel. From it the waters of life are flowing.] The Aquarian principle makes it possible for there to be many *different* things working together and united together as one. This is what makes them true 'parts.'

The Virgo principle governs qualitative apartness. It makes it possible for things in the material world to be different from one another.

As with all the signs it too has an active manifestation, and when involved with a process is more clearly seen as a law of nature. As a dynamic principle it is a law of divergence—one that was recognized by the biologist and paleontologist Pierre Teilhard de Chardin as one of the laws promoting evolution. [2] This law is not yet recognized as such in mainstream biology or physics.

The reason that the Virgo principle is a major factor in evolutionary processes is because diversity aids evolution. It is when many different parts act together coherently that there is a chance for them to unit as a true whole. And it is precisely the diversity that provides the need for them to find a *higher* point of sameness. In other words, they must reach up if they are to find a point of unity.

As an active principle of qualitative divergence, this type of being makes it possible for there to be different individuals, different levels of being, different dimensions, different types of qualities, and even different realms (physical, emotional, mental). Virgo unites its two parent principles (Leo and Pisces) in a neither-nor fashion (neither x nor y). As the polarity partner to Aquarius (and likewise a child of the Leo and Pisces principles), Virgo has the character it does based upon its combining its two parent principles, but in a different way from that of Aquarius. It is individuality (made possible through Leo) impressed upon the undifferentiated sameness of the substrate (as reflected in Pisces) that produces qualitative differentiation in the world. It is the wholeness of an individual with its distinctive boundaries that makes it possible for parts to have distinctive boundaries, i.e., to be different and set apart. (What is a whole at one level of being is a part when experienced at another level.) Individuality is the means to qualitative differentiation in the world. And apartness (Virgo) is merely the flip side of its polarity partner, Aquarius. It is part-whole duality that produces qualitative differentiation or apartness.

It is in seeing these two 'children' as offspring of their 'parents' that makes it possible to see why light has wave-particle, part-whole duality. In having a nature that is both universal (Pisces) and individual or particular (Leo), spirit contains within itself the grounds for the part-whole duality that exists in light. Like spirit, the material world itself is a duality-within-unity. Everything is composed of light, and light has wave-particle duality. The wave aspect serves as the wholeness of a system, and the parts are what matter consists of. But, as with spirit, the part-whole duality is not a real duality. In a hierarchically constructed universe a part is a whole, depending upon the level of experience. (Ultimately everything is a part of the universe as a whole.) This is a phenomenal way to express the fact that spirit is both a universal substance and a principle of individuation.

All these subjects will be discussed more in depth later. But one very important thing that looking at these interrelationships reveals is what was touched upon in the previous section when discussing the -/+ principle Capricorn as it creates spatial boundaries. The -/+ principle Virgo in this domain creates qualitative boundaries—that which we speak of as qualitative differentness. These principles make it clear that the world is created in such a way that there need be no empty spaces to separate one thing from another, qualitatively or spatially. To create distinctions merely entails the union of two levels, but with a drop in perspective from one level to another.

Even relative to part-to-*part* relationships at the same level of being, when something changes from one quality to another, the two different parts merely switch from one type of being to another type (another part or facet of the whole). This is the relationship of otherness or part-ness and does not involve any separation. If Being is what is, and the material world is composed of types of being, then there should never be empty spots in a system. The reason this is important is because these are the parts of our body. Later we will be talking about what happens when parts come apart, as they do at death, and partially so with any wound or illness. These conditions entail lack of wholeness, which is an unnatural condition.

In summary, these four types of being acting through the qualitative domain make it possible to have individual wholeness, part-ness (the part-to-whole relationship), apartness (differentiation), and universal undifferentiated unity.

(Up) – **Leo**: individual wholeness (*higher-level, 'superior' being*)
(Down) – **Pisces**: universal undifferentiated unity; qualitative sameness (*lower-level, 'inferior' being*)
(Up and Down) – **Aquarius**: qualitative part-ness; parts as they are parts of a whole (*connections between inferior and superior levels of being)*
(Neither Up nor Down) – **Virgo**: qualitative apartness; parts as

they are apart (*qualitative differentiation*)

In addition to these laws producing wholeness and evolution towards ever higher states of wholeness, the four laws in the qualitative domain also produce the so-called 'inverse-square' laws. Spirit is a power as well as a substance, and is behind all manifestation of power in the universe. When two systems interrelate, a force center is generated. Leo governs spirit as an individuated expression of power. With a material force there is a specific influence, and it arises at the point of, or center of, the impact. Because the quantitative domain interpenetrates with the domain of power and the qualitative domain, that specific influence is greatest in magnitude at the center or point of generation, and then fades gradually in an ever-widening circle outward towards a condition of irrelevance in terms of the original influence. Leo rules the center and point of the greatest value of the impact. Virgo provides for the dividing line between qualitative, quantitative and causal relevance, verses that which is irrelevant—the circumference of the circle so to speak. Pisces governs that which is undifferentiated and is meaningless relative to the specific causal influence.

Inverse square laws are a fundamental form of value present in light. Like the parts-to-whole dichotomy, center-peripheral relevance produces direction in the universe. They are evidence that there is value in nature.

In the astrological category system, the natural direction of processes in all three of the domains is revealed through their classification via the 'elements.' The primarily positive (fire) signs lead the way. In the kinematical domain Aries governs that which is to-the-fore while Scorpio rules behind-ness. Aries is the purely positive (fire) sign in this domain. It rules motion as a 'going forth.' Motion is not only a going forth; we move forward in time. [The symbol for Aries is a ram charging forward.] In the spatial domain space expands instead of shrinking (and individual extents jut out into space). Sagittarius is the primarily positive (fire) sign in this domain and rules growth and extending-ness. [The sign Sagittarius is symbolized by an arrow (f).] The tip of the arrow attests to its leading role. It is natural for space to be an extent and to expand. Leo is the purely positive (fire) sign in the qualitative domain. It governs individual expressions of spirit. It is natural for the number of individuals to increase in the universe, and for those individual that do exist to increase in the expression of their individuality—to evolve.

The existence of directed movements in nature presupposes something absolute (non-relative and other than material being) towards which things are headed, in other words that they are headed towards something other than other material parts. In the astrological cosmology it is spirit or absolute Being-Unity that produces these directed movements. Systems are seeking more being and more unity. Relativists claim that there are no directed movements in nature, and that is a consistent stance for them to take since they believe there are no non-material realities. However, if the world shows evidence of there being directed movements, the relativist position may not be the correct one.

The Twelve Types of Unity

The twelve types of being are concurrently twelve types of unity intended to promote wholeness in the three domains that all systems manifest.

(1) The Kinematical Domain and Unity: Change (♏) and motion (♈) are serial in nature. Serial-ness is made possible through the Scorpio and Aries laws. These types of being are what is meant by a process. A process is continuous, with no breaks in its nature. *Serial-ness is a kinematical type of unity.*

There is a tendency to think of motion as something that bodies manifest externally as they move about in their external world. But motion is essential to the meaning of what a 'part' is within a body as well. Scientists now speak of the internal parts of systems as composed of processes. As this relates to wholeness, the internal parts of systems are naturally intended to flow and interact in a manner that expresses kinematical wholeness.

The third type of being in this domain is Newton's third law of motion. It is made possible through the Libra (♎) Law. Its symbol is a scale beam. In France Libra is called 'Balance' (from the word 'balance'). *Balance is a form of unity*. In electromagnetic phenomena, the attraction (and constant interaction) between positive and negative poles of electric and magnetic forces is an expression of this form of unity—which is what we call *polar unity*.

The fourth type of being in this domain produces the unity of stasis—non-change of amount or degree of motion. [This is Newton's first law, or the Taurus (♉) Law.] Standing still or stasis is non-change, which is *continuity through time—a form of unity* producing state.

(2) The Spatial Domain and Unity: In this domain, the Sagittarian (♐) type of being makes possible *the continuity of extent, which is a form of unity*. An 'extent' is by nature continuous. (If there is a break in continuity, then there are two extents.) The Sagittarian form of differentiation makes it possible for the space-time continuum (that which we call 'space') to be continuous. It is also responsible for the spatial wholeness of individual systems.

The second form of unity in this domain is the Cancer (♋) type of being. Cancer produces *the type of unity we speak of as 'inclusion.'* We verbalize these names for different types of unity but do not think about what they mean. Everything in the inward part of a body is included within its nature.

The third type of being in the spatial domain is Gemini (♊). It produces *the type of unity we refer to as spatial 'connections.'*

The last type of being in this domain is Capricorn (♑). It

makes spatial boundaries possible. This would seem to entail division—the opposite of unity—but in fact merely involves a relationship between levels of being. In a hierarchically constructed universe, 'within' and 'without' are a matter of the level of experience. Material boundaries exist because what is a whole at one level is a part at another, and the whole, because it is a precise individual, makes it possible for parts to have adequate boundaries. Boundaries are a result of part-whole *relationships*, and a relationship is a form of unity.

(3) The qualitative domain and unity: The first type of being in this domain, Leo (♌), makes individual *qualitative wholeness possible, which is a type of unity.* It does this through *centralization, which is a type of unity.*

The second type of being, the Pisces (♓) principle, produces *the type of unity we refer to as qualitative 'sameness,'* and universal *'at-one-ment.'*

The third type of being, the Aquarian (♒) Law, produces *the type of unity we refer to as order and (dynamically) coherence.* It is the unity of many different parts existing together and acting together as one. This form of unity is the relationship of *part-whole connectivity.* It produces hierarchy in nature.

The fourth type of unity in this domain is the Virgo (♍) principle. This produces qualitative differentiation which, as with the spatial separation produced by Capricorn, is a form of apartness. Like spatial boundaries, this would appear to not be a type of unity at all, but in reality Virgo is a polarity partner to Aquarius, the principle of part-ness. It is the relationship of wholeness to part-ness, and between levels of being that makes spatial boundaries (Capricorn) and qualitative distinctions (Virgo) possible. As was mentioned in regards to Capricorn in the spatial domain, in this domain Virgo expresses a *relationship,* and a relationship is a form of unity.

Even relative to part-to-part distinctions, things that are different are not essentially separated. They are merely different *facets* of one another, and different *parts* of a higher level whole. (Facets/parts are a type of unity.) Although the material world allows for differentiation, because that multiplicity entails merely different levels and different facets, even qualitative differentness is a form of unity.

[*When it can be shown (as has just been done) that each of the types of being (each law of nature) is also a type of unity, this is evidence of the connection between unity and being. When, furthermore, these are shown to be twelve types of unity producing three types of wholeness (as has just been elucidated) that is evidence that the material world is intended to be whole. Perhaps even more importantly, because (via the Leo principle) these laws are shown to be centered upon us, it is clear that the laws of nature are intended to produce individuals, and that those individuals are intended to be whole.*]

This completes this chapter on the category system. These are the laws of nature that are expressed via light. Every photon has the capacity to manifest these twelve basic types of being in our three dimensional world. The twelve it does manifest are the 'colors' we enjoy as we paint the landscape of our own nature and that of our collective world.

The astrological category system is a powerful system of thought that documents the basic archetypes underlying the material world—the same that are the means for mental concepts as well. The power and scope of this category system can be seen in every aspect of its nature.

To begin with, there is the self-evident and simple character of the basic types of being that are closest in nature to the archetypes. Motion, change, inertia, causal exchange, spatial extent, spatial connections—these are fundamental to our world experience.

In addition, there is the ability to connect them to the laws of nature. Furthermore, the category system shows how the laws of nature in a particular domain are related to one another, making it possible to understand each one better.

The cosmology that emerges from this category system is extremely rational and coherent. As will hopefully become evident as this book unfolds, the paradigm it makes possible fulfills all the criteria for truth that were given in Chapter 1.

Another strong point present within this category system is its 'innocent' nature. Astrologers do not know they are working with a philosophical category system, and one with an underlying cosmology that can overturn the interpretation modern physics has of the universe. Astrologers have not been interested in the philosophy of science or in cosmologies. Astrology has always been used as a practical rather than a theoretical discipline, and used specifically to explain circumstances at the human level of being, while scientists conversely have been interested in the laws of atomic interactions. Astrologers, for instance, are aware that people with the principle Pisces prominent in their make-up are not interested in being distinctive individuals. They prefer to lose themselves in a crowd. What astrologers are not aware of is that this is the second law of thermodynamics, and as a principle is an expression of spirit as the undifferentiated substrate—the opposite of spirit as a principle of individuation. The astrological cosmology has to be extrapolated from astrology as it is used. The innocence of astrology's 'rulership' system is evidence that the underlying cosmology is there to be discovered. This is not an invented metaphysical system.

Astrologers know very well that astrology works. It works because it is real and true. The category system is there just beneath its use in practice with human beings because it is real and true as well. And that highlights another strength that it possesses. One day astrology will be studied scientifically. When it is seen that it works in the way it is currently used relative to human beings, that will provide empirical verification that the category system itself is valid. This category system possesses an inherent capacity for proof.

In addition to being self-evident, rational, innocent, and empirically verifiable, another thing that is a major strength in this category system is its incredible comprehensiveness. And by that I do not mean just that it is a complete system relative to the twelve basic types of being it speaks of—in other words, that everything possesses these. In addition to the basics like extent and motion, there are millions of different kinds of things in the world—some of them quite complex. It is necessary in a (good) category system to be able to account for, not just the most fundamental types of being, but literally for anything and everything. No scientist or philosopher has ever attempted to do this, but it is done every day in astrology. Due to hierarchy in nature, and because life is simple as well as complex, no matter how complex something is, it is composed of the twelve basic types of being. In other words, its complexity is due to the combination of these simple types of being, and specifically in a layered fashion. In addition, no matter how complex something is, it has an overriding ruling principle. For instance, Gemini rules spatial connections. In the human level world it governs bridges. No matter how complex a particular bridge might be, its ruling principle is Gemini. That is because the essence of it is to connect two places in space. Even people have a ruling principle. It is their sun sign, and denotes what quality they lean the most towards in this lifetime. Any particular person, place or thing is composed of all the laws of nature. But what is involved in a ruling principle is an overarching ray of light that defines the essence of the whole. The complexity comes about in terms of levels of being, each of which are simple themselves and involve the twelve basic types of being. (This was touched upon somewhat at the beginning of this chapter, and will be discussed more in depth in the chapters to follow.) This category system can, not only account for the simples in our world like mass and motion; it can account for anything and everything, and in a very rational and simple manner.

Over and above all these pluses, the most important thing about this category system is its capacity to show the interrelationship between the material world and spirit. This set of fundamental principles is the basis for formulating a new spiritual paradigm to replace the old materialistic one, and the beauty of it is that it does it through matter and its laws. And it is not just that all twelve of these laws can be shown to be types of being and unity (thus for identity purposes supposing the existence of pure absolute Being-Unity). But rather that it is a category system with specific laws that provide evidence (via material processes) that spirit and soul are real substances even though they themselves are metaphysical.

These laws not only show that spirit is real, they show that it has the two expressions spoken of (as a center in individuals and as their substrate). Hence these principles show the manner in which we are related to this spiritual substance. They indicate that, as individuals, we are the means to creation. And by that is meant, not just that we have a spirit through which we can create, but that we are the creation that we create. Our bodies are composed of these twelve basic types of being. They are actualizations of the potentials in our soul, and that soul is a piece of the substrate. Our spirit, our soul and our body are all three spiritual in nature.

It is for these reasons that I said at the beginning of this book that it is possible to have answers to spiritual questions—answers it was never before believed were possible. With this category system the laws of physics—matter itself—has answers to spiritual questions.

From this point forward this book will entail taking this category system, and from it constructing a spiritual scientific cosmology—a new paradigm. The place to start is at the beginning—with spirit.

Chapter 3

Spirit

If something called 'spirit' exists, and we are indeed to undergo a paradigm shift, we need at the very least to be clear about what is meant by the terms 'spirit' and 'spiritual.' When someone is asked to describe what 'spirit' is, they either have never tried to think at all of what it might be like, or at best have visualized it as energy (which actually is matter in a different form) or as some kind of vague, ethereal (but still basically material) substance. It might be a good idea, before discussing the role of spirit in a new, spiritual paradigm, to have a clear concept in mind for the term being used.

The signs of the zodiac show clearly that the material world is composed of twelve basic *types* of being, and furthermore that these derive from a source that is pure absolute being. In this category system absolute being is referred to as 'spirit'; and that makes sense since spirit has been linked to absolute, infinite and eternal realities. Nevertheless, this connection between spirit and these three qualities needs to be explored.

The connection between spirit and matter also needs to be clarified. If the material types of being emerge from or are composed of spirit, then that is contradictory to what has been traditionally thought about spirit, which is that it is non-material in nature. So what is this thing called spirit? Is it material or is it non-material? For that matter, what is matter? Is there something spiritual about matter?

And what about our ability to know what spirit is? To know that spirit exists and what it is like, we would need to have something about our own nature that was spiritual in essence. For knowledge to take place there has to be something in the knower that is the same as that which is known. If matter is derived from spirit, and especially if it is composed of spirit, then we could come to understand through matter, including our own material bodies, what spirit is like. Even if matter is not spiritual in essence, we still could have something spiritual about us (e.g., a 'spirit') by means of which we could experience spirit and understand what it is like. As far as that goes, it is not necessary to choose one or the other means to knowledge. It could be that we have something about us that is absolutely spiritual in nature—and not material at all. It could also be true that matter is composed of spirit.

In order to provide for a clear concept, it seems to me that the best place to start might be with trying to understand and define spirit in a pure sense first. If spirit exists in a pure, non-material form, then some of the most basic attributes of matter could conceivably be used as a guide to understand what pure spirit is not.

We are a material system and we have a number of different qualities, two of the most fundamental of which include our capacity to move about and our spatial attributes. Relative to spatial attributes, material systems have spatial extents. And whether looking at a quantum, an atom, or we ourselves, these extents are finite in extent, i.e., limited. If what is spiritual is non-material, then it would by definition not be limited. What is not limited or finite would presumably have to be infinite.

In addition to having a spatial extent, another major material quality that all systems possess is the capacity to move and change. Something that is not material is not going to move, and what does not move does not change. (This is also why spirit cannot be looked upon as energy, which moves and, in the form of kinetic energy, is motion itself.)

Nor would spirit possess any temporal qualities. Time is a material attribute closely related to motion and change. What does not move, cannot move through time.

Time can be spoken of in terms of a system coming into and

out of being as well as in terms of time's involvement with motion-change. What has nothing to do with time cannot come into temporal being, pass 'through' time, and then pass out of temporal being. Spirit, because it does not move or change and remains forever as it is, would seem to need to be either outside time altogether (atemporal) or else be everlasting, i.e., go on and on in time forever. The latter does not work because time (even everlasting time) involves motion-change. What is outside time but exists would be eternal in nature.

In addition to the primary attributes of spatial extent and motion/change, matter has all sorts of miscellaneous qualities. The various qualities that qualify matter entail being different *types* of being. Where there is a split into types via differentiation there has been a separation. It is not a spatial separation; but it is a qualitative separation, and hence is still a type of division. What is not extended cannot be divided. Pure spirit would not possess qualities as we know them.

There are other ways to see that spirit would not possess qualities. Quantum physicists claim that the world is composed of quanta. These are viewed as processes of energy interactions. These interactions are the qualities that matter possesses. In other words, not only do systems relate externally, internally they are a conglomerate of dynamic interactions—a nexus of interrelating motions. Even when considering the essential nature of 'motion', kinetic energy is a 'going forth' relative to something else. In other words, it is a relationship. It does not matter whether matter is visualized as a set of dynamic relationships or as a set of stationary relationships, the very idea of a relationship entails two or more things relating to one another, and this entails division. Pure spirit would be neither extended nor divided, and hence would not possess any material qualities.

What has no qualities is no-thing, i.e., no material thing, but not necessarily nothing. The symbol that represents our concept of nothing in mathematics is zero. When symbolizing no-thing in spatial terms, geometricians refer to that which has no length, breadth or depth in our three-dimensional world as a point. (Actually a 'point-*instant*' since space is now looked upon by physicists as being temporal in nature as well as spatial, i.e., as a space-time continuum. But, when describing spirit, the 'instant' of the point-instant must be visualized as truly atemporal, and not just an infinitely small bit of time.)

In addition to not possessing qualities, pure spirit would also be unqualified in the sense of being non-relative. When relationships make something what it is, then that object or state of affairs is relative in nature. Nothing material stands alone, but rather relies upon something else for its nature. Einstein's special theory of relativity states that each thing in our world depends upon *every* other thing in the world to make it what it is. Spirit on the other hand, if it exists, is not material, and hence is not relative. By definition, what is not relative is absolute.

In summary, if spirit exists in a pure form, it is best defined as something that is infinite, absolute and eternal.

These three attributes can be determined by looking at what pure spirit is not, i.e., non-material. But what might pure spirit be in a positive sense? First of all, if spirit exists, it has to have being. To exist is to be, and being is just what is meant by existence. Spirit is not a type of being. These are what qualities are. When you strip away the 'type' from types of being, what is left is just pure Being. Spirit, if it exists, would have to be just pure, absolute, infinite, and eternal Being.

[All this does not mean that spirit would not be involved with the world of matter. In a materialistic paradigm, what is meant by 'being' and by 'existence' is the nexus of material qualities that exist. These are types of being. And the question can be asked, "How do types of being get their being? How is matter related to spirit?" And we will ask that question too as soon as spirit is first defined in an absolute sense apart from matter altogether.]

In addition to being pure absolute Being, spirit may also be just what is meant by pure absolute Unity. Zero in being non-extended is a point of absolute unity. But zero is not nothing. And as with Being, it is somehow involved with the world of matter. Spatially speaking, it takes a zero point to unify the internal parts of any system. It also takes a zero point externally to unify separated systems. Zero point must be real if there are real connections in the world, and what is real exists.

If spirit is involved with unity, it cannot *possess* unity. Matter possesses unity—a number of different *types* of unity actually. These were discussed in Chapter 2 in regards to polar unity, spatial unity, etc. If spirit is non-material in nature, it would not *possess* unity, or express a particular *type* of unity. It would only be pure, absolute Unity.

Furthermore, if spirit is just what is meant by Being and Unity, these cannot be two separate things. For spirit, these would not be qualities in the normal sense of the word. Being and Unity would have to be the same thing for spirit.

Let us try out this hypothesis. Let us say that pure spirit exists, and define it as absolute, infinite and eternal Being-Unity.

Having created a foundation by setting forth a clear definition for 'spirit,' the next step is to figure out if such a thing really exists. There are two ways of finding out the truth about this. One way is if we possess this thing or 'substance' in a pure, absolute form. Another is if our matter is derived from it or composed of it. We can then know that it exists and what it is like *through* matter.

Let us explore both these routes, but reverse the order right now, considering the possibility of the material road to understanding first in **(A), 'Understanding Spirit through Matter,'** and then come back later to **(B) 'A Direct Experience of (Pure) Spirit**.'

(A) Understanding Spirit through Matter

If spirit is Being, and matter consists of types of being, then matter may either be derived from spirit in some way, or consist of spirit. We will use the latter, stronger hypothesis—that matter is composed of the substance of spirit—and you will see why it is possible to do so in a minute.

If matter is composed of spirit, and spirit is both Being and Unity, then it is to be expected that matter will have both being and unity—and it does. Relative to the Being aspect of matter, it goes without saying that matter possesses being. It would not exist if it did not.

According to the AC, matter consists of a number of different types of being. It is the thesis of the cosmology that emerges via the astrological category system that there are twelve basic types of being in our world, and that these produce every type of material quality we experience.

These twelve types of being are also twelve types of unity. Together the twelve types of unity produce three types of wholeness in systems, referred to previously as three 'domains'—the kinematical domain, the spatial domain, and the qualitative domain. Each of the three domains produces its own type of wholeness.

If it becomes possible to see that matter is composed of types of being that are concurrently types of unity, then that is one indication that matter is composed of spirit—which is Being-Unity. But there is another way to arrive at this conclusion as well. Because matter is in essence Unity as well as Being, it is fruitful to go deeply into the subject of wholeness, which is a material form of absolute Unity. It is via qualitative wholeness that it is easiest to see how spirit produces matter. Unfortunately, science does not know anything about the existence of the qualitative domain. The only two domains currently recognized in physics are the quantitative-spatial and the kinematical domains. Surprisingly, one of the best ways to make the case that matter is composed of spirit may be to utilize some current data emerging in math and science in regards to the wholeness connected to these two domains. Let us consider spatial wholeness first.

Spatial Wholeness

One of the primary qualities that systems possess is that of having a spatial extent. It is geometry that specializes in the subject of spatial extent. Because something spiritual has no material qualities, it would have no extent. Geometricians refer to something with no extent as a dimensionless point—variously a point-instant, which has no extent in either space or time. It is not manifesting as a world line in any of the dimensions, and is zero relative to all the spatial coordinates—even time.

Unlike a point-instant, zero is a mathematical concept—but one that is closely related to the geometrical concept of the point-instant. Zero is no thing, and at one time it was treated as nothing, but that has been changing. Mathematics, like science, is in a state of evolution. It has taken a long while for us to grow from the use of zero as a mere place holder to an understanding that it is something. (John Barrow in *The Book of Nothing* gives an excellent history of the evolution of our concept of zero.[1])

One of the tasks that zero accomplishes within the real world is to connect separated systems (at zero point). Zero may be no (material) thing but, if it is not something real, then we may not be able to account for real connections in the world. At point zero unity takes place, not nothing. What is real exists. What exists has (or is) being. We have spoken before of the relationship between unity and being.

As with external connections between separated systems, the internal parts of a single system also need to be connected. It is in looking at this type of connection that the Being as well as the Unity of zero can most clearly be seen. This is also how it is possible to begin to visualize how matter is composed of spirit. As this relates to the material attribute of spatial extent, one of the most important discoveries in modern mathematics has been the solving of the problem of how a continuous extent can be continuous. This branch of math is called the 'mathematics of the continuum.'

Mathematicians, when visualizing the spatial attributes of our three-dimensional world, see the three possible dimensions of length, breadth and depth in terms of lines, surfaces and cubes. When trying to understand through geometry what a spatial extent is, it is helpful to look at the simplest type of extent, and that is a one-dimensional line. In geometry a line, if it is to be seen as continuous and unbroken, needs to be looked upon as *composed* of dimensionless points—in fact, an infinite number of them.

The first time I thought about the possibility of a line being composed of dimensionless points was while reading a book by a mathematician-architect by the name of Claude Bragdon, and I heartily recommend two of his books to any earnestly enquiring reader.[2] The new mathematics of the continuum is now beginning to formalize ideas of this sort, but Bragdon writes very plainly and simply for the beginning explorer.

To see a line as composed of points is to see it as composed of no-things, and that is not an easy thing to grasp conceptually because the extent of a line seems so solid, but it makes rational sense. First of all, the term 'extent' carries the connotation of wholeness within it. By definition a spatial extent is a whole. If there are any breaks in an extent, then there are two extents. Secondly, if the points of unity within the extent did not have being, they would not be real and the union could not occur. Thirdly, if the parts of a line or extent are seen as themselves extended and the point only as something that unifies them and gives them continuity, then the 'extended' parts would have continuity to their extent, and that would have to be explained. What this comes down to is seeing the parts of the line as themselves non-extended, but as capable of producing the line. It is when the nature of the point as the infinitely small is expressed to an infinite degree that a material extent is produced. 'Matter' then becomes the *relationship* between two expressions of spirit. The astrological category system indicates that spirit has two manifestations, both of which are infinite. We experience one manifestation as an infinitely 'tiny' point-instant when spirit is busy creating a world. But spirit also has another manifestation as an infinite potential—the substrate. Wholeness—a material extent—can be seen as the expression of the infinitely tiny to an infinite degree. It is this that makes it possible to begin to visualize matter as composed of spirit.

This is also a way to see that Unity is Being and Being is Unity. In looking at the spatial extent of a line, it is not just the wholeness or continuity of the extent that is dependent upon the point. It is true that the point produces the continuity; but the continuity is also the extent, and the extent is the being of the line. In other words, the unity is the being.

There is currently a controversy in both physics and mathematics as to whether we should quantize over minute quantities or over dimensionless points. Describing particles as dimensionless points is a correlative of the mathematics underlying relativity theory. (The ideal particle is infinitely small; the ideal field is infinitely large.) A view of the world as composed of point-instants is gaining in popularity due to the breadth of its explanatory nature—including its capacity to account for wholeness in nature. Relative to the real world, this is the question of whether matter arises from no-thing or begins with small extends (tiny quanta, tiny bits of process, tiny strings, a Plank's length, etc.). Quantizing over the point-instant entails a reversal of previous atomistic paradigms that have involved quantizing over 'infinitesimally' small (but still extended) extents. (Even quantum theory is still a type of 'atomism' in this respect. A quantum can never be smaller than 10^{-33} cm.) If this preference for the point-instant continues in mathematics and areas of physics like relativity theory, matter may come to be looked upon as composed of no-thing. Our most basic material attribute—that of spatial extent—will have to be seen as dependent upon and composed of that which has no extent!

The famous mathematician Georg Cantor (1845-1918) is today the recognized authority on the mathematics of the continuum, or what is sometimes referred to as point-set theory. The mathematics of the continuum confirms the somewhat more intuitive perception of geometry that it is two perspectives on (or two types of) infinities that are behind the wholeness of a system—behind the very composition of its extent. The truths of geometry can be visualized easier because geometry is closer to physical reality in that it studies spatial extents. However, the nice thing about mathematical confirmation is that math provides 'proofs' and more extensive ways to manipulate ideas. It was Cantor who gave a mathematical proof that wholeness and the infinite are integrally related. A point on a line can be represented by a number. If every point on a line is symbolized by a number, Cantor showed that it takes an infinite number of numbers to produce a continuum.[3] Like Albert Einstein and his theory of relativity, Cantor is one of many people who must be looked upon as midwives to the new paradigm. We will talk a little more about Cantor's life in Chapter 4 and how difficult many aspects of his career were because he did not believe in the (atomist, materialist) manner of quantizing over tiny infinitesimals but rather had faith in the reality of the infinite—both the infinitely small and the infinitely large. Although his work was criticized in his lifetime, his proof is now universally accepted.

Unlike physics which deals with physical matter, all this talk about lines in geometry may at first seem far removed from science and the real world. Our world is not one-dimensional like a line. Nor is it two-dimensional, consisting merely of surfaces. It is three-dimensional, and is composed of three-dimensional world-lines. However geometry is the branch of mathematics that is closest to physical reality—spatial attributes in particular—and it was Claude Bragdon who helped me see that the dependence of our cube world and even of higher worlds upon the infinite remains as dimensions progress upwards. It takes an infinite number of infinitely 'small' points to make a line whole. It takes an infinite number of infinitely 'small' (narrow) lines to make a surface whole. And it takes an infinite number of infinitely 'small' (slices of) surfaces to make a cube whole. Thus what happens each time a system ascends to a higher dimension is that spirit as an infinite potential is again accessed.

[The fact that it takes an infinite number of points to make an extent whole means that any truly continuous extent apparently has to have the attributes of a world. That is because only a world is unbounded and hence unending in extent. What spirit expresses as its substrate pole (infinite potential), an individual soul must express as unending-ness. We will talk more about how a soul is a kind of 'world' in the chapter on '**Soul**.']

Under the current materialistic paradigm it has been very difficult for the existence of anything infinite to be recognized as real by either mathematicians or physicists. This is true of both the infinitely small and the infinitely large. It is only recently that the mathematics of the continuum and relativity theory have helped establish the reality of both types of infinities, including their interrelationship. That is why the current evolution of thought in mathematics and physics is so important to the construction of a new, spiritual paradigm.

Kinematical Wholeness

As with the wholeness involved with spatial extent, motion is serial in nature. It has its own type of wholeness. Although he provided a proof of the need to quantize over dimensionless points, Cantor was not the first to devote his life to this idea. The German philosopher and mathematician Gottfried Leibniz (1646-1716) was a co-inventor of calculus. The purpose of calculus is to find a zero point between two moving processes. Integral calculus studies rates of change between two or more variables.

At some point, two changing things that are being related to one another must find a common ground where they meet at zero point.

The ancient Greek philosopher Zeno (490-430 B.C.) is famous for his thoughts on this subject. You may have heard about his story of the race between Achilles and the tortoise. (This and some of his other paradoxes in regard to zero point are given in the Appendix.) When there are two moving objects going in the same direction at different speeds, one does and must pass the other. In other words, there is a zero point. And yet, if you think of the distance between the two as ever decreasing material extents, you can keep halving the extents forever and never get to zero point, since an extent always has extent. (In quantum mechanics, there is no zero point. As was noted, the smallest length a material extent can have is Planck's length, 10^{-33} cm.)

It is the same with motion itself as it is with two moving, changing things finding a common ground. If motion is not internally continuous and serial in nature, then our experience of motion occurs in jumps. That is how our original moving pictures were made—through thousands of still pictures being spliced together so that the jumps or leaps were not detected. It is how cartoon figures appear to be animated for us in the movies. (Quantum mechanics also pictures motion as taking place in jumps or quantum leaps.)

The problem of the meeting ground between two related changing things is handled in calculus by a concept called 'convergence to a limit.' The idea is to get as close to zero as you want to get *for all practical purposes,* and then to forget the difference. We know that we do reach zero point. This can be proven negatively through a 'reductio ad absurdum' argument. (If a mathematical or logical conclusion shows an inconsistency than it must not be true, and is 'absurd.' [4]) However, to prove in this negative way that something has to be true (which we already know to be true), does not prove in a positive way how it actually comes about.

To have invented the calculus was a wonderful thing. It was a powerful tool needed in the evolution of our science. But it made Leibniz dramatically aware of what was still lacking and still needed. Leibniz believed that we must postulate the real existence of dimensionless points if we are to understand the domains of both space and motion. And it is easy to see why the problems of a continuous spatial extent and the continuity of motion are interrelated problems.

Under quantum mechanics particles have now become viewed as tiny bits of process. Under the previous, atomistic paradigm these were seen as tiny bits of extent. Because energy is a force as well as a substance, and also because processes seem more connected than individual extents, there has been a strong movement (via quantum 'field' theory) to see process as the fundamental attribute of matter, and spatial extent as a quality derived from forces between particles. This is how quanta are viewed in quantum field theory. This view has been especially useful because quanta are tiny bits of energy, and kinetic energy is just what is meant by motion.

Despite the convenience of seeing extent as a derivative of process, this merely transfers the question of continuity from space to motion. Quantum theory begins with a given, i.e., that motion exists, but does not say how it can exist. Quantum physicists are currently using the wave-form of particles to explain the continuity of both motion and extent (particles have wave-particle duality). The problem is, they also concurrently deny the existence of the wave-form because it is so metaphysical in character (cf. Chapter 4, '**The Role of Science in the New Paradigm**').

Actually, neither relativity theory nor quantum theory answers this question relative to the meeting ground between the infinite and matter. Relativity theory acknowledges that its ultimate particle (a dimensionless point) does become a field with an extent. (Relativity theory is a field theory.) However, it does not say how this is possible. Quantum theory begins with a tiny field and so begins with process (and hence with extent) as a given, but does not explain how motion or extent can be continuous, in other words, how there can be such a thing as 'extent' or 'motion' in the first place.

The mathematics of the continuum explains what continuity is and its dependence on two types of infinities. What it does not explain is how something infinite can wind up divided and thus less than what it is. A point is infinite—the infinitely 'small.' But it takes an infinite number of points to produce a spatial extent. This suggests that something infinite can have something outside itself, i.e., other infinite points. Since multiple points are all themselves infinite, the infinite would be divided into an infinite number of things, and that makes no sense.

This problem with understanding how something infinite can wind up being divided and hence presumably finite is the same problem that exists when trying to understand how something absolute can wind up being relative. A point (if it has real being) is absolute being, rather than a type of being like motion or extent. When something is 'absolute' (e.g., absolute zero) it is unconditioned, unqualified, and non-relative. What is non-relative (and infinite) does not relate to anything external to itself. If the types of being depend upon absolute being for their existence and their nature, which is relative, and if the types of being are relationships (and yet composed of point-instants), what is the 'relationship' involved? How can something that is non-relative (and not even separated in the first place) relate? This question of how spirit divides, multiplies and then interrelates (apparently with itself) is the question of how the material world comes into existence and, perhaps even more importantly, it is the question of how matter can still be spiritual in nature.

The AC provides a solution for these paradoxes but, without this, philosophers have been unable to do so. Leibniz, as with

Aristotle, epitomizes a process of taking reason so far, and then hitting a wall. What the astrological system of thought indicates is that two kinds of infinities produce an extent, and furthermore that these are just two different manifestations of one spiritual substance. The point-instant is (always and ever) a part of the substrate and does not have to be divided, multiplied or rejoined—rather the 'separation' that produces a point is simply an individual perspective on the whole. It is matter that produces the individual perspective. Matter surrounds a point in the substrate. All points *are* the substrate and these are not in and of themselves divided.

Leibniz came to believe that the point-instant was behind the very composition of matter. In the process of trying to understand how the world developed from the point-instant his philosophy evolved into a 'monadology.' He called the point-instant a 'monad' which comes from the Greek 'monas' ('unit' or 'one').

A monad in philosophy is an indivisible unit—the smallest thing possible. It is also something from which everything else derives. [The Italian Renaissance philosopher Giordano Bruno (1548-1600) used the term monad to describe both the universe as a whole, and the creative essence behind all things.]

As soon as Leibniz came to the conclusion that his monad was infinite and hence contained the world within it (in potential form if nothing else), it would have been philosophically advantageous for him to have rationally concluded that there was nothing external to this monad, in other words that there was only one monad. Instead he said there were multiple monads. The problem Leibniz had with his philosophy is that he could not logically separate or link his monads. In terms of separation, unlike the AC, he had no way to conceive of matter as being due to a relationship between two different manifestations of spirit. He had no concept of spirit as the substrate—only as a point-instant. Thus he could not see how to link his monads.

As if the question of multiple monads were not a serious

enough issue, if a monad is seen as containing the material world within itself, there is the question of whether it contains the world in actual or potential form. Leibniz believed, like many absolutist philosophers before him, that pure Being had to be actual in Being—that it could not be just a potential. Although quantum physicists tend to be relativists rather than absolutists like Leibniz, they are currently grappling with precisely the same problem. The wave-form as a field potential is seen as the source of material particles like photons and atoms—but, in the standard Copenhagen interpretation of quantum reality, a field potential is seen as not real. But how does something real come from a potential if the potential itself is not seen as real? [In the AC the wave-form of a photon is real (reference Chapter 4 on '**Science and the AC**'). Spirit as the substrate is real, but it is a potential for the photon in particle form, and for all the material forms built up from photons.]

Leibniz's specialty was motion, although he was interested in the character of extent as well. If the point-instant is seen as absolute being (as opposed to a *type* of being), and as the only real thing, then extent and motion may be perceived as lacking being. Because he saw no way to connect his (separated) points, Leibniz decided that motion could not exist. He came to the conclusion that only what possessed absolute Being (the non-extended 'place,' and non-temporal 'instant' of the point-instant) could be real. [The same could be said of extent—that the *connection* between points needed to make an extent was simply not real.] Leibniz decided that motion as we experience it involved a continuous creation of point instants in a serial manner. Motion entailed one point-instant going out of existence then another coming into existence in a new place at a new time—in other words, in the next instant—the creation of the next point (and the next instant). But the motion itself he believed was not real, i.e., the link between the points. (For this reason he referred to them as 'windowless' monads.) What is absolute does not itself move, and if points are not linked, there can be no motion as we know it.

In the AC, motion becomes real again. Motion is a relationship. The signs Leo and Pisces show clearly that spirit has two manifestations. One is as the substratum, which is everywhere and everything (in potential form). It is in the material world that spirit becomes a point-instant. But spirit never ceases being the substrate. It never ceases being every possible 'where' and every possible 'thing.' For this reason, in the AC, the point-instants do not have to be reconnected. The divisions are only from the perspective of matter. Matter *is* the multiplication and the division. Matter *is* the relationships and the relativity. Motion and extent are due to spirit relating to itself.

As for problems with whether spirit is an actuality or potentiality, as was noted, spirit is always an actuality. It is Being and cannot not be. However, spirit (*from the material side of creation*) is a potential. It is the potential for an infinite number of material types of being. Those that are actualized in a system are actual matter. The world of matter with its material relationships is real. Spirit is not delusional. Matter is spirit really perceiving itself and relating to itself in terms of a set of its own potentials. Matter is not only composed of spirit, it is composed of a spiritual relationship. Matter is an ongoing love affair between the two poles of spirit.

In mythology this polarity of spirit has been expressed as the mother and father aspects of God. [In philosophy the mother aspect has been called variously the receptacle (as per Plato), or prime matter/the substrate (as per Aristotle). The father principle has been likened to a seed implanted within the receptacle/womb of nature. That seed can be visualized as a point.]

The question of how spirit as no-thing can multiply and divide, the question of how something absolute can produce something relative, the question of how something infinite and non-extended can produce a finite extent, and the question of potentiality-versus-actuality are all important and major philosophical questions that any good cosmology must answer. This section on '**Understanding Spirit through Matter**' is actually just the beginning of the philosophy that emerges from an examination of the astrological archetypes; and the ideas will be further developed in the next four chapters. The purpose of the current discussion is simply to suggest that we cannot explain the nature of several of the qualities that matter possesses without giving spirit the ultimate role in their production. Hopefully a case has been made for that hypothesis. (Relativity theory and the mathematics of the continuum are themselves very strong hypotheses.)

If only one material quality could be shown to be composed of spirit, that would be *a* proof of the thesis that our material world is a spiritual one. We have just discussed two qualities. And these two are among the most fundamental attributes that matter possesses.

Furthermore, space and motion may not just be very fundamental qualities that matter possesses. They may be the basis of all the other qualities we experience. Everything in our world is composed of light. Light in turn is composed of, and differentiates in terms of, motion and extent. Relative to motion, light is energy and, as kinetic energy, *is* motion. Relative to space, there is no 'space' external to light. As this relates to what is called the space-time continuum, this *is* a field of light. As for the manner in which light differentiates, electromagnetic vibrations have a wave-like nature. Different types of electromagnetic vibrations have different lengths or extents to their respective waves. The quality of a particular wave depends upon how many wavelengths pass a given point in a given period of time. This is spoken of as the wave's 'frequency' which blends the two factors of motion and wave-length. These are the two attributes that determine the quality of an electromagnetic vibration, and everything in our world is composed of these vibrations.

To be able to describe these two material attributes in terms of the point-instant and the substrate is to be able to make a case that matter is composed of spirit.

(B) A Direct Experience of (Pure) Spirit

I like looking at the requirements for *spatial* and *kinematical* wholeness because these help me see that matter is composed of spirit; however, it is in looking at *qualitative* wholeness and the principles in that domain—Leo in particular—that I am able to visualize spirit as, not only the basis of the composition of matter, but also as continuing to exist within systems in a non-material or pure state as well. Spirit is not just the substance of the parts of a system. It is the means for them to be a subject as well. It is in the center of a subject that the spirit in a system remains in a pure state, and it is via that center that spirit qualitatively unifies the types of being making up the material body.

The qualitative domain has to do with how all the qualities in a system—and in the cosmos as a whole—are unified, despite the fact that each quality is different. The various laws of nature which give the cosmos its character are each quite distinct. Yet, as a whole, they interact correctly, fit together like pieces of a jigsaw puzzle, are interdependent upon one another, and in short are integrated integrals. Even though each law is quite different, if one law ceased to exist or even changed by a fraction, the universe would fall apart. In order for them to work together as one, they need to be qualitatively unified.

Contained within the cosmos are all types of systems, from we ourselves, to atoms—even to the fundamental system at base of all systems—the photon, which is the unit of light itself. These systems are themselves orderly internally. An atom may be tiny, but its parts must interact coherently if it is to remain in being.

External and internal types of order are not necessarily distinct from one another. The external world presents as a system, and (as has been hinted at) a system may be a world.

Science now looks upon the matter that systems are composed of as just what is meant by the laws of nature. Because systems are created from these laws, and the laws of nature are naturally orderly and coherent, this may be one of the primary reasons that systems have the capacity to manifest the incredible order and coherence that they do.

The arrangement of the qualities or parts within a system is what is referred to as qualitative unity. The question then arises as to what makes this qualitative connectivity possible. Whether considering the laws of nature as the parts of the world, or the parts of systems as the laws of nature, the order in the universe is a material expression of a form of unity in the world, and one that needs to be explained. Something is behind this form of wholeness, just as something is behind spatial and kinematical wholeness—and it may be the same thing.

You may have noticed in the pages just past that the term 'different' has been used quite a bit. The opposite of the term 'different' is 'the same.' In order to think on qualitative wholeness, it is necessary first to distinguish it clearly in the mind from spatial wholeness. For two things to be unified spatially they must be connected spatially. For two things to be unified qualitatively they must be connected qualitatively. We use the terms 'different,' 'similar,' and 'the same' to indicate whether or not one thing is qualitatively connected to another. That which is 'similar' is like the concept of a spatial 'infinitesimal.' It never quite gets to zero point. It is when things are the 'same' that they are truly (absolutely) qualitatively unified.

The very idea that sameness entails an absolute factor gets back to the role of spirit in producing qualitative unity. When material things are different but (somehow, somewhere) unified, it may be necessity to postulate something non-material as the unification factor. The laws of nature are not unified in terms of any one particular law. Since they are not, and since these laws are just what is meant by matter, their unification factor may indeed be non-material.

Qualitative unity entails sameness, but different things are not the same, and the very differentness of the involved parts of a world or system intensifies the mystery of how they are unified. Closely related to the above is the fact that the unification factor behind the order in the universe is not visible. Visibility is something that occurs relative to the material parts aspect of experience. Atom-to-atom detection of any kind is a relationship between the eyes (which are material) and other bits of matter. The very fact that the unification factor is not visible may be an indication that it is not material. Conversely, if the unification factor is spiritual in nature, we would not expect to be able to see it.

Although the laws of nature are not the same relative to each other, what unifies them would seem to need to be the same as them in some way. The one thing that is the same about all the different types of being is their Being.

In the section on spatial wholeness, we spoke of the need to see the infinite as both absolute Being and absolute Unity. This is true of the qualitative domain as well. 'Sameness' is absolute Unity. If absolute Unity is absolute Being (a real substance in and of itself), there actually is a common element that connects the types of being.

The four astrological signs in the qualitative domain show how a system is unified qualitatively. Most importantly, the principles in this domain show the role of spirit working within a system in a pure, absolute form as it produces qualitative unity for the system. The four types of being in the qualitative domain were introduced in Chapter 2, and are called Leo, Pisces, Virgo and Aquarius. Pisces and Leo are the two primary or 'parent' principles. No law of nature is spirit. But each law manifests spirit in its own material way. Of the twelve, Pisces and Leo come closest to expressing what spirit is in a pure sense. These two archetypes express the two poles of spirit. Pisces rules spirit as a universal, undifferentiated substance that is not separated qualitatively into this or that type of being. It also rules spirit that is not separated into the personal spirit of this or that person. In distinction to Pisces, the principle called Leo shows what spirit is when it is expressed through an individual. This is spirit when it is to be found within a system—still in a pure state—but directly involved with the parts within a particular individual, giving *that* system its Being and its Unity. We will come back to the subject of the substrate aspect of spirit in the chapter on '**Soul**' since soul is a piece of the substrate. It is spirit as it manifests in a pure form within a system (the Leo principle) that will be the subject of the remainder of this chapter.

In the AC, Leo is a universal law of nature. It produces processes of centralization in systems and states of centralized order. Processes of centralization and centralized order are evidence of something doing this, and it is logical to think of the 'something' as existing in the center of a system. In the AC that 'something' is called spirit. It is a power as well as a substance.

Because in the AC Leo is a universal law of nature, it must be seen as existing in every system. In other words, every material system has a spirit.

Despite the presence of this power and substance in every system, the remainder of this chapter will concentrate on how pure spirit manifests within us. The purpose of this is so as to provide a personal form of 'proof.' Each of us knows what we possess, because we directly experience these things. By looking closely and honestly at some things within our own nature that are manifestations of pure spirit it becomes possible to experience a very personal type of proof that spirit is real—that we do in fact possess a spark of this divine substance within ourselves.

In the next couple of pages, while trying to understand what pure spirit is as it manifests through us, we also need to know where it is, and why it exists where it does. Although Plato came to a decision, via philosophy, that there had to be absolute being (and absolute unity) 'somewhere' in order to give being and unity to things, the problem he wrestled with was, where on earth it existed. He finally decided that it did not and could not exist in the material world, because it was not material in nature. He decided the Form of absolute Being was in a higher realm somewhere. Plato did not consider the possibility that we ourselves might have absolute Being within us as our spirit. Leo not only confirms that it is not 'out there' somewhere; it indicates that it is inside of us. Leo even says precisely where it is—that it exists in the center of our material body. From antiquity Leo has been said to rule, not just spirit, but also centers. Each sign has a corresponding planet, but Leo is unique in that the sun corresponds to this sign. It is not by accident that the sun is at the center of the solar system. Our own body has a sun at its center which from antiquity has been called the 'solar' plexus and is ruled by the Leo/sun principle. The solar plexus is a chakra which is a source for energy to come into being within the physical body. Everything that relates to centers and centralization is ruled by the Leo/sun principle. Leo is said to rule the heart, which is the center of the circulatory system. Relative to human personality traits, people with Leo prominent in their nature are said to have a strong need to be the center of their family unit, their work-place, or whatever. To be center-stage is said to come naturally to them. (This sign is primary in the character of actors.) Leo rules centralized types of governments, and was said to rule kings and queens a thousand years ago.

I have read many books on astrology, but one of the most interesting authors I ever came across was an obscure astrologer by the name of William Davidson. He was a holistic physician and had used astrology to diagnose problems with his patients. To my knowledge he never wrote a book other than pamphlets derived from lectures on medical astrology, but the American Federation of Astrologers did publish an article by him in the *Journal of the American Federation of Astrologers* that I found very helpful when I was trying to understand the philosophy behind astrology. The article was an attempt on his part to express the essence of each sign in terms of one or two words.[5] The myriad of things said to be ruled by one particular sign are all closely related. Davidson believed that it was helpful to try to express the essence of those relationships in terms of a key word or two. He pointed out that the key word for Leo is 'centers' and that the essence of Leo as a *dynamic* principle is in terms of 'centralizing.'

Davidson did not talk about Leo ruling spirit as other astrologers have. And even those astrologers who have pointed out that Leo rules the spirit in a human being have never attempted to explain why a concept like spirit might be linked to centers. That is the purpose of philosophy. It is important to, not just accept or trust the rulership system, but rather to test it and see if it is logical.

To begin with, whatever is creating the qualitative unity between different parts needs to be itself united with, rather than apart from, the parts (i.e., internal rather than external). But if it is 'internal' in the sense of being a part, and being placed where any one of the parts are, it will be material and will be a *type* of being. There is only one 'place' that is within a system but that is not a material place. That is the center. This can be illustrated through a graph of a coordinate system. The three axes or 'ordinates' possible in a three-dimensional world are joined at the center, via zero point. They represent types of being.

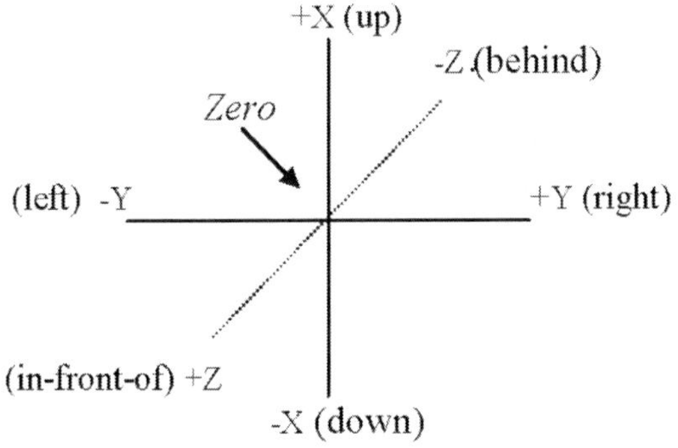

Relative to *what* spirit is, 'what' is related to 'where.' When something is no-where spatially it is no-thing qualitatively. Space is a fundamental type of quality; and the key words for the types of spatial directions correspond to the simple basic types of being. We indicate where something is in space through simple qualitative (and relative) terms such as being 'above' (some point of reference) versus 'below' it, or to the 'right' versus to the 'left,' or 'in-front-of' versus 'behind' it.

A center is not only unqualified in the sense of having no qualities, it is also non-relative. A center is the crossing point between above-versus-below and up-versus-down. It has no comparative and is not defined in terms of relationships. What is not relative is absolute.

The center of this graph symbolizes infinite being as well as absolute being. The types of being are extended, and what is extended is finite. A center as zero point is not extended and what is not extended is infinite—in this case infinitely small.

A center is where infinite and absolute being can naturally reside and be what it is. When trying to understand where spirit as zero or no-thing might exist—and especially what it might have to do with order as well as ordinates, it makes sense to see it as existing at the center of a system, and seeing a system as a set of material parts that interact coherently in terms of *centered* interrelationships.

The graph shows how the center gives both being and unity. The being or nature (the meaning) of the ordinates is derived from (and stabilized by) the center from which they emerge. The center also unites them. [In terms of spirit as a substance or power, this illustrates its ability to give being (and be the source of the types of being), and also to give unity.]

The rulership of centers by Leo holds up to reason. The rulership system arose because astrologers over the years kept track of the things they noted through astrology.

Leo is said to rule, not just centers and the spirit in a person, but other things as well. It is possible, by looking at what it rules, to see the ways in which we experience pure spirit in our own lives. The answers are surprising because spirit in a pure form apparently manifests in us in a number of very common ways that we take for granted without any realization of how truly spiritual these things are.

Some of the things Leo is said to rule include our 'life' and our 'will.' We can put these things to the test of reason to see if they seem to have something about them that appears to be non-material in nature, and to concurrently see if these things also have the capacity to give material being and material unity to the parts of us. If it appears that these things need to be absolute in order to do what they do, we can know through them that spirit really exists. We can also learn through them what spirit is like—what we experience it as.

(1) Pure Spirit and 'Life'. The term 'spirit' has been connected to the term 'life' (e.g., the 'spirit of life'). 'Death' is often spoken of as the obverse of 'life.' Life is connected to being. When a human being dies, they cease to be. In a material sense, for a person, to 'be' means that they are a qualitative whole in their own right. In a spiritual sense, Being may be that which connects the types of being—the material parts of a system. At death the connection between the parts ceases, and they drop down to a lower level of being. It takes a little time for all the parts to disintegrate and actually separate, but at the moment of death, the level of Being really has dropped to the molecular level. While the system was living, something unified the parts and gave that being its being. When people die it is said that their spirit departs. The very suddenness of the instant of death is one indication that something has been suddenly withdrawn.

Currently the concepts involved with our materialistic, mechanistic scientific paradigm do not include a concept of 'life,' at least not in the sense that there is something non-material in a system, either unifying the parts or sustaining them in being.

Astrology indicates otherwise. In astrology, as noted, the Leo-sun principle, in addition to ruling spirit, is also said to rule that which we call 'life,' which would suggest that life is indeed a spiritual thing.

Although their ideas are not currently accepted in mainstream science, there has been a branch of biology that has been around for years known as 'vitalism.' This school of thought entails the belief that the life of a living system is something above and beyond the various material parts. Pierre Teilhard de Chardin (1881-1955) was a well-known biologist who was born in France but lived and died in the U.S. As a biologist, Teilhard's exploration of the world of living creatures led him to become a vitalist. He was intensely interested in the tight qualitative order that living systems possessed. He saw a real difference in the way the parts of a rock are related to one another versus how the parts of even a very simple one-celled organism are interrelated. In a rock, the parts are externally stuck together, and not commonly interconnected. One piece at the front of the rock will be connected to near neighbors, but not to parts on the other side of the rock. The parts of living systems on the other hand are united to one another by something that is not in terms of any one part, but yet is common to and somehow connected to all the parts. They are centralized—and the higher on the scale of evolution living systems exist, the more different kinds of parts there are that are centralized. Hence they possess ever greater centralization or, as Teilhard expressed it, ever greater 'centro-complexity.' [6]

Other biologists had of course noticed that living systems in particular are highly centralized, but it was Teilhard who suggested that centralization was evidence of a real power and substance at the center of systems that caused them to be centered, and made it possible for them to be a coherent, orderly, qualitative whole. Teilhard looked upon whatever was at the center of systems as some type of non-material substance called 'life.' He also visualized it as a force—a dynamic principle of centralization—and called this substance and force 'spirit.'

In astrology Leo rules states of affairs in matter directly and immediately involved with the effects of pure spirit, such as the centralized coordination of the parts of the body. Leo also rules material parts that have evolved to facilitate this centralization. Leo is said to rule the heart and the spinal column, including the brain stem. These are dramatically involved with centralized coordination. In a materialistic paradigm, these material parts are considered to be the only things needed to centrally govern the parts. No spiritual explanation is thought to be necessary. In the AC, these parts are a means to centralization, but they are not the ultimate cause. The vitalistic and mechanistic schools of biology need to be integrated with one another.

When trying to understand what life is, it is important to include, not merely different schools of thought in biology, but also philosophy in this process as well—especially relative to what philosophers would call the power aspect of the principle of 'life.' Philosophers are very interested in causes, and not just the ordinary 'mechanical' causes that science specializes in. To use a graph of a coordinate system to symbolize the relationship between centers and coordinates only poorly expresses the fact that spirit presents as a force bringing about centralization, giving the parts an active form of unity. Coherence is a *dynamic* expression of order within parts. The parts of a true system are 'center-seeking,' and spirit itself is apparently the goal or 'end' of these centripetal behaviors. When philosophers speak of a force that produces behaviors that are directed towards some 'end,' they call it a 'final' cause (variously a 'teleological' cause or force). In the thesis being formulated, spirit may be looked upon as a final or end ('omega') cause drawing matter back to spirit—to Unity—and specifically to the center of a system, and in this way producing material wholeness.

There are many, many states of affairs in our world that

indicate that there are end causes. So important is an examination of these causes relative to understanding what spirit is, an entire chapter is dedicated to causation—Chapter 6.

Although many philosophers have supposed that there are teleological causes in the world, they have not thus far thought about the possibility of these existing within systems—at the center of a body. They usually have conceived of these as 'out there' somewhere, and so have run into philosophical difficulties in their theories when trying to account for the 'where' and the 'what' of final causes (as Plato did with his Form of absolute Being). Astrology (with Leo, and its rulership of centers) is a tremendous help in this respect.

Another reason for including philosophy in this attempt to understand what life consists of is because the base concept needs to be re-defined. The term 'living' system needs to be set into a broader context to include things not normally thought of as living. When trying to understand what the 'life' of a system is, it is important to look at what it does. Biologists currently believe that there are systems that are living and systems that are not, e.g., a one-celled organism is living while an atom is not. Per the AC, if the spirit in a system centralizes the parts, and if spirit is a principle of life, then when a system has independently centered, coordinated parts, we need to look upon it as possessing a primordial form of 'life.' An atom is a centered system with coordinated parts. Hence it has a spirit. [Because it has part-whole (wave-particle) duality, it also has a soul.]

Thus far while talking about 'life' as it relates to qualitative wholeness, we have been concentrating on the Unity aspect of pure spirit. But there is a Being aspect as well. In addition to unifying the body, whatever exists at the center of a system also apparently has the power to actually give being to it and not just in the sense that it keeps it alive by unifying the parts. What astrology teaches via the Leo principle is that the center of a system is a place where energy comes into a being and is disbursed throughout the body, actively maintaining it in being. For thousands of years there has been an interest in centers in humans as places where energy arises in their body. This subject is present in the concept of 'chakras' in Hinduism and Buddhism, which are seen as billowing wheels or vortices of energy rising up within a physical body from a creative center.

The Eastern discipline of yoga entails actual practices intent upon increasing the flow of energy supposedly arising from these places. In both the East and West there are many 'energy-medicine' practices and schools of thought that utilize these Eastern concepts (e.g., Reiki, acupuncture, tai chi, qigong). The theory behind their claims is that there are centers in the body where energy emerges in a centripetal or out-from-center manner. The goal of these practices is to help open blocked centers so that the energy needed to feed and sustain a system's life, and to heal damaged parts, can flow unimpeded into the body.

According to what the AC reveals, energy is being created from spirit. The world is in a state of constant creation. And creation is not occurring 'out there' somewhere, but rather at the center of every system. The photons being created feed the systems they arise in and sustain it in being. These also help it to grow both spatially and qualitatively.

The external world grows as well. We are creative beings, and not just in the sense of creating ourselves. When two or more systems interrelate externally, these relationships are mediated by photons. Photons *are* relationships and *are* energy. A totally *new* type of relationship brings a totally *new* photon into material manifestation. (Leo is said to rule a person's creative potential.)

If it is spirit that is sustaining a system in being and constantly infusing more energy into that system and its external world, then it is a first ('alpha') cause as well as an end ('omega') cause. Given what Leo indicates relative to the being as well as the unity of spirit, and also given Leo's rulership of creativity, spirit can best be looked upon as alpha and omega, producing centrifugal (out-from-center) processes in matter as well as centripetal (in-towards-center) ones.

The creation of more energy is the creation of more being, and the more being that spirit constantly creates entails both a quantitative and qualitative more-ness. For something to become more quantitatively it must grow. For something to become more qualitatively it needs to get *better*.

Teilhard spoke of spirit as having the power, not just to centralize matter, but also to what he called 'superiorize' it. He was a paleontologist as well as a biologist, and as a paleontologist he was able to examine the evolution of living systems as evidenced in fossil records. Teilhard came to the conclusion that the principle or power of centralization that existed in systems was concurrently an evolutionary force. Evolution, he believed, was not due to a Darwinian ('survival of the fittest') type of process, but rather to spirit. He saw that the more highly evolved a system was on the hierarchy of being, the more complex it was, and specifically the more centralized and centro-complex it was. When systems evolve, they take on more being, but not just more in a quantitative sense. They become qualitatively superior.

Spirit not only unifies a system internally, it also acts to unify systems with other systems. When two or more systems interrelate in a cooperative and coherent manner they have the potential to (together) form a new higher level of wholeness. Systems evolve and 'superiorize' in this way. They rise up to ever higher *levels* of being. Atoms unite with other atoms and become molecules, molecules unite together and produce one-celled organisms, and on and on. Qualitative superiorization produces hierarchy in nature. Each higher level of being entails a higher level of unity. Atoms are wholes, but when they unite and become a part of us, what is produced is a higher level wholeness, and a higher level being. The more layers a system is composed of, the more superior and centro-complex it becomes.

Teilhard believed that spirit was the power behind these evolutionary processes in matter.

At about the same time that I discovered Teilhard and his ideas, I also ran across Dr. Davidson's articles and was intrigued by the fact that he used two terms to categorize states and processes belonging to Leo—'centralizing' and 'superiorizing.' Davidson did not manufacture keywords or test them out philosophically. He was just trying to concisely interrelate all the things that each sign (in this case Leo) had been said to 'rule' for thousands of years.

[Leo has traditionally been said to rule, not just the central in place, but also the superior in place—all things that are above other things or at the top of some system. In the human body it is said to rule the crown of the head. In addition to that which is superior spatially, it also rules that which is superior qualitatively—the evaluative sense of something being qualitatively higher than some other thing in a system of gradation. I mentioned previously that Leos like seeing themselves as the center of their family (work-place, or whatever), but they also tend to see themselves as superior to others. In a more general sense, in atoms and all material systems, Leo governs processes of matter becoming qualitatively higher in nature and hence 'superiorizing.']

The key words that Davidson used for Leo were almost exactly the same as Teilhard used to express the essence of what he believed spirit (as both a substance and a power) did. As a scientist, Teilhard's ideas arose from his experiences as a biologist and paleontologist—and one who had no interest in astrology. Davidson was not a paleontologist, nor a philosopher, and he did not think in terms of how centralization might produce superiorization. He was merely interested in formulating a rational category system, one in which everything Leo ruled could be stated in terms of one or two simple words. Despite their separate ways of arriving at their conclusion, when the ideas of these two men are put together, they support and augment one another.

There is evidence, via biology and paleontology, that evolution is a process that is constantly occurring in our world. If this is a *natural* process—one caused by a natural law—then it is rational to suppose that an impetus towards greater Being and Unity exists within all systems. In astrology, Leo is one of twelve basic laws in our world. The AC indicates that evolution is a natural process present in every true centralized system.

To summarize, both being and unity are commonly linked to the concept we call life. Furthermore, the ideas connected to that which we call 'life,' and 'living' are not passive ones. Life presents as a power to actively unify systems and actively sustain them in being, while concurrently pressing them on towards greater degrees and levels of being and unity. When something gives a system its being as well as its unity—and when spirit has been defined as Being and Unity, it may be correct to see the life of a system as its spirit. The parts of systems are types of being. What unifies all the types would not itself be a type of being any more than the laws of nature are unified in terms of one law of nature. Pure, non-relative, absolute being, as the agent of unification, is a natural hypothesis. An absolute nature in turn makes it rational to connect it to the center of a coordinate system. When living systems show evidence of being centralized, and centers can be philosophically linked to being absolute, the simplest and most coherent explanation for the evidence may be a postulation of something absolute at the center of a system that has the power to give it its current being and unity, as well as providing an impetus to reach for ever higher levels of being and unity.

(2) Pure Spirit and 'Awareness.' Webster's defines spirit as "the agent of vital and conscious functions in man."[7] In astrology, in addition to the vital or life principle, Leo is also said to rule awareness, and specifically *individual* awareness. [Pisces, the polarity partner to Leo, is said to rule universal consciousness (corresponding in some ways to what Carl Jung called the collective unconscious). In addition to universal consciousness, in a gray world where we are not as aware as we should or could be, Pisces also rules loss of awareness as occurs when we relegate what should be conscious to our unconscious mind, or what occurs when we sleep, are in a coma, etc.]

If the astrological rulership system is valid, then if awareness is examined closely, it will show evidence of being different from matter, including evidence of giving both Being and Unity to the parts of the system it exists within. Before discussing awareness, it is relevant to make an important distinction between the mind and the awareness.

There is much scientific evidence surfacing which indicates that the mind is more than just the brain—that in fact mind may not be a physical substance at all. Memories, for instance, are not localized in any one part of the brain. The brain appears rather to be a transducer allowing memories to be re-called from some sort of 'mind' that is beyond the physical brain.[8]

Due to all the latest research emerging in neurophysiology and even in quantum physics, some people are beginning to look upon mind as a spiritual substance. If mind is not physical in nature, it is natural on the surface of things to see it as spiritual. That is partially because spirit has been seen as something different from matter, but even more so because spirit has not been adequately defined. Once the concept of spirit is clarified, it is clear that mind and spirit are two different things.

When the nature of mind is considered carefully, it presents as material—as some sort of mental 'body'—although not necessarily a physical one. (The terms 'physical' and 'material' can denote different things, with 'physical matter' and 'mental matter' both being species of the higher, more inclusive term 'material.') The reason that mind presents as material is because thoughts possess the three material 'domains' that were spoken of previously. First of all, they have a qualitative aspect or domain to them. In other words, there are different kinds of thoughts, and thinking entails linking these things together. They also have a kinematical domain. They move. This is what a thinking 'process' is. They even have 'spatial' extent—albeit of a type that has to do with mental 'space.' We arrange them, some thoughts are more extensive than others, some minds are bigger than others, and so forth. In a similar manner as with physical bodies and their environment, there is a 'within' and a 'without.' We have a mind that is a mental body, and it thinks and interacts with other minds within a mental 'plane' or world of thought.

Astrology makes it clear that mind is material—but a different kind of matter than that of physical matter. The Mercury-Gemini principle for instance rules, not just physical spatial connections; it also rules the capacity of the mind to make mental connections, and has been said to be the primary ruler of the mind. The Uranus-Aquarius principle rules, not only the way the central nervous system connects all the different physical parts of the physical body; it also rules the ability of the mind to connect many different kinds of things together in thought—especially in new and innovative ways. The Jupiter-Sagittarius principle (which rules physical spatial extent), not only makes it possible for physical things to expand in extent, it makes it possible for the mind to take what is known and expand upon it by means of an ampliative or inductive inference or judgment. Thus in astrology this principle is said to rule judges and futuristic predictions based on what is currently known about a situation. It is this sort of thing that makes it clear that the same archetypes that produce our physical experience also produce our mental world, but also that these are two different things. It is not difficult to consider the world of mind as a different world, or at least a different plane within a world—especially now that the physical world is looked upon by quantum physicists as merely a nexus of energy relationships, or interactions between processes.

To claim that the mind is material does not mean that there is nothing spiritual about our mental life. As with the physical body, our mental life appears to entail spirit, and not just as a derivative expression of spirit through the types of (mental) being, but also in regard to a direct expression of pure, absolute spirit. For instance, mental order entails qualitative wholeness, and qualitative wholeness in turn is evidence that something is centralizing the parts. If mind is a body, that makes sense. As to what spirit is or what we call it in the mental body, these different types of bodies in a person are still the bodies of one person, and they need something to unify them. Spirit itself is not a divided substance. And yet it is possible to give spirit different names as we experience it in different ways, such as Unity versus Being. We may give it different names when we experience it in different bodily ways as well. Whatever is at the center of the mind is probably appropriately called the life of the mind. But I suspect it is also that which we call 'awareness.'

One way to know that the awareness is different from processes of thought is due to the fact that we are aware of our own thoughts, just as we are aware of all the processes in the physical world. In other words, if we are aware of thoughts, then awareness is different from thoughts.

When we take in percepts, or reason, we need some sort of base from which to do this. We perceive ourselves as an ego *center*, thinking things mentally, as well as doing things physically.

(Because people have confused the concept of spirit with that of mind, and that of mind with awareness; and because they have used the term 'consciousness' for this muddy blend, I prefer to use the term 'awareness' rather than 'consciousness.')

When the stance is taken that the awareness is spiritual in nature, then some of the attributes of awareness become simpler to explain. For instance, the fact that we grow in knowledge, just as we grow and evolve physically is one piece of evidence that the source of that growth may itself be an unlimited 'alpha' source.

Another thing about awareness that makes it appear to be spiritual in nature is that it has both a Unity and Being aspect to it, and of a type that is centralized. Relative to unity, it stands at the center of our physical perceptions, taking them in and unifying them for us. Intellectually this unification aspect entails making sense of our thoughts as well as of our physical percepts. Qualitatively speaking, this is about giving them order.

The awareness is our means to Being as well as to Unity. It gives being to both the external world and to us. Relative to the external world, things come into being (for us) when we become aware of them. There may, for instance, be other worlds, but they have no being (for us) until we become aware of them. Likewise, as we now know through quantum theory, it is our awareness of particles (including atoms) that makes them change from a field potential (wave-form) into a material state and thereby come into being. In this case physicists are claiming, not just that they come into being for us, but that they really come into being when and only when we are aware of them. (The wave-form of an atom is a set of potentials. When we become aware of the set that pertains to it, it actualizes and becomes matter.)

This makes of the awareness a source of creativity. The things being created are material types of being. In focusing our awareness on certain things, if quantum theory is correct, we are creating our external world, and apparently ourselves as well. [If this is true, it has many ramifications for us. The soul, for instance, is a set of specific potentials. This would indicate that we must be aware of our potentials (and believe in them), if they are to be actualized.]

It makes sense to perceive of the different types of being as created *of* (the substance of) pure being as well as created *by* pure being. Qualitative order in the world then also begins to make sense. Order, whether in the physical world or in our minds, presupposes specific kinds of connections between the different types of being involved. It entails many different things fitting together as parts of a whole. If spirit is the source of the types of being, it is logical to see each type as a perspective on pure *absolute* Being. If each thing perceived is a perspective on or 'part' of absolute being that explains why they fit together to produce order and coherence in our thinking, and in our physical world as well. Relative to the latter, it is why we can speak of the laws of nature as 'lawful' and as appropriately interactive with one another.

If the awareness (as with our life principle) is spiritual in nature, then it is absolute and infinite in being. One important question this immediately raises is how individuals can arise who are individual in character, and hence only a part of the universe, and who have a limited awareness as well as a limited physical being. What the AC indicates is that the individual life spirit in our physical body is individual only because the physical matter of our parts surrounds something that in essence is not limited. It is the same with the awareness. Our individual awareness is limited because and only because of the material bodies (mental and physical) that surround our individual spirit. (From antiquity the Pisces principle has been said to rule psychism. That is because we are in contact with the substrate as well as having a personal center, and hence have access to a universal form of awareness.)

In the material world we utilize our awareness to be aware *of* ourselves, and aware *of* the things around us. We are most aware of things in our own small world. But, if the awareness is spiritual in essence, it is not necessarily always so confined. If the awareness is in and of itself neither extended nor divided (and hence not limited) then, although it might normally focus on our own little affairs in the immediate present, it might not always be restricted in this way. It could conceivably have a much wider range of possible activities whereby we could access other things going on in the world in different places, the thoughts of other people, even things that have happened at other times. For those who have experienced intuitive perception, psychism, glimpses into the future, remembrance of past lives, etc., these are evidence that there is something about the awareness that is not obeying normal space-time limitations.

Buddhists believe that the awareness is infinite in essence, and that it is possible for us to deliberately expand our own awareness towards a state of infinite cosmic consciousness. Buddhists also believe that the awareness, like the life principle, is related to centers. It is a tenet of Buddhism that yoga practices can, not only increase the flow of physical energy coming through the chakras, but also increase the degree of awareness we possess—ultimately to an infinite degree.

If the awareness is absolute and infinite, it is also eternal in nature. Buddhists claim that it entails the eternal 'now'—that pure awareness is outside time altogether. They believe that, when we achieve absolute awareness, we will be a part of the eternal now. I believe that it provides for the now for us this very moment. Everything material is in a state of constant flux. Without a point of absolute zero in us we could never be in a precise place or a present time—the 'here' and 'now' that we commonly experience and so little appreciate or examine. [Relative to the 'here' being a zero point, this is the problem of the integral calculus spoken of previously relative to how something material converges to a limit (zero point).]

In addition to our experience of a 'now' there may be another way to realize that the awareness is eternal. The French philosopher, Rene Descartes (1506-1650) was fascinated by the fact that we cannot doubt that we exist. The relationship of awareness to absolute, infinite and eternal being can be seen in its ability to perceive, not just our material being, but also our Being. The known must be in the knower, and it must be the same in some way. We can doubt that one or all of our parts exists, or even the entire material world (and we can certainly doubt our potentials), but we can never doubt that we, the 'I am' exists. This is because we would have to be aware of the doubt, and to be aware is to exist—to Be. We cannot be aware of not existing. Our pure Being (spirit) is not a type of being, and the contemplation of pure Being by pure Awareness, not only is a measure of their identity, but also of the absolute and eternal nature of both. Spirit cannot not exist. Pure being cannot not be. The awareness of pure Being *is* Being and also cannot not exist or cease to exist. To put it in a slightly different way, if our Being is something more than just the types of being, and if we are aware of this, then the awareness itself is absolute.

To summarize this section, if the awareness is different from thought (if we can be aware of thinking), and if it gives unity and being to both our mental concepts and physical percepts, and if it shows evidence of not being limited in normal space-time ways, then we need to consider whether or not it might be spiritual in nature. (Then too, in consideration of the use of astrology to formulate the AC, there is also Leo's traditional rulership of the individual awareness as well as of centers, spirit and the life principle.) If the awareness is a spiritual substance, then our sense of somehow being an ego center is philosophically justifiable and appropriate—as are the things the awareness seems to be able to do. Spirit in a pure state may not be just pure absolute Being-Unity, but rather Being-Unity-Awareness.

To bring together sections (**1**) and (**2**), just as pure 'Unity' and pure 'Being' present as the same thing, if 'life' and 'awareness' are both manifestations of spirit, they should show evidence of being the same thing. In section (**1**) the life principle was spoken of as being a creative force as well as a unifying one. One of the most important aspects of quantum theory is that it presents evidence that the awareness does not just take things in; it creates our world. That which keeps things in being or can make things come into being is a principle of life.

(3) The Seat of Thinking and Reasoning: In a materialistic paradigm thoughts are seen as produced only by the brain. As was noted, the AC sees both the mind and the physical body as 'material'—the one physical matter and the other mental matter. But, whichever way the thoughts are visualized, there may be no way to adequately explain how we conceptualize and put all of our thoughts together without positing something absolute at the center of the mind. Thoughts are the 'food' of the mind and ultimately become the 'parts' of the mind. The parts of the mind need something to unify them, so that they have an appropriate place. Each perception and thought itself needs to be characterized and categorized. We need something within us to tell us where everything fits. We need a touch-stone in order to reason well, and we need it even to form simple concepts.

We think and perceive by relating things—putting them into categories and seeing how they are the same and how they are different. The various types of being are different and, if they are to find any type of qualitative order or unification within our mind, they must in some way be the same as well as different. If something unites the types of being, it is presumably not a type of being, and hence is not material in nature. Pure Being presents as the ideal source for this unity, precisely because it is not a type of being and yet is Being; thus it is something that every type of being shares in common. Because it is absolute it is even the source of the very concept of qualitative 'sameness.' To be the 'same' is to be absolutely unified qualitatively, and anything differentiated into a type of being is not the same when perceived as different. On the other hand, if anything was totally different from everything else, there would be no way to conceive of it. Different things may be apart qualitatively and yet, if they are not a part of some larger picture, we cannot understand what they are. Different types of things are best understood when they are recognized as perspectives on some whole, in other words, with each type of being visualized as an expression of, or perspective on, pure Being. Pure Being can unite the types of being because it is the source of the types of being and hence contains them all in potential form within itself.

If spirit is real, it makes sense to see pure Being as existing in the center of us, which is the one point within us that has no characteristics and hence is not a type of being. It is possible to think of this means to our judgments as being somewhere 'out there' (as Plato did when he saw the Form of Being and Unity as existing in the world of the Forms). However, it makes much more sense to perceive of pure Being as existing in the center of us. If spirit exists in us as real Being and real Unity, and specifically as a dynamic force capable of actively unifying and making sense of the types of being, then some of the things that we seem to be able to do, such as conceptualizing and categorizing things in our mind, have a logical explanation—and a simple one.

[In terms of touchstones, consider again the coordinate system given previously. These six simple qualities (up, down, right, left, behind and to-the-fore) can each be intelligible concepts, not just because they have opposites, and not just in terms of the other two paired coordinates, but even more fundamentally because of the center of all these, which center is both their source and place of unification.]

(4) Judgments of What is True: In addition to general conceptualizing, we are able to make judgments of that which is real and true versus that which is not. What is real has being. For something inside of us to know that particular types of being in the material world are real and possess being, it may be necessary for this touchstone to, not only be inside of us (since the known must be in the knower), but also for it to be Being. It would not just *have* being as the various types of being do. If this source within us judges the reality and being of all the different types of being, it needs to be able to include them all in its cognitive abilities, and would hence be more than one particular type of being.

In addition to knowing what is real and true in the physical world, something inside of us also needs to unite the two worlds of mind and physical matter. We are aware of the physical world and aware of our mental world. When these two are united with one another in our mind, and are *correctly* united, we say that we are judging correctly or seeing things 'truly'—in other words, as they actually are in the physical world. (Because reality is being, and both the physical and mental worlds are real, this ability to unite mental and physical realities also includes the above noted capacity to judge when something possesses real being.) In speaking of this ability to judge real being and to correctly link physical and mental realities, philosophers speak of this road to truth as the 'correspondence' theory of truth.

(It is not merely philosophers that are interested in these correspondences. Psychologists are as well—but from a health standpoint. The 'parts' of our mind cannot be whole when we cannot do the above processes well. When the mind is not whole, it becomes sick. Psychiatrists label a problem in this area as 'paranoia,' 'schizophrenia,' and other conditions that interfere with our ability to discern what is real.)

In addition to the correspondence theory of truth, philosophers also speak of the 'coherence' theory of truth. The correspondence theory focuses on recognizing true Being in the world. The coherence theory focuses on recognizing Unity. Our mind orders the things it thinks about. We know that what we are conceptualizing is true when the various aspects of a complex concept (such as a cosmology) fit together like the pieces of a puzzle. Thoughts (and the minds that think them) are not just amorphous conglomerates of qualities. They possess order. Nor is the mind passive. It seeks order, coherence and wholeness just as avidly as the physical body does.

The types of being possess coherence because they are parts of, or perspectives on, pure being. When we reason well, the thoughts in our mind are all coherent. In order to see the coherence between the various types of being, the awareness needs to be the source and ground of their unity.

To reason well and to be logical is to further the wholeness of our own mental bodies—and of the collective knowledge of the human race as well—which is what we call our 'science' and 'mathematics.' When everything fits well in our heads, we embody truth. There is a life of the mind as well as a life of the body; and spirit, as the 'spirit of truth' presents as behind that type of wholeness as well.

Trying to understand how we know what is real and true brings up the subject of falsehood. It stands to reason that, if truth entails types of being that are real and are coherent with one another, then falsehood entails types of being that either are not real or are incoherent with a larger picture of reality. And it could be that what is false is both these things. Relative to incoherence, when the pieces of what we think about or perceive do not fit together, they do not cohere and there is true separation between the parts. These are places of non-being. Unity is Being.

We think of that which is untrue as primarily a mental thing, but the physical world may also be false if it has lack of being or incoherence in it. And the material world certainly does have some relationships in it that are very incoherent and that hence lack cohesion between the involved 'parts.' This is especially true of relationships at the human level of being. Plato's famous myth of the cave compares our world to a shadow land that reflects reality, but in a distorted way (*Republic* 7:514-517).

Despite the distortion, Plato nevertheless believed that the material world had a measure of reality because it did at least *reflect* the real world. Buddhists on the other hand believe that the material world is pure illusion. I do not think so. I agree with Buddhists that pure Being exists. But we disagree relative to the material types of being. Per the AC, the types of being are just what their name implies. They are perspectives on absolute being, and the only way they would be illusions, or even have a measure of unreality is if they were not true expressions of absolute being. And then their incoherence would betray them as such. The overriding beauty of the world speaks for the truth of its existence. What is beautiful (and good) is not an illusion.

Nevertheless, there is falsehood in the world. People have compared falsehood to darkness. The world is composed of light. The types of being are like 'colors' or perspectives on light. These types of being are the laws of nature and are naturally coherent when created correctly. Because the world is composed of light, light is what is real and true. But because of our ability to create, and even create incoherently, this has made of our particular world-of-living-creatures a land ultimately composed of light, but hindered by shadows. This is why it is possible to speak of the world as currently being somewhat gray.

And it is not just the external world that is affected in this way. We have wounds as well—dark spots that keep us from being whole. We are not true to our real selves—our true potential—when we do, think or feel things that are not clear and correct expressions of the laws of nature that we utilize to create ourselves and our world.

When the mental and physical realms are perceived as merely different types of bodies, truth and reality can be seen as the same thing. For our minds to be whole, they must encapsulate the truth.

Furthermore, the mental and physical realms interpenetrate, and the mind is the stronger of the two bodies. The mind is a set of beliefs. What we believe to be true in the mind has the power to make it come into being in the physical realm. This 'mind-over-matter' state can be a good thing. But in a somewhat gray world we have the capacity to believe what is not true. It is not just what people do that makes a world what it is, it is also what they believe. (This is why paradigms are so important. They entail a collective belief system.)

And as with the external world, this is true of our own

character as well. It is not just what we do to ourselves but also what we think and come to believe about ourselves that creates who and what we are. As a probability field our soul is a set of specific potential that we have the capacity to actualize. To the extent that we believe in real and ultimate truths about ourselves and our potentials, the probability increases that these will be actualized and (most importantly) will be 'real' and 'true'. Conversely, if we believe shadow things about ourselves (things that do not add to our true being and coherent nature) these things are very likely to be actualized as well (and our real potentials temporarily put on hold).

Many books are now surfacing that address the issue of the effects of 'positive thinking' and 'visualization' on the creation of our internal nature and external world. The real secret behind positive thinking and visualization is belief. We create reality, not just through focusing attention on something, but by believing it is real. For something that is not in the physical realm as yet, to have faith is to believe in what can be physically and what already is mentally. There is magic in belief, and true magicians who were not just practicing slight-of-hand tricks, may have known how to affect matter in this way.

In summarizing this section on truth, it is possible to see via theories of truth in philosophy that truth entails being and unity. The thoughts of the mind and the attributes of the physical world are produced by *types* of being. To judge their real being requires a touchstone of Being within us. To judge their coherence likewise requires a touchstone of Unity within us. This touchstone cannot be any one particular type of being or unity. Judgments of truth require something absolute within us. This judgment seat or touchstone fits the criterion for that which was defined as 'spirit.'

(5) The Judgment Seat, Goodness, and 'Conscience.' As a judgment seat, our awareness apparently judges, not only truth, but also Goodness. It has thus far been suggested that it is possible to experience the Being of spirit in a pure sense, and that some things that we commonly speak of as existing, such as life and awareness, may be examples of its presence in us. Our so-called 'conscience' may also be a manifestation of spirit.

When trying to figure out what a 'conscience' might be, there are several things to be considered: (**a**) Is there such a thing as right and wrong, or good and bad? (**b**) If there is, how do we make judgments about these things, and with what? And finally (**c**): Where might this particular judgment seat exist?

Relative to the first question (**a**), to say that something is good or bad, or even to make a value judgment of which of two things is better, we must assume that there is value in the world. Our current materialistic, relativistic scientific paradigm denies that natural value exists in our universe. We use the term 'conscience' in common language, and believe that we have one. But, if there is no value in nature, and one state is no better than another, then there is no need to have anything in us that can make judgments of this kind. Anything that we do is no better than anything else. Everything is equally okay.

In that branch of philosophy called 'ethics' and 'value theory' there has historically been a school of absolutist philosophers (Plato, Aristotle, Kant, Spinoza and others) who have strongly believed that there is something called 'the good.' Because they are absolutists, they believe that, in order for the good in our world to be what it is and to be known as what it is, there must be an archetype behind it. Because the good in the world is so generic and there are so many different types of good things—but most of all because it seems to be involved with unity and with existence or being—the archetype of the Good (for Plato and the Neo-Platonics) was believed to inhabit a high place on the hierarchy of archetypes.

The absolutists are opposed by the relativists (Nietzsche, Hume, Sartre, Kierkegaard, and others). In ethical relativism there is no definite right or wrong (just as in epistemological relativism there is no real truth of the matter). Because that which is absolute has historically been related to God as well as to universals and absolute Forms, Nietzsche's famous way to express the ethical relativist viewpoint was, "God is dead!"

Plato believed that the very highest archetype was the Form of the Good. He identified this with the Form of pure Unity. He believed that the Form of Being was a slightly lower level Form, but the AC does not concur with this idea. Absolute Unity-Goodness would not exist if it did not possess Being. More importantly, and more to the point, there is an identity factor between these three things. That which is good is just what is meant by Unity. It is 'good' to be unified. That which is good is also just what is meant by Being. It is good just to be and exist. If Unity and Being are the Good, then spirit needs to be referred to as absolute Goodness-Unity-Being.

In coming to some answer in regards to (**a**) whether or not value exists in our world—even bad versus good—I hesitate to digress into value theory per se. Value theory and ethics is an entire domain itself within philosophy, and the purpose of this chapter is really (**b**) to determine if we have a conscience and what that might consist of within us. I am going to digress for a few pages anyway, even at the risk of some degree of continuity of thought. Plato put the Form of the Good at the highest level of the Forms, and he said ethics was the highest and most important area of philosophy. He was right. If we come to understand through a new paradigm that we are spiritual in nature, and do not understand what happens when we are not being true to our own Good nature, the rest is meaningless.

Of all the things that the materialistic paradigm has encouraged, a relativistic perspective in regards to value theory has been the most harmful. From this view 'good-better-best' and 'good-versus-bad' are simply names we give to things. The purpose of science and philosophy is to help us find the truth. Their role may be more important than we realize because, if a paradigm is wrong, and we believe it and have confidence in it, we help create that which is not true, and in that way foment darkness in our world. The ethical relativist perspective is wrong. There are things that are right, and things that are wrong. Relativists have claimed that good and bad are contextual, i.e., that they change in different contexts. Every material context is different, but there are things that will further our unity and our being. This is true of the unity and being of our collectives as well. From a material perspective that means things that will further our wholeness and that of the groups we belong to. In a gray world, we sometimes cannot choose pure goodness; but we can choose the lesser of two evils—positively stated this means choosing that which will preserve and facilitate the greatest *degree* of wholeness.

Every individual in our world is naturally intended to be whole. A world of clear, coherent light is the natural state of things. The astrological category system attests to this. Our external world is also intended to be a coherent, orderly whole-in-the-making. When looking around it is possible to see that this is currently not the case.

The grayness in our world is a serious issue. If we are to change this, we need to understand what this thing called the Good is and why the fullness of goodness is not being expressed in the world.

The 'bad' things we encounter bring sorrow and pain. The good things bring health and happiness. People cannot consciously and fervently work to eliminate bad things if there is no decent definition for these things—if they do not even realize that suffering and unhappiness are not a normal, natural part of the world. There are many people who believe that we need darkness in order to have (or to appreciate) the light. According to the AC on the other hand we have 'colors' for contrast. The parts of an individual are different types of being and can be looked upon as colors. We do not need lack of light for contrast.

Just as important towards promoting contrast is part-whole duality. Part-whole duality and hierarchy in nature together create a natural means for comparison, and badness is not needed for contrast. In terms of part-whole duality, wholeness does not need unwholesomeness for a comparative. Parts and wholes provide a natural contrast and, although wholes are over parts, the parts are not 'bad.'

In terms of hierarchy in nature, different systems are at different levels of evolution. A human being exists at a higher level of being than an atom and is hence far more complex. To say a system with more levels, and existing at a higher level of being is good, while one with less layers and at a lower level is bad, is like saying a fourth dimensional world is good while a three dimensional world is bad. Systems (and worlds) occur in stages, and each is good. The 'grades' provided by dimensions and layers produces a natural contrast (and a better and better future to aspire towards); but none of the levels are 'bad' so long as each is whole at its own level of being.

That which is bad is not natural or normal, and it need not exist. It is very difficult in a gray world to believe in the priority of goodness over badness, much less that the bad can be totally eradicated—even to the elimination of death. But we need to know the truth of this, and to understand precisely and clearly what is going wrong. Each of the domains manifests its own type of wholeness, and hence consequent unwholesomeness when the laws of nature in that domain are not correctly expressed. Relative to the quantitative (spatial) domain, when we lose part of our body (or even just cut its surface), we learn very quickly that it is bad not to be spatially whole. Relative to the kinematical domain, motion is serial in nature. It has its own type of wholeness. It is good for any motion to flow freely and continuously with no breaks to its serial nature. Wherever there are breaks in a medium, these cause friction. This is 'painful.' Where there is friction impeding and distorting the flow of motion, there is loss of useful energy. (In order to survive we then need to confiscate energy from other systems.) Friction in the kinematical domain makes the second law of thermodynamics in the qualitative domain act in a negative way as well, degrading things qualitatively. This is not good. The qualitative domain is actually prior to the other two because it has to do with coherence and with wholeness per se. Its relationship to value is why it can be spoken of as the 'qualitative' domain.

There is a great deal of incoherence within us. It is the reason we age, sicken and die. The incoherence externally between individuals produces child abuse, family conflict, divorce, wars, tyranny, injustice, exploitation, economic inequities, and ecological crises. It is very difficult in a gray world to believe that we can overcome and eliminate these things. Nevertheless we can! But first we have to know we can and believe we can. There is enough evidence available right now to know the truth of this.

At the beginning of this book the reader was advised that it was possible, through the philosophy that emerges from the astrological category system, to know who we are and what the purpose of our life is. The heart of these answers is in this section. Our reason for being is to be, and that means to be whole. To be whole is to be healthy and happy. And it is to be everlasting—precisely because Unity is Being. Furthermore, the purpose of life is not just to be whole. It is also to increase in wholeness everlastingly via the hierarchical concrescences that have been mentioned previously. There is no end to the potential dimensions in light. Wholeness is what Goodness is. We are the expression of Being, Unity and the Good. In order for this purpose to be realized, we must be Good.

In addition to the overriding purpose of expressing the Good through simply living, there is another purpose that we have right now—a duty actually—and that is to eradicate anything that is not good in ourselves and our world. We need to bring ourselves and our external world back to a natural state of health and wholeness.

The signs show clearly the validity of the things spoken of here. To begin with, they speak the truth via their weighted nature. The positive signs are prior to the negative ones. (As was noted, the negative signs are not 'bad.') The negative signs support the positive ones from behind, below and within so to speak. The positive and negative signs work together to produce value in the world; and the heart of value resides in spirit and wholeness. All twelve of the principles show themselves to be types of being and unity intended to promote wholeness. Wholeness in the qualitative domain is prior. The four signs in the qualitative domain show how the two different manifestations of spirit produce a qualitative whole, and specifically one with part-whole duality. Aquarius in that domain makes it clear that parts, in order to be parts (and reflect the character of pure Unity), must be coherent. 'Parts' are 'matter,' and coherence (lawfulness) is the natural state of parts. Aquarius also makes it clear that it is levels of light and levels of creation that produce contrast in the world—not light versus darkness. Leo and Pisces rule the high and the low and hence levels of being. Aquarius rules part-whole duality and hierarchy in nature. Part whole duality is a primordial expression of levels, and produces a natural weighted contrast in all systems. As systems evolve and coalesce into ever higher hierarchical concrescences this produces hierarchy in nature.

The second way in which it is possible to know the truth about these things is through looking at the world. If we never research astrology—if what it says about the nature of the world and the reason for our creation is never validated via astrological research—it is still possible to know the truth through the world itself.

First of all, there are weighted contrasts everywhere. Motion is a going forth. Time moves forward into the future. Inverse square laws exist. Space is expanding.

Secondly, everything in the world is directed towards unity and being rather than towards disunity and nonbeing. The laws of nature are not types of non-being producing lack of order and unity.

The world as a whole is a system and it shows plainly that it tends towards wholeness, both qualitatively and spatially. Spatially speaking, the saying, 'nature hates a vacuum' is a deep truth rather than a trite phrase. In terms of the qualitative domain, the strength of the laws of nature—their coherence, order, and resistance to anything that opposes their nature—is evidence of their holistically weighted character.

Individuals are prior to external worlds, and individuals want to be whole. In living systems, every time a wound heals in the body, that is a personal proof of the weighted nature of experience.

Not only do systems want to be whole, they want to achieve ever higher levels of wholeness. They are constantly moving towards greater being and greater unity. Atoms are uniting into molecules, molecules are uniting into mega-molecules, these are uniting into one-celled organisms, and on and on. Everywhere there are systems merging with one another in hierarchical concrescences in what appears to be a universal effort to promote ever greater levels of being and unity. Evolution is a weighted process. Scientists are ignoring evidence of evolution and of hierarchy in nature everywhere. These provide proofs of natural value in the world.

If there is true evolution in our world, if it is a natural (non-Darwinian) process, and if it is a universal law of nature and not just something living systems manifest, then this is evidence of the existence of a principle of goodness working throughout nature moving things towards *betterment*. Its end is the Good.

We need to understand evolution better. Thus far it has only been studied relative to living creatures. If it is universal then every centralized system is evolving. Pierre Teilhard de Chardin spoke of the relationship between centralization and superiorization, and we need to understand that relationship as it speaks of a universal principle of betterment present within all things. Currently Darwin's explanation for evolution is in vogue, and that is understandable since, in a materialistic paradigm, there can be no natural principle of evolution. If there were, the world would be purposeful and heading somewhere. Directed movements are antithetical to a relativistic philosophy.

Darwin's theory that evolution is due to 'survival of the fittest' may have something to say about how things work for living creatures in a gray world that is lacking wholeness; but overall it has never been an adequate explanation for evolution. Cooperation is a much stronger thesis, and that is what spirit leads systems to engage in. Cooperation leads to order and coherence. These lead to unity, and unity supports the need of systems to stay in being and increase in being. Unity is the way of life.

It is not merely the lack of a spiritual paradigm (with its directed movements) that makes scientists unwilling to see evolution as a natural process occurring throughout creation. It is the second law of thermodynamics. This appears to be moving things, not towards greater centro-complexity, but rather towards an undifferentiated, non-individualized oneness.

The second law of thermodynamics is as little understood as is the law of gravity. In astrology, Pisces rules the second law. A study of this sign is especially helpful towards understanding order since this law is said to promote entropy or disorder. Systems create via the archetypes. Due to free will, they can create in ways that are not right. It is hazardous to utilize any of the principles in a way that is incorrect and hence incoherent; but when this particular principle manifests negatively, it can be very destructive. That is because the purpose of creation is to produce qualitative differentiation and individuality. The Leo principle is primary in this process. The Piscean principle is supposed to be a helpmeet; but goes the opposite direction if it is abused.

Because of the way the second law of thermodynamics acts (in conjunction with friction between systems), it can be used as an argument against natural evolution. The thing to remember is that this law is said to lead to less qualitative being or order only in closed systems wihin a closed universe. The archetypes indicate that every system is open at the center. There is constant creation going on. Nevertheless, in a gray world there is friction when there is incoherence between parts. It is because of friction that motion does not manifest clear (serial) wholeness and is hence not perpetual. Due to the interpenetration of the domains, friction in the domain of kinematics makes the second law act in a negative manner. But it is precisely the role of friction that makes it possible to know that this is not a natural way for law to act. Order and coherence is an expression of lawfulness. Incoherent frictional interactions between systems and coherent (lawful) interactions between systems are opposites. Coherent interactions between the parts of systems are smooth and 'lawful.' Coherent interactions are anti-entropic. The matter in our world is composed of the laws of nature. These lawful interactions are naturally coherent. When they are incoherent, this is unlawful and this is the source of friction. Friction and the negative effects of the second law of thermodynamics are unnatural. (We can eliminate these from our lives, precisely because they are unnatural).

Any frictional interaction (within or between systems) does waste energy and does move the world one step closer to disorder rather than towards greater order and greater ordered complexity. Lack of order at its furthermost limit is undifferentiated sameness. This is the substrate. (Pisces is a principle of sameness.) Despite what occurs when we do things that are unlawful and wrong, the astrological category system shows clearly that the substrate (and the Pisces principle) is the back, supporting side of spirit as a principle of individuation and subsequent differentiation (Leo). Pisces is not the opponent of Leo. All of the negative signs are helpmeets. When many different systems reach up for a higher level of wholeness via a hierarchical concrescence, they do so by sharing something in common—something that is the same—a common purpose. This is how a new higher-level group soul is born. A new set of potentials is actualized from out of the substrate and begins a process of actualization. A probability field is a piece of the substrate. To call the substrate pole of spirit the *sub*strate is to miss the point that creation occurs in layers (and dimensional steps), and that the substrate is a field potential above us as well as below. (Not merely the new set of potentials, but also the new center for the new higher-level soul *is* the substrate. The two manifestations of spirit are one.)

Another way in which the Pisces law helps produce ever greater centralized complexity in systems rather than disorder has to do with what scientists refer to as the 'principle of least energy expenditure.' It has long been known that there is something in nature that makes systems choose whatever path will be easiest and simplest. Leo rules qualitative more-ness (highness), Pisces rules qualitative less-ness (lowness). The qualitative domain interpenetrates with that of the other two domains (amount of energy). When atoms (and people) join together with other like systems in orderly and mutually beneficial ways they need *less* energy to exist as a whole. The second law of thermodynamics has to do with less-ness, but it does not oppose order (or evolution towards ever higher states of order). That is because the Piscean form of differentiation encourages relationships that *save* energy, and cooperative interactions save energy. The Piscean principle hence encourages cooperation in its own way, and furthers superiorization as surely as Leo does.

The negative effects of friction and the second law of thermodynamics not only wreak havoc in our external world, they also cause our internal parts to wear out. There are a number of scientists who believe that this type of entropy is unnatural. Biophysicist Mae-Wan Ho presents a convincing scientific and mathematical argument that living systems are naturally counter-entropic. According to Ho, systems experience entropy only when their natural order and coherence is compromised—when something happens to them to damage their natural wholeness.[9] Because Ho is a biophysicist, she is very interested in the correspondences between atoms and living creatures. It is the biophysicists that may someday help us see wholeness as a natural, sustainable presence in all systems.

The world is moving in a particular direction, and it *is* towards greater unity; but it is towards greater differentiated unity. Individuals constantly join with other individuals at the same level of evolution as together they collectively reach up to a higher level of being. A system becomes more complex *and* more 'universal' by encompassing other systems. In this way both Unity and Being are increased. Spirit does not create just to destroy creation. The existence of part-whole duality and hierarchy in nature makes it possible for there to be differentiaton within unity; and this is what creation is all about. Wholeness is a material way to express unity, but specifically as differentiated unity.

Once wholeness is understood as a material form of Unity, it is possible to understand what unwholesomeness is, including its relationshp to non-being and that which is bad. Unity, Being and the Good are correlatives. Relative to Goodness, in the material world wholeness is that which is good. That which does not promote wholeness is bad. Relative to Being, badness (luckily) has a measure of unreality about it. I say 'measure' because, although there is no ultimate non-being, there can be non-being at a particular level. When there is a wound in a body there is no non-being in the wound. The molecules are real and are present in every spot. It is the level of being that is wrong. In a wound the molecules are not at the (cellular) level that they can be. When we have a wound, we are not realizing our potential (the form that is in our soul). The molecules in a wound are not true 'parts' of us (as cells). That is how we come to miss a 'part.' There is still matter there it is the level that is wrong. And it is incoherence that prevents the correct level from manifesting as it should.

Non-being does not exist. To know the truth of this we have but to look at the universe. Every 'spot' within it is filled with energy or matter. Even dark space is light in wave-form.

To know the truth about the relationship between Being and Goodness (not to mention the power of the Good), we need to ask ourselves why there is anything at all and not nothing—why there is being and not non-being. (Because being is a power as well as a substance, this is also why all wounds will be healed. Nature hates a vacuum—qualitatively and spatially. All power resides in being, and *correct* being, not in non-being or incoherence).

In addition to looking at the Being and becoming that is everywhere, we also need to ask why there is value everywhere, from ordered complexity to movements in a specific (positive) direction. The simplest atom is tremendously complex and orderly. The complexity and beauty of life everywhere is a testament to the power of Goodness, Unity and Being. In terms of directed movements, it is antithetical to the spirit of truth to say things like "Well, maybe someday the world will start shrinking instead of expanding," or "Well, maybe someday time will go backwards into the past," or "Well, maybe someday the power of a force will be greatest at the circumference around the point of contact and lessen as it moves in towards the center of the contact," or even "Well maybe someday there will be no order." The world is made in terms of these weighted contrasts and either we are going to respect empiricism as a guide for truth or we are not. Astrology is a wonderful touch-stone, but we only need to look at the world to know the truth of the things mentioned in this section.

Wholeness is the material expression of pure Being-Unity, and it is through the contemplation of all these things that it is possible to know the truth about creation—to know with certainty that we are meant to be free of disease, pain and death. Despite the obstacles presented by the bad things in the world and by the negative effects of the second law of thermodynamics, a new, spiritual paradigm is coming. Because it views the world as arising from absolute Being-Unity, it will necessarily entail seeing the world as naturally good, and the individuals in that world as intended to be fully whole and only Good. A spiritual paradigm can help people achieve faith in the possibility of fulfilling their potentials—both individually and within their collectives. Along with faith there must be work, but the end result is nothing less than paradise. Paradise is not a place, but rather a state of being.

Returning now to the goal of this chapter and to the second subject **(b),** If there is value in nature there must be something within us to advise us when one activity, circumstance or state is higher or better than another. And that brings up the third question **(c),** where this moral compass might exist. Plato was unable to say where the Form of the Good existed. And yet Plato himself was aware that the known must be in the knower. There has to be a point of sameness in order for there to be any type of relationship, including the concepts that lead to 'knowledge' and 'understanding.' The knowledge of the good is no exception to this rule.

If we have something within us that can judge that which is good, then that judgment seat needs to be the Good so that identification can take place. If there are many types of things that can be good, then the standard to judge needs to be, not in terms of any one particular 'type,' but rather undifferentiated and independent, and hence absolute. If we are to assess and judge 'good-better-best' and if the standard to do so is absolute, then our spirit is our means to judge what is good. Furthermore, the place of this touchstone must be seen as in the center of us, and we have talked about why that is so. Furthermore, if that standard is in the center of us, it is not a different judgment seat from the one that judges truth and reality. Nor is it different from the one that we use in conceptualization and reasoning processes—thus our conscience may be able to intuit that which is good, but also able to think it out.

Thinking is a material process, so the question can be asked as to what material state of affairs might assist in the process of using our conscience. And that too the astrological category system makes clear. Absolute Being and Unity is expressed materially through wholeness. For this reason, wholeness can be looked upon as a natural source of value, and as a measure, standard, and guide for determining that which is good.

We have talked about how spatial wholeness is dependent upon the point-instant; but that type of thought process in philosophy is a little removed from our day-to-day experience of wholeness, the latter of which is of vital and immediate importance to us. It is good to be whole. If we lose part of our bodies (or part of our bodily function), this value dimension really comes home to us. The two things we value the most are our being and our unity. To reiterate, it is good to be whole—and just to exist and to be. If our parts come apart and we cease to be whole, we cease to be.

Relativists have insisted that there are no absolute standards for that which is 'good,' but there are. The fact that it is 'good' to be whole (and to be) is strong evidence of the existence of a definite, absolute standard for goodness—not only residing in spirit, but also manifesting in the world via wholeness—the material expression of spirit. We either exist or we do not. We are either whole or we are not.

We are able to judge when our physical bodies are whole by whether or not they are all there and coherently interactive. We are able to judge when our mental and emotional bodies are whole as well. And we are able to judge when our family 'bodies,' national 'bodies,' and worldly 'bodies' are whole. We know by the degree of cooperation, coherence, peace, love and other signs of unity that exist. We also know by the degree to which they flourish and remain in being.

The judgment seat that is our conscience is the same judgment seat that is utilized to judge wholeness as Unity and Being for any purpose. And as with any assessment of wholeness, it depends upon the spirit within us—on absolute Unity-Being.

We believe that we have a conscience just as we believe that we are alive and aware, are an individual, can judge what is real, etc. It is better to have a cosmology that, not only supports our belief, and accepts the existence of a 'conscience,' but can also explain what it is. Such a cosmology is far better and more comprehensive than one that cannot.

*[As was mentioned in the section on truth, all of these manifestations of pure spirit go together and act together precisely because they are all the same spark of pure spirit. Because spirit is a power as well as a substance, and because of what has been gleaned thus far, it is possible to know that this judgment seat is not a passive one, but rather a creative power of 'life.' In other words, not only is our spirit a factor in the judgment of the good, it plays a role in its creation as well, as is discussed more in depth in the next two sections on '**Creativity**' and '**Will**.'*

Furthermore, in terms of our focus (awareness) and belief structure (the power of belief), this means that we need to focus on and have faith in the good in ourselves and in others. Naturally it would be a form of blindness not to see the wounds in ourselves and our collectives. To deny wounds is to be unable to work towards healing them. However, to not see these merely as tissues that lack life, and to not see the real goodness, potential wholeness, and true reality of the person (and people) behind the wounds is blindness. We need to be careful how we judge both ourselves and others, as we help create negative conditions in our world when we do so.]

(6) Pure Spirit and Creativity: The first question to be asked in terms of 'creativity' is whether or not new energy-matter is needed to create, or if we can just reshuffle things around and put them together in new and 'creative' ways. A child knows that, if he has only so many plastic Lego pieces to build with, eventually he will run out of pieces. Furthermore, if these are distributed among four children and only one-fourth are his, he will run out of pieces even sooner—unless of course he confiscates the pieces of energy-matter belonging to the other children. No matter what he does, he will run out sooner or later and his 'creativity' will end.

We believe that we have the capacity to create circumstances in our life. The archetypes confirm that there is true creativity—that new energy proceeds from the center of every system constantly. If this is true, then new matter is being created. Since energy-matter is material in nature, something producing energy-matter would have to be non-material. This subject was touched upon in Section (**1**) on the life principle as a creative (alpha) process. Since our current materialistic paradigm allows for the existence of nothing that is not material in nature, creativity is denied. From a materialistic viewpoint, matter being created would have to come from nothing and that is illogical. The AC agrees with that. But it is not illogical for something material to come from no-thing, i.e., from no material thing, but from something that is real and yet not a type of being—and that is what spirit has been defined as.

Recently it has been observed that particles do pop into existence from no-where. (Cf. the so-called 'zero point field' phenomenon in physics.[10]) Thus far the claim is that particles also pop out of existence, thus preserving the *amount* of energy in the universe. If the coming and going can be measured someday, and the coming exceeds the going, that would be a proof of the existence of creativity. As far as that goes, any coming from no-thing is creative, even if things go back to no-thing. (Even ordinary motion is creative because, in order to be serial, it accesses the substrate.)

No-thing is by definition no material thing. What is not material is not finite; hence the source of what is created would not itself be finite. What exists but is not a finite type of being can only be absolute being. If we have the potential to create, then the source within us needs to be infinite and absolute. The place where something infinite and absolute can exist and be what it is, is at the center of a system. Leo rules centers, and from antiquity Leo and its house (the fifth house) has been said to rule the creative potential in a person. That rulership stands to reason.

[*If we are creating, we are creators. Each of us needs to ask ourselves if we truly believe that we are free to create the circumstances of our lives. Either we are or we are not.*]

(7) Pure Spirit and 'Will.' Creativity is closely related to the concept of free will. We believe that we do have a will and that it is free to choose what we will do given various alternatives. (The subject of will gets into that aspect of spirit which reveals it to be a power as well as a substance, hence the term 'will power').

To posit the existence of free will does not necessarily mean that the concept of 'law' and lawful interactions is damaged—that the laws of nature are violated. The types of being *are* the laws of nature. And these *are* the parts of systems. The parts of a system can be arranged in any number of ways and still be whole. There are correspondences between the different laws of nature. When we see formulas like $E=mc^2$ which relates the doing power of energy to mass, or $E=hf$ which relates the doing power of energy to a spatial concept (frequency), it is possible to know that the different aspects of creation can be expressed in different qualitative ways without violating law. The parts of a whole are perspectives. To shift from one perspective to another is not to break the law. This is what even ordinary change (e.g., going left to go into the bedroom instead of straight ahead) is all about. We know that the parts of systems manifest unbelievable variety, and that the wholeness of a system is yet achieved, so long as there is coherence and viability between the parts.

In addition to making things change, and manifest one type of being rather than another, there is also true creativity. If that were not so there would be no future with its accompanying growth. Growth and creativity of any kind is a miracle that can only exist because an individual spirit has access to the quantum potential. Growth and creativity entail the realization of new potentials.

Whether creating something entirely new or merely switching from one type of being to another (or even, as in the case of motion, re-activating a particular course), the substrate must be accessed. This is easy to see in terms of truly new creation, but it is true of switching types of being as well. There must be a movement back to no-thing, and then a re-emergence as a different (or even similar) type of being materializes. Whether this source and crossing point is referred to as a zero-point field, the substrate aspect of spirit, or the quantum potential, it needs to exist and yet not be a type of being, but rather the source and potential for anything, and that is what absolute Being is.

Furthermore, if there is such a thing as free will, it does not entail just freedom. The emergence of quantum theory in modern times has at last brought us a scientific means to explain what 'freedom' is. A photon, atom or molecule in wave-form is a field potential that is not limited in the way the same system is when it is in particle (material) form. Atoms and molecules (and people) have a field potential that is not definite in nature and hence not prescribed by the laws that gross matter is composed of. We have many potential paths that we could conceivably follow.

Nevertheless, even if we have infinite potentials available to us, but we cannot choose between these alternatives, how are we truly free? Real free will entails a director with the power to choose. If we have a free will, we cannot merely posit the existence of freedom in experience, as per quantum theory. We must also posit a consciously aware director with the power to choose. And if we have a director, it would seem to need to, not only have access to an undifferentiated field, but in some way be related to it, so that it can then actualize any type of being necessary in its creative process. In astrology, for thousands of years, Leo has been said to rule the will. The astrological category system indicates that the will is an individual creative center accessing the quantum potential (of which it is a part). This rulership stands to reason.

[Although a directed change, or even an entirely new process, does not necessarily entail violation of law, spirit is nevertheless Unity, and if what is created does not further personal and world wholeness, it is not lawful, because law has to do with the coherent interconnection of parts—relationships that produce wholeness. The parts of our world and the parts of our bodies must all fit together in lawful ways.

The laws of physics are naturally orderly and coherent. Therefore any incoherence in a world is evidence that natural law has been violated. The laws of nature exist at many different levels, including at the level of human beings. The world of living creatures is currently manifesting considerable incoherence. The source of this would appear to be ill will and bad intentions. (An 'intention' involves an end or final cause and entails a conscious visualization of the outcome and a deliberate aim for that particular outcome). The initiation of an act is a creative first cause. When we are aware that the end will not promote wholeness, and we still initiate an unwholesome act, that is wrong. Will can break the laws, but it obviously should not. That is what conscience is all about. It knows when something is lawful and when it is not, when it will promote wholeness, and when it will not. If the end furthers wholeness, then it can be said that the will has been exercised in a way that entails 'good will.']

(8) Pure Spirit and Happiness: In section (**5**) on '**Conscience**' the point was made that, if we are to have values, we need a touchstone of absolute Goodness, and we need it to exist in us. The relationship of goodness to wholeness was also mentioned. It is good to Be and to be whole. In that section pleasure was discussed, but the relationship of Unity-Being-Goodness to happiness was not elaborated upon. It would seem as though conscience would be the last thing connected to pleasure, since we normally think of it as related to duty rather than to pleasure, but such is not the case. Our physical body feels good when it is whole. There is pain and discomfort only when it is not. It is the same with the emotional body, and it is the emotional body that deals in 'happiness.'

The Leo-sun principle is said to rule happiness and fun. Colloquially we speak of people being 'sunny' when they are happy by nature. A happy person is perceived as radiating light and energy. (The concept of radiation in turn has to do with centrifugal processes of centers actually producing light and energy.)

All the things that Leo rules are immediate modes of spirit. Happiness has been said to be an emotion, but it may be something a little more than that, and its rulership by the Leo principle is a clue to this effect. In order to understand what it is, it is important to understand first of all what the emotions are.

As with mind, the emotions appear to be material in nature. We have different kinds of emotions. They move (and move us). They even have a spatial domain to them. Some are large; some are small, etc. Mind needs to be conceptualized as material in nature, and the emotions present as the same. And yet, the emotions are different from the mind, as anyone knows who has tried to be reasonable with a person who is being excessively emotional. Nor are they physical in nature, although I do suspect they are more closely related to the physical body than to the mind. Perhaps this was the first body to separate off from the physical body, or perhaps they are not yet as distinct as the mind.

The close relationship between the physical body and the emotions can be seen even in the words we use to express emotions. Thus for instance we speak of physical 'feelings' (sensations) and emotional 'feelings.' We know the difference, and yet they are close. As with mind, our materialistic paradigm has tended to find the basis of the emotions in the physical body. I do not agree. I believe that we possess three different kinds of material bodies—physical, mental and emotional.

A 'body' is nothing more than a nexus of relationships. As portions of these ways of relating change, they can branch off. An example is the evolution of different species. Even though the mental, emotional and physical bodies a person manifests entail different kinds of bodies branching off in an individual, rather than different kinds of physical bodies branching off in a group or species, the evolution of species nevertheless presents as an analog. In the physical plane, different species branch off as they become more and more different physically. At a certain point in time different groups become so different that they can no longer mate, and are truly separate species. At the same time they all remain part of the world of living creatures.

When a bundle of new ways of acting branches off within an individual united by one soul, the bundle apparently remains a 'body' for the individual that is the owner of the conglomerate of relationships. In other words, the different types of bodies are still the vehicles of the same person. Although separating into distinct bodies, they continue to interpenetrate and interconnect because they are linked by one principle of life and awareness, and because they are also linked by the same soul. In other words, similar expressions of one particular archetype expressed in different bodies are all linked by a common archetype in a shared soul. For this reason, the emotional body is still able to affect the physical body and so forth.

If we are to better understand these bodies, it is important to recognize their differences and their individual distinctions. One reason the mental and emotional bodies have not yet been recognized as distinct and real may be due to the 'physical' emphasis of the current paradigm. There is a denial of the reality of anything that does not consist of (and interact with) physical atoms. We need to consider the possibility that there are mental and emotional photons, and that we use these to construct our mental and emotional bodies as surely as we do our physical bodies.

For any 'body,' there is an internal world and an external one. Groups of beings that have similar bodies share similar external worlds or realms. If there are physical, mental and emotional bodies, it is rational to presume that there are physical, mental and emotional realms. The possession of mental, emotional and physical bodies (and their collective realms) is a tenet of the so-called ancient 'gnosis' or esoteric wisdom. Many people in the past have claimed that we possess these bodies and that these domains exist. It makes sense to me and the more so because I use astrology to clue me in on which directions of thought are profitable and which are not.

When an astrologer looks at a person's chart, they are looking at a probability field. Astrologers can look at the chart and see it now in terms of the physical plane, now in terms of the emotional plane and now in terms of the mental plane. For instance, Capricorn is an archetype in the spatial domain and rules that which we experience as spatial limits. In the physical body, Capricorn rules the epidermis of the skin, since this is the outermost area of the skin, and the skin itself is the outmost boundary of the body. In our external social world, this archetype makes it possible for us to be set apart from other people. In the emotional plane of experience Capricorn rules the emotion of loneliness, which entails a feeling of being isolated from others. The emotion of loneliness is different from the physical reality. What astrology indicates is that these states of affairs utilize the same archetype, and yet involve truly different bodies and planes of experience.

If the emotions are material in nature—a body—then this body will have a wholeness aspect, and will be connected, as with all the bodies, to the spirit. Relative to wholeness, as with our physical and mental bodies, the emotional body is also apparently intended to be whole. At base of all the various types of feelings and emotions, a general feeling of 'pleasure' and/or 'happiness' is paramount. And at base of pleasure and happiness resides a state of wholeness. When something is wrong or missing in our physical body we feel 'pain.' When something is missing in our emotional body, or threatens its wholeness, we experience emotional pain or 'unhappiness.'

It is the same with the external groups that we participate in as wholes-in-the-making. To feel disconnected, unloved and alienated from other people involves a lack of group wholeness, and is a primary source of unhappiness in the world. To be united with other people is to find happiness. What makes us feel good and happy in the material world is either maintenance of current wholeness, or an increase in the extent of a sense of inclusion and unity. And at the back of any wholeness is spirit—the impetus behind greater being and greater unity. Wholeness reflects spirit; and the existence of wholeness—whether mental, emotional, or physical—necessitates the postulation of spirit. In a spiritual paradigm, spirit needs to be perceived as actively involved with producing emotional as well as mental wholeness.

If the emotional body is a body, it has a pure form of absolute being at the center of it just as the physical and mental bodies do—and this would have to be the same spirit as resides at the center of the other bodies or nexus' of relationships. At the same time (as with the difference between 'life' and 'awareness') we would experience it slightly differently in accordance with the type of matter surrounding the spirit. Some of the expressions we use for pure spirit in the emotional body are also used for the other bodies. The emotional body is *living*; and thus 'life' is involved with it just as it is with the physical body. In addition, we are *aware* of our emotions just as we are aware of our physical body and the thoughts of our mental body; and thus 'awareness' is a part of this body as well.

At the same time, each body also seems to allow absolute being to shine through in special ways. For some time, certain oriental and eastern religions have claimed that there is something connected to that which we call 'happiness,' compared to which material happiness is a pale reflection. They refer to it as absolute Bliss. If they are right (and we do apparently connect wholeness to happiness and to that which is good), spirit may not be just pure absolute Being-Unity-Goodness-Awareness. It may also be pure absolute Bliss, and actually the basis of that which we call happiness. Buddhists believe that we can increase our state of awareness through yoga practices and that the goal of life should be to denounce the material world (which they believe is an illusion) and become one with absolute Awareness, which then becomes for us a state of absolute Bliss. The material body and individual ego are viewed as a source of unhappiness. They speak of this state of absolute awareness and bliss as 'nirvana.'

I disagree with Buddhist philosophy in regards to both the non-existent or illusionary nature of the physical world, as well as its negative nature. I think the material body is real so long as it is coherent. It is the incoherent parts or wounds that create pain, and the only non-existence is due to a lack of the correct level. What I do think Buddhists are right about is their concept of relating happiness to awareness, both of these to centers, and both to an absolute substance.

Buddhists believe certain artificial practices open the centers. I suspect that the opening of centers also occurs naturally when we change how we think, feel and act in regards to ourselves and others, so as to repair the wounds in us and in our collective world of living creatures. In this way we 'find' happiness—

because we do not have to find it. It is not 'out there' somewhere. It is in the center of us, and naturally ours when we find the fullness of life that we are meant to have.

(9) Pure Spirit and Love. Spirit is the power of Unity. It does not just unite the parts of an individual. It is a power that also draws separated individuals to unite together into collectives.

Love is an attractive force and, as with any manifestation of spirit, it has different names in different bodies. The processes produced by spirit drawing systems together in the external physical world are due to what is called the force of gravity. In the emotional realm, this power is experienced as love.

From antiquity, Leo has been said to rule, not only life, awareness, will and happiness, but also love. Astrologers look to Leo (and Leo's house, the 5th) to see when a person is likely to fall in love. By relegating the rulership of love to Leo, what astrology appears to be telling us—and in a very innocent (unconscious) manner—is that love is spiritual in nature—that it is the substance from which the world is created, and that it is a power capable of unifying people as well as moving them towards a better, higher state of being.

Relative to Unity, love is certainly a unifying power. Love connects person to person. When we are loved, we feel emotionally whole. We even have an intuitive sense that it originates from the center of our being. For centuries we have been told, not only that love is perhaps the greatest power in the universe, but also that it originates in the heart. Leo rules the physical heart. The term 'heart' used in a symbolic sense involves an intuitive, symbolic reference to the core or center of our being. (If there is a spirit of truth, intuition itself may begin in the heart, where we instinctively know the truth about things.)

In addition to its unifying power, there is also a Being and Awareness aspect to love as well. Never do we feel more alive or aware than when we first fall in love. It is also tremendously uplifting. It brings out the best in us. If it *gives* unity and being, it may very well *be* Unity and Being. Teilhard de Chardin believed that the spirit of life that centralized individual systems also unified groups of systems, and was what we call love. The AC (with the help of the sign Leo) affirms this view.

[One indication that love may be a manifestation of pure spirit is the connection it has in the human mind with concepts of God. In the Psalms—a holy book for Christians, Muslims and Jews—God is said to actually be love. In ancient Greece, Eros (love) was said to be the first and oldest of the gods.]

Pure love may indeed be just what is meant by spirit. If so, and spirit is also the substance from which matter derives, then the material world is literally created from love.

To summarize the ideas presented in this chapter on '**Spirit**,' if we do not posit the real existence of pure absolute Being in the world, it is not possible to explain in a simple and coherent manner so many things that we do regularly experience. These range from the reality of a spatial extent to all the basic, commonplace characteristics of systems—from the unity, coherence and centralization of their parts, to the evolutionary processes that we see them engaged in. Science cannot currently explain what love is, what awareness is, what the will is, what life is, what causes psychic phenomena—and on and on. Science under the materialistic paradigm denies the existence of value in nature, even while it is manifest all around us. If we can explain all these things with ease once the existence of pure Being becomes part of a world paradigm, why on earth would we choose an awkward, incomplete paradigm? (And to explain them 'with ease,' is meant simple explanations that are concurrently coherent.)

In discussing these things, it was possible to see that they involve giving Unity as well as Being to our material parts, which was one of the tasks set forth at the beginning. Most importantly, in order to give a truly adequate explanation for these things that we do experience, it was necessary to posit the existence of an absolute factor within us. The known must be in the knower. There needs to be something infinite, absolute and eternal within us in order to clearly and simply explain our experiences of these things.

Although in this chapter absolute Being was given the name 'spirit,' other names for the same thing were also mentioned. Plato talked a great deal about the nature of pure, absolute being as it related to qualitative unity. It was basic to his theory of the Forms. But Plato did not call absolute being 'spirit.' It is the same with spatial unity. Leibniz called his ontological unit a monad. Geometricians do not call the point-instant 'spirit,' and mathematicians do not think of zero as a spiritual substance. Needless to say, physicists would not dream of using the term spirit (much less 'love') to explain attractive forces in nature like gravity. On the other hand, Teilhard did call the power and substance that he believed caused centralization and superiorization in systems 'spirit.' In addition, he related this to God. The main thing is to be faithful to the description involved with the term—its character as absolute, infinite and eternal Being-Unity. Leibniz believed in what philosopher's call the 'identity of indiscernible.' This simply means that when two things are the same in their character, they *are* the same thing. The monad, the point-instant, the Form of absolute Being, spirit—these things are the same thing.

When trying to understand this metaphysical substance, it is important for the term spirit to be used. It is also important for spirit to be conceptualized as something at the center of us. The materialists and logical empiricists have made scientists (and philosophers) supersensitive about the term spirit. This divisiveness needs to end. Rational examination and spiritual examination do not contradict or threaten one another. And although, as has been noted, matter has sometimes been seen as diametrically opposed to the nature of spirit, when the paradigm shifts, there is no essential difference between the two. We need a concept of spirit in order to rationally and scientifically understand ourselves and our material world.

Chapter 4

The Role of Science in the New Paradigm

According to Jewish, Christian and Muslim tradition, when God created the world, he said, "Let there be light!" All material things are composed of light. If all of the atoms in the world disintegrated, physicists claim they would dissolve into a shower of photons. Because light is the substance from which everything is created, it must contain within itself everything in potential form, including the fundament laws of nature. For this reason, physics boils down to trying to understand what light is. The basis of the new paradigm in turn involves understanding the nature of light and its relationship to spirit and soul.

There are currently three theories of light utilized by modern science—electromagnetic theory, quantum theory and relativity theory. Each adds something to the other and hence there is a certain complementation between the three, but there are also inconsistencies between them. Each has something to give to the new paradigm as well; and in return a spiritual paradigm has the capacity to reconcile some of their differences.

The Nature of Light

Electromagnetic Theory

Electromagnetic theory was the first of the three modern theories of light to arise. It conceives of light in terms of waves, and describes them as composed of electric and magnetic components, which are their 'parts.' In addition, different types of waves are said to have different frequencies. As different waves with different frequencies relate to one another, they do so in terms of differing internal phase interrelationships. As with the electric and magnetic components, these different phases within a particular electromagnetic wave are also parts of the wave.

Electromagnetic theory is relevant to a spiritual cosmology for two reasons. First of all, in this paradigm there are true individuals—wholes with parts. Electromagnetic theory is a field theory, so it does not oppose quantum field theory, but there is a greater emphasis on the fact that waves have internal components, and hence on the individualistic factor. What has parts is a system or whole.

The second reason that electromagnetic theory is important to the new paradigm is due to the role of astrology. Electromagnetic theory helps show how phase and frequency differentials in waves produce qualitative distinctions. The twelve signs in astrology are the qualities possessed by our world and they too manifest in terms of phase and frequency differentials.

The electromagnetic waves of which the universe is composed are not waves *in* a medium as water waves are. Rather, they are themselves the substance of creation. In addition to being the substance of our bodies these waves are also the means for us to interact with other systems in the external world. When the interaction is with the molecules of our eyes, we call these waves 'light.'

Light is a subclass of electromagnetic waves. The atoms of our eyes see electromagnetic waves only in the range of about 400 nm to 700 nm. [A nanometer (nm) is equivalent to a billionth of a meter.] Although light as a visual experience is a subclass of electromagnetic waves, the more inclusive term of 'light' is generally used for all electromagnetic radiation, and this book will do so as well, since 'visibility' by the atoms of our eyes is just a specialized form of detection occurring at a particular level of being. An electromagnetic ray of any range, occurring at any level, can theoretically be detected. Any detection is a form of 'seeing.'

Electromagnetic waves have a spectrum of wave-lengths that range from miniscule lengths compatible with the sub-world of the nucleus of an atom, to the lengths of ordinary electromagnetic vibrations, to huge lengths involved with radio waves, and possibly even astronomical lengths involved with the gravitational force. The atomic-level world of atom-to-atom interactions is the range of waves that we know the most about, and these are spoken of as the electromagnetic force per se. Those at astronomically higher or lower levels are referred to as 'electromagnetic-like.'

Electromagnetic theory arose before quantum theory, and is a wave theory. In quantum theory on the other hand, light occurs in a particle form called a photon, after the Latin for light ('photos'). Later, quantum theory was combined with electromagnetic theory to produce a quantum field theory. This worked well with the notion that light has wave-particle duality. The effort to synthesize these two theories has been only partially successful. Although the idea of light existing in a wave-form is needed to explain such concepts as wave-particle duality, the connection between otherwise separated particles, and even the vibrational character of light, in quantum theory the wave-form is said to not be real, whereas in electromagnetic theory the waves are real.

Light, whether perceived as a wave or a particle is an elemental system. It has three basic attributes—the 'domains' mentioned in Chapter 2. One domain entails energy as motion and 'doing power.' Light is energy ('radiant' energy). A second domain has to do with the fact that light has spatial attributes. Although the 'space' of an electromagnetic vibration is a little different from the spatial extent of a gross material object, even a ray of light has a specific extent to its wave-length. The third domain has to do with light's qualitative nature per se. This has to do with how all of its qualities are linked into a coherent whole, making it a primordial system.

Relative to the domain of energy as motion and doing power, a ray of light has the capacity to move through space. In addition to itself moving, when light is absorbed by a material body, it gives that body energy—the power to move. This capacity of bodies to do things is called 'work.' Electromagnetism is spoken of as a 'force' because it has the ability to affect other things and to facilitate work. In quantum electrodynamics, when two complex bodies interact, they are said to do so via photon exchanges. The donor of the photon acts as a causal agent upon the recipient, giving it more energy and hence more capacity to move and do things. Such interactions are called 'mechanical' causes. Newton's laws of motion are laws of mechanics.

[Science currently speaks of electromagnetic energy and kinetic energy (the energy of motion) as two different *forms* of energy. What the AC indicates is that radiant energy is a simple system that possesses the *quality* of kinetic energy, as do other systems as well.]

Light is a force; but it is also a substance. It is in terms of its being a substance, that it has spatial attributes and other characteristics. Relative to its spatial attributes, light does not actually move *through* space. There is nothing external to light. Light *is* space, or rather its spatial attributes are what we call space, whether this is the spatial extent of an individual ray of light or the universal space-time continuum.

Because everything is composed of light, every type of individual material spatial 'extent' that exists in the world derives from how it begins in a photon. In terms of its wave nature, one wave-length is a fundamental unit of space.

The wave-length of a ray of light is a factor of its quantitative domain. How far something extends into space is a measure of how 'much' it possesses internally. Spatial extent is a geometrical primitive for the principle of 'how much.'

The two domains are related to one another. In electromagnetic theory the concept of 'frequency' (f) combines the two. Frequency is a measure of the number of complete waves passing a set point in a certain unit of time. Green light, for instance, involves 600 trillion waves per second.

The third domain of light has to do with its qualitative character. This domain is prior to the other two in the sense that it treats of a ray of light (or photon) as a system or whole *with* parts.

It is the other two domains that actually produce the parts or qualities. Electromagnetic theory explains qualities via frequency variables, and frequency is a product of spatial extent and motion. The quality called color provides an example. Each color in light has its own specific frequency.

Quantum theory looks to the particle nature of the photon rather than its wave nature to express the three domains. In terms of the quantitative and kinematical domains, it describes the photon as a quantum or packet (a quantity) of energy (doing power). In terms of qualitative differentiation, quantum theory again looks to the particle aspect. Thus quark theory attempts to explain the laws of nature in terms of twelve basic particle regularities.

There are mathematical correlations between these two different ways of looking at light in terms of either the particle or wave aspect. The amount of energy in a single photon is related to its frequency (and hence to its wave length) via a number the frequency can be multiplied by called Planck's constant (symbolized as h). The formula is $E=hf$. A unit of energy is equal to h times the frequency of the wave. [Planck's constant is equivalent to 6.62×10^{-34} joules. (A joule is the energy used to lift 1 kg. a distance of 10 cm.)].

Because the archetypes begin in the wave-form, I prefer to see the three domains in terms of waves. This is particularly so in regards to the qualitative domain. I like to visualize the archetypes in terms of colors, and these in turn in terms of frequencies within waves. The concept of color may actually have a wider application than is currently believed. Each atom in our world can be identified via the colors it emits and absorbs. As this idea relates to the AC and how the laws of nature arise in light, the primary colors, I suspect, arise from the elements; in other words from a primal red (fire signs), primal blue (water signs), primal green (earth signs), and primal yellow (air signs). When a ray of light (which presents as possibly a two-dimensional system) interacts with matter (or actually becomes three dimensional matter), the simple elements apparently expand into wave lengths that express the basic types of being. These also can be experienced as normal colors, such as those emitted and absorbed by an atom. Astrology makes it clear that the differing frequencies in electromagnetic vibrations translate into the fundamental types of being.

Light actually has a continuous spectrum with no particular colors singled out, and yet the astrological category system indicates that there are twelve basic potentials for qualitative differentiation in a three dimensional world. Thus the question can be asked as to why there are certain wave-lengths that predominate—even why there are limited types of being (limited colors).

What the AC indicates is that the qualities of a world increase as its dimensional expressions increase. (In Chapter 2 the types of being were correlated to the spatial dimensions.) The number of dimensions a particular world is currently manifesting, not only express the level of its evolution, the type of space produced conversely may be a factor in limiting the number and character of the types of being that can be produced from light. In our world, atoms are our matter. Hence the type of space these manifest determine the wave-lengths that are appropriate for our three dimensional world. Electromagnetic waves must obey certain wave-length demands within an atom if they are to remain strong (everlasting) and not cancel themselves out. It is electrons that are responsible for the qualities of atom, and hence for the qualities we experience in our world through atoms. An electron is a standing wave and its 'orbital' relates the diameter of its wave (its wave-length) to the circumference of the atom.

The circumference of the atom limits the actual cycling motion of a wave that is captured within an atom. The wave needs to fit the atom. There is a constant of nature that relates h to π, the latter of which in turn relates the diameter of a circle to its circumference. The constant is called $\hbar$ ('h-cross'). The equation $\hbar$ takes both h and π into account and is sometimes itself referred to as Planck's constant because it is found in so many quantum calculations. The equation $\hbar$ is equivalent to $h/2\pi$, and hence, as noted (via π), relates h to the diameter of a circle.

The frequencies (orbits) that are allowable for electrons within atoms occur in multiples of $\hbar$, i.e., in terms of $1\hbar$, $2\hbar$, $3\hbar$, etc. There can be 1, 2, 3… (full) wave lengths in an atom but they must be full so that the wave does not interfere with itself and cancel itself out.

Because electrons are responsible for the qualities of atoms—and hence for the qualities we experience in our world through atoms—those electron orbital possibilities may one day be related by physicists to the basic laws of nature—perhaps even to planetary orbital possibilities. Chapter 9 (**The Language of Astrology**) discusses this subject more in depth.

As was noted, scientists hope someday to find the basic laws of nature in a few particle regularities. What looking at the solar system (or even electron orbitals) teaches is that there is a need to look more closely at the wave nature of light rather than (or in addition to) its particle nature.

Quantum Theory

During the twentieth century, a new way of explaining the nature of light came about called quantum theory. What was different about this new theory is that it viewed light as being dual in nature—capable of manifesting two different forms—one as a particle or photon, and one as a wave, depending upon its circumstances at the moment. It is the wave-form that may help explain what the wholeness of a whole is, and consequently what soul is.

As a photon, light is a discrete, localized particle with a definite nature, such as being a green photon or a red photon. When in wave-form, on the other hand, a photon does not possess these normal physical qualities and is actually quite metaphysical in nature. In its widest character light in wave form is a field potential for the photon. It makes it possible for a photon to potentially be anything anywhere. In a slightly more restricted sense, the wave-form of a particular photon serves as a 'probability field' for that particle. As a probability field it is like a Book of Life for the photon. Light is a tremendous carrier of information and it records everything that any particular particle has ever been in any previous incarnations in material form. In doing so it helps determine where a photon will most likely next materialize, and what kind of photon it will most likely be when it does. The probability field is still not definite in nature and only determines these things in a loose, probable way. A photon that should be green could conceivably wind up being blue, purple, or even colors that we have no conception of. It could appear in Timbuktu or on the other side of the universe. It could even manifest as a meson or an atom, but it will most probably be a green photon reflecting off of a particular green wall because that is what its current external interactions and internal characteristics as recorded in its probability field guide it towards. This guidance along with relative looseness is why it is called a probability field. There is guidance, but also freedom.

The wave-form of the photon cannot be seen and will never be seen. Any form of detection makes it immediately turn into a particle. Because the wave-form can never be seen, many physicists think that it really does not exist—that it is just something about the photon that makes it be what it is when it does materialize. Others believe the wave-form does exist, and that we need to expand our understanding of the world to include a basic reality that is not physical in nature. Those who say the wave-form does not exist must explain how a non-existent can hold the character of a photon when the photon itself does not exist. Those, on the other hand, who believe the wave-form does exist, must explain what it is. If it is not physical, by definition it may be metaphysical. Its characteristics are certainly metaphysical.

It is now known that atoms also have a wave-form nature. Like a photon, they too are particles, but of a much more complex type. Their wave-form also acts as a probability field that determines the where and what of their nature—both where they will (most likely) exist in the external world of atoms, and what type of internal structure they will (most likely) have. The wave-form is quite individualistic and specific in its guidance in the sense that a carbon atom will most likely be a carbon atom, not an oxygen atom. In terms of this internal qualitative guidance, its probability field is what philosophers speak of as a 'formal cause.' The wave-form of an atom determines what type of form it will have.

As with atoms, molecules are particles as well, and are known to possess wave-particle duality. There are huge carbon molecules that have an incredibly complex internal structure. Both where they exist in the world of atoms and what their internal structure is like is controlled by their wave-form, which guides their materialization.

It is the thesis of the AC that people possess a probability field. Chapter 6 on '**Spirit and Soul as Powers**' looks at the thoughts of some biologists who believe that living creatures possess some type of field that has formal causal efficacy upon the parts of their bodies. These biologists do not speak of this field as a probability field because quantum physicists currently believe only atoms and molecules have individual probability fields. In constructing a new paradigm, it may be possible to make the case that this belief is in error.

A photon, in either wave or particle form, is behind the construction of even complex systems. The wave-form of an atom or molecule is not different from the wave-form of a photon. It is just light in wave-form. In particle form, the photon is behind the 'parts' aspect of atoms and molecules as well. Physicists believe that if all atoms and molecules ceased to exist, the energy would still be there but in the form of photons.

In the new paradigm being formulated with the help of astrology, three concepts will be linked. First of all the wave-particle duality of light will be combined with the notion of part-whole duality. The probability field of an atom, for instance, will be looked upon as the wholeness of the atom. The part-whole duality of an atom in turn will be linked to the concepts of soul and matter. The wholeness of an atom will be looked upon as its 'soul.' Its parts (nucleons, electrons and ultimately photons) will be looked upon as its 'matter.'

If in the AC the concept of soul is linked to probability fields, then obviously soul cannot be an emergent quality that comes into being when a system becomes living or human, but rather one that exists in any true system. If soul is not an emergent property, then it is possible to learn about it through physics—in particular through the photon, the most elemental system of all, and especially in terms of how the photon is visualized via quantum theory.

In order to understand quantum theory, it is helpful to look at how ideas about light in general have developed over the years. This highlights the philosophical questions that have occurred, and it also gives an overview of how knowledge of any kind often vacillates between opposing ideas.

A major historical controversy in regards to light was whether to see it as a particle or a wave. Closely connected to that question was whether the light field of the world was to be looked upon as a unified whole or divided into discrete units. As this relates to our view of matter, for centuries the material world was seen as divided into and built up from tiny bits of matter with a miniscule bit of extent. With the advent of relativity theory, scientists began to see the world as a unified whole. (In quantum field theory it was easier to visualize particles as connected in a fluid manner when they were described as dynamic processes.) Included in the divided-versus-united issue is the question of the real existence of individuals as such. If a photon, an atom, or a person is only a concentrated area of energy in a universal electromagnetic field, they are not independent wholes.

Another historical controversy was over the question of whether to see light as a metaphysical or physical substance. Massive objects can be seen. Light is invisible until there is an interaction with matter. Energy cannot be seen except in its relationship to objects. At first there was a problem with trying to understand the difference between energy as a particle and matter. When energy and matter eventually became seen as two different ways of looking at the same essential substance, this problem evolved into trying to understand the difference between energy and the wave-form nature of light which (unlike a photon, its particle-form) is truly metaphysical and can never be empirically seen or detected. Since philosophers (and scientists) have been divided over whether anything 'metaphysical' is real, this historical question is closely related to the real-versus-unreal controversy in regards to the modern view of the wave-form of light.

The wave-vs.-particle, unified-vs.-divided, material-vs.-non-material, visible-vs.-invisible, real-vs.-unreal questions entail very closely related concepts. For that reason, the very fact that in our lifetime light has begun to be seen as concurrently both a particle and a wave is a positive sign that the other controversies will be reconciled as well.

Our history of trying to philosophically and scientifically understand light goes back as far as the ancient Greeks. Aristotle had described light as something invisible that could be seen in terms of white light (or color), but only due to its relationship to something else, i.e., to solid bodies, to water or to air—what we would refer to as the three states of gross matter—solids, liquids and gases. The process of 'seeing' light (and seeing color) he thus explained as due to the relationship between matter and light (*On the Soul*, 418a).[1] According to Aristotle, although light was not visible, it caused visibility. Light not only was involved with visibility, it also had the power to move things (*On the Soul*, Books I & II). The fact that light was not material was evident in the fact that it was in itself not visible until it met up with matter, but also according to Aristotle because it was involved with an unseen, everlasting part of a living being—their 'soul.' Per Aristotle, it even made up an everlasting part of the world. It was the uppermost, unseen surrounding shell of the world.

With the advent of the Renaissance, classical mechanics was born and with it new theories of light. Rene Descartes (1596-1650) believed that there was an invisible substance between bodies called the 'ether.' This was a very fine type of matter. The ether was everywhere. Light was simply a disturbance that propagated through the ether with different wave-like periodic motions.

The period of classical mechanics is sometimes called Newtonian mechanics due to the contributions of Isaac Newton (1642-1727) towards understanding motion-change and mechanical forces. Newton presented the first particle theory of light. He believed that small particles moved (in a straight line) through the ether after being emitted from a light source.

In the early 1800s an Englishman, Thomas Young (1773-1829), was able to prove that light demonstrated interference effects, which occur only with waves. Young showed this through the double-slit experiment, wherein a light ray (with multiple photons) is split by being aimed towards and progressing through two separate slits. If light is composed of particles, the screen on the opposite side should show two areas of light where the photons hit in two spots corresponding to having come through two slits. Instead, two semi-circular waves come out the other end of the tiny slits, similar to the way water produces waves when it passes through two separate openings. As with water, the two separate but ever widening semi-circles of light then run into each other on the other side and either augment or interfere with one another's waves, depending upon whether the crest of one meets a crest in the other (constructive interference) or the crest of one meets a trough in the other (destructive interference). With light, constructive interference increases the brightness of the collective meeting ground, while destructive interference darkens the collective effect. What happens with this experiment with two slits is that a pattern of bright and dark bands is generated on a screen at the end of their path. The double-slit experiment proves that light is a wave.

James Clerk Maxwell (1831-1879) added to the view of light as a wave by combining known ideas about electricity and magnetism with that of the nature of light. He was able to show that light waves were a type of electromagnetic wave. Maxwell described a single ray of light as consisting of two transverse waves of electricity and magnetism at right angles to one another, constantly changing and constantly affecting one another through their interacting undulations, creating self-motion in this way from within. (A changing magnetic field produces an electric field, and a changing electric field produces a magnetic field.) Maxwell did not yet do away with the notion of the ether that had been so important to the classical period of Newtonian physics. He merely combined this with electromagnetic theory and related these to light.

Max Planck (1858-1947) and Albert Einstein (1879-1955) then turned light back into a particle again when they both provided simple and better explanations for what happened when a heated body radiated energy (Planck) or when certain metals absorbed energy (Einstein). The former was spoken of as 'blackbody radiation,' and the later was called the 'photoelectric' effect. One dealt with emission of light, the other with absorption of light, but both furthered the particle theory. In the photo-electric effect, for instance, Heinrich Hertz (1857-1894) had noticed that certain metals emitted negative electric charges in packets when a beam of high frequency light was shined on them and they absorbed this energy. (Later these electric 'charges' were named 'electrons.') Einstein saw that the experimental data could best be explained by visualizing the absorption of light as occurring in chunks. (Later the most fundamental chunks were called 'photons.') Each photon of a specific energy amount ('quantum' of energy) caused the emission of one electron with the same amount of energy.

Despite the distinction between particle and wave theories, and his assistance with a particle explanation for light, Planck saw that the *frequency* of an electromagnetic wave (and hence the wave length) was related to a definite *amount* of energy—the amount being the packet or quanta. [As noted previously, the relationship between the frequency of the wave (f) and the amount of the energy of the quanta (E) is in terms of what is called Planck's constant (h), i.e., $E=hf$.]

Light very soon *became* the ether, or rather the notion of a fine material substance was done away with, and the idea of energy in the form of an electromagnetic wave arose. (With this the energy-matter dichotomy was soon to be eliminated.) At the same time as the concept of the ether was eliminated, whether radiant energy was a particle or a wave was still very much in question. It was known that electromagnetic waves were not only waves, but invisible waves, and light was also invisible until it interacted with matter, so this tended to point to light being a wave. But when it interacted with matter, it seemed to be a particle.

At the same time as all this was going on, the view of what an atom consisted of was changing as well. Ernest Rutherford (1871-1937) had proposed a picture of the atom that included a dense nucleus with electrons rotating around the nucleus like planets. In his model, an electron was seen as radiating light continuously (light of all wave-lengths) as it rotated around the nucleus.

Niels Bohr (1885-1962) constructed a better picture of an atom as having electrons in different orbitals and as emitting or absorbing light radiation when and only when an electron jumped from one orbit to another. In Rutherford's model, the electron would continually lose energy, and as it did so would prescribe a smaller and smaller orbital. Eventually it would fall into the nucleus and the atom would cease to be an atom. In the Bohr model, the electron absorbed light when it jumped to a higher energy level and emitted light when it jumped to a lower energy level. Thus the electron could be perceived as a stable part of the atom and not as something that would eventually spiral into the nucleus.

In classical physics the notion of angular momentum was used to describe orbiting planets, and Bohr realized that the units of Planck's constant were the same as those used in classical physics to describe angular momentum. 'Frequency' has to do with any cyclic rotation, be it a planet or the complete cycle (circuit) of an electromagnetic wave. Pi (π) is the ratio of the circumference of a circle to its diameter. If this is related to Planck's constant (h), the relationship is ($h/2\pi$) or what is symbolized as $\hbar$. Planck theorized that, if there was a chunk nature to the energy of an electron, then the radii of an electron out from the center of the atom could only be $h/2\pi$, or $2(h/2\pi)$ or $3(h/2\pi)$, etc. [In other words only in terms of the fundamental units $1\hbar$, $2\hbar$, $3\hbar$, etc., that relate a fundamental unit of space to a fundamental unit of motion-change (the doing-power aspect of energy)] In this way Bohr saw that there would be only certain allowed orbital radii in terms of the energy chunks. Although the Bohr model quantized the electron, it still utilized the concept of orbital radii and waves.

In the classical view of matter vs. energy, scientists had pretty much come to the conclusion that energy was a wave, and matter consisted of particles—essentially electrons, protons and neutrons. A French scientist, Louis de Broglie, published a paper in 1924 which suggested it was possible to combine wave and particle explanations relative to matter in the way that it had been suggested relative to light. If an electron could be described as a particle, it could also be described as a wave. If an orbit is a wave, and the wave ends meet when it goes around once, then what happens each time is that the wave reinforces itself through constructive interference. In this way it is stable, since even partial destructive interference would lead to the wave eventually canceling itself out. Even if the orbit entails two waves, so long as it involves two *complete* waves, the ends will still exactly meet, and there will be no destructive interference, and again a stable orbit. There can even be more than two complete waves, so long as the waves are in terms of whole units. The equations Bohr worked out in 1913 and those de Broglie published in 1920 were the same.[2] The one was derived from a quantization of angular momentum, the other by visualizing the electron as a wave, and capable of a stable orbit because there were complete waves meeting at the ends.

(If one thinks of an electron as a wave, not merely the chunk nature of energy limits the orbitals, but also the very nature of three-dimensional space. As noted, there are only certain ways in which standing waves around the nucleus of an atom can fit and can have ends that meet. Presumably higher dimensions would have more possibilities.)

In addition to the stability of the atom, another thing that the Bohr model helped explain was the line spectra of atoms. Light has various potentials for different frequencies (different colors). The spectrum itself is continuous. No particular lines within white light are singled out such as 'red' or 'blue.' Yet, when light is shone on an atom, particular lines are singled out. A hydrogen atom emits and absorbs only certain frequencies of light, which colors in turn define its nature. An oxygen atom emits and absorbs only certain frequencies of light, which define its nature as oxygen rather than hydrogen, etc. In the Rutherford atom, the electrons would have to emit a continuous spectrum. The Bohr atom explained the line spectra of atoms. Specific (divided) orbitals produce specific (divided) colors.

DeBroglie had the idea that an electron could be both a particle and a wave, depending upon how it was viewed. Visualizing light as energy had made it possible for the notion of the ether, as a rarified type of matter, to be eliminated from theory. But it had also divided the world into energy vs. matter. The Bohr atom helped show the essential unity of the world again. The difference was that, with the ether concept, light was seen as a rarified type of matter, whereas with the DeBroglie concept, matter could be seen as composed of radiant energy and electromagnetic in essence.

Another thing that helped show the ultimate identity between light and matter was the continuing use of the double-slit experiment in new ways. As instruments became more refined it was possible to shine, not merely a beam of multiple photons at two slits, but a single photon. The photon registered on the screen at the end as a dot, showing itself to be a particle. But then, as more photons were fired through a double slit, they did not land in the same place, but rather in a pattern that produced the aforementioned wave pattern of light and dark bands. Each individual photon was clearly both a wave and a particle. It was a wave when it went through the double slit (i.e., because it went through both slits). But it was a particle when it got to the end. At the end, the wave pattern entailed multiple separated dots but—precisely because these were in a wave pattern—it proved that each photon had been a wave in the middle and capable of going through both slits at once.

Firing single photons through slits merely helped clarify what Young had already discovered in terms of the wave-particle duality of energy. But what helped show the identity between energy and matter was the capacity to fire tiny material particles through the slits. Electrons were fired through double slits, and eventually whole sodium atoms, and they acted the same way as photons. And it was not just that the double-slit experiment confirmed that, like radiant energy, electrons and atoms were waves; it confirmed that both energy and matter were both particles and waves. It was one thing to see that light had these strange characteristics; another to see that atoms, the very foundation of the material world, also did. The new emerging information both eliminated one dichotomy and produced a new one. It helped eliminate the distinction between energy and matter, but produced a new dichotomy between a wave on the one hand (of some sort) and energy-matter on the other hand.

At the same time as all this was going on, the matter vs. non-matter, visible vs. non-visible historical vacillation took on a new twist. Thanks to the double-slit experiment it was discovered that whether a photon is visible or invisible depends upon whether it is in particle or wave-form. When in particle form, it is visible as light. When in wave-form a photon is invisible.

The visible versus invisible question had always been related to the material versus non-material question. We see material objects. We see radiant energy only when it interacts with matter, and then it becomes a particle or photon. To accept the fact that light is invisible when not seen is not too difficult of a concept. But when it was discovered that an atom could occur in wave or 'particle' form (as an atom), it became evident that atoms also had the capacity to take on a non-visible, non-material form.

As if visualizing matter as capable of occurring in an invisible form were not a difficult enough idea to incorporate into the new type of world view, the role of the observer became a fundamental concept of quantum theory as well. Whether a photon is visible or invisible depends upon whether it is being observed or not. In a way this makes sense, since 'light' is the fact of an observation. The problem is that this is true of atoms as well.

The way this knowledge came about was through switching the point of observation in the double-slit experiment. When this is switched from the screen at the end to the two slits in the middle, a different area is being observed, and something very different happens. When in particle form, even light is an 'object' so to speak, and has the characteristic of specific place. It is localized. In the first case, when the screen at the end is the point of observation, the photon (or atom) becomes a localized particle or little dot on the screen, just as you would expect a particle to do. [Then successive photons, electrons or atoms arrange themselves in a wave pattern with light and dark bands—the proof, not just that the particle is a wave as well as a particle, but also that it is an (unobserved) wave in the middle of the journey when it goes through the two slits, and a particle at the place of observation at the end of the journey.] In the second case, when the slits are the focus of the observation (and are then one end of the journey so to speak), the beam becomes a particle at that point, the point of observation. In the first case (when the double slit is not the point of observation), since the particle is in wave-form and the wave is spread out, a photon or atom can go through two slits at once. When on the other hand an atom or photon is observed at the point of interaction with the double slit, it becomes a localized object, only goes through one slit, and the wave effects of constructive and destructive interference on the screen at the very end then do not take place. This astounded physicists, because it meant that it was the process of the observation that decided whether light would be a wave or a particle.

Bohr stressed that the observer must be conscious, since it is observation (by us) that turns the wave-form of an atom into a particle. He believed that only living things were conscious. (He also believed living things were different because they did not have a single, individual wave-form like an atom or molecule.)

{There is a better way to interpret this characteristic of quantum behavior, and that is to explain the required 'difference' as being due to different levels of true wholes. And it is not just that there are different levels of wholes and different levels of parts, but that a part is a whole, depending upon the level of interaction and observation. Parts are what we call 'matter' and photons and atoms are the 'matter' of our world. It may be precisely our awareness that make them become parts of *our* world and hence '*material*-ize.' [That does not mean that the atom is not an individual or whole in its own right (and involved in a world of its own); just that it may only become visible to us (and hence 'matter' to us) when we interact with it and utilize it as a *part* of our world.]

As for claiming that awareness only exists in so-called living things, Bohr in this way presupposes an 'emergent' hypothesis with no explanation of how or why awareness emerges as systems evolve. If atoms were not aware of their internal and external physical environments, they would not respond to internal and external stimuli. Furthermore, they respond *as a whole*. These have centered cognition. (It is when awareness is seen as a factor of only a mental body that this interpretation is likely to take place.)

And as for Bohr's belief that living things do not have an individual wave-form nature, how would he know? These would have to be fired through a double slit to determine this. It is simpler and more comprehensive to see all true wholes as aware, and the awareness to be a factor, but the *level* of awareness the deciding factor as to whether something is a particle or wave, a part or a whole, matter or soul.}

The visible versus invisible, material versus non-material question was also always historically related to the real versus unreal question. There is nothing like seeing to assist believing. Even the saint 'Doubting Thomas' wanted to see and touch before he believed in the reality of something. Modern science has a materialistic bias in its empiricism. It is believed that we can never prove the existence of something metaphysical. The mainstream of modern science decided during the twentieth century that anything non-material (and this included 'spirit' or 'soul') was either unreal, or could never be proven to exist. Therefore, it was thought best to leave such subjects alone. Little did science know that the foundation of quantum theory (which in turn is the basis of how we explain the world) was going to depend upon utilizing the concept of light in wave-form which is neither energy nor matter and will never be visible.

Given that science had embraced empiricism and its emphasis upon physical observation, it was natural in the historical development of a theory of energy-matter for light in wave-form to come to be seen as unreal. It was accordingly decided that the wave-form did not really exist. A photon or atom really existed only when it was being observed. When it was not being observed, it was only a 'probability' and had no real existence. The probability is simply *something* about the photon or atom that makes it materialize where it does and be what it is when it is observed, and then becomes real and exists. (How something that does not exist can carry a probability is unexplainable by this theory, and hence irrational.)

Coming to see the probability as a nonexistent wave actually was due to a number of concurrent developments in the field of physics. An Austrian physicist, Erwin Schrödinger (1887-1961) developed the 'wave mechanics' theory of quantum behavior. These equations concurred with Maxwell's equations relative to electromagnetism. What Erwin Schrödinger intended as a description of the behavior of a real wave, was utilized in quantum theory to express a 'probability.' A wave is more spread out than a localized particle. What is spread out is less 'definite' and hence may be less real.

Another thing that contributed to seeing the wave-form of light as unreal was the very idea of a 'probability' and a 'potential.' It was discovered that a photon (or atom) in wave-form has a potential to materialize anywhere in the universe and be anything. When it materializes it is specific in being and specific in qualities. A potential can be thought of as something that is not real, since it can 'potentially' be anything. A probability is somewhat more limited than a potential in that it entails a set of increasingly more likely possibilities; but a set of probabilities is still not definite in character or localized in space like a particle.

In addition to the spread out nature of a wave-form—the indefinite nature of a probability, and the potentially infinite nature of a potential—there is also something inexact about the qualities of a particle. A German physicist, Werner Heisenberg (1901-1976) helped elucidate an inherent uncertainty in regards to two quantum characteristics. This uncertainty is related to which state is being measured or observed. At the quantum level of being, it is not possible to know precisely where something is and how fast it is moving at the same time. When place is accurately observed and measured, the motion is fuzzy. When motion is accurately observed and measured, the place of the particle is fuzzy. A portion of this uncertainty is due to the fact that we change things when we measure them, but even more basically this uncertainty may be due to an inherent inability to look at two levels at once—in this case a characteristic of the object as a whole, versus one of its material qualities or 'parts.' Motion is a characteristic of matter and belongs to the 'parts' aspect of experience. 'Place' is where something is individuating, and is more closely tied to the reality of a centralized (whole) object.

(In quantum theory place in space and motion are spoken of as attributes that have 'complementariety.' This means that particles do have these two attributes, but we cannot get a clear perception of both at the same time.)

It was not merely the uncertainty principle that Heisenberg contributed to quantum mechanics. In 1925 he developed the first coherent description of quantum mechanics called 'matrix mechanics' which describes quanta as energy interactions rather than tiny bits of matter with an extent. Describing particles in terms of the quality of motion-change combines well with the notion of uncertainty. When something is perceived as a dynamic interaction, it is constantly changing and is never definitely this or that.

All this fits with relativity theory. In relativity theory, what something is (its nature or character) involves how it is relating to and interacting with every other 'part' in the world. Thus objects are not very definite in and of themselves.

A German physicist Max Born (1882-1970) worked in collaboration with Heisenberg on the matrix mechanics version of quantum reality. It was Born who got the idea that the wave theory Schrödinger had used in his version of quantum mechanics could be used to describe a probability wave.

It was seeing the world as affected by probabilities which dealt a real blow to the previous notion in classical mechanics that the laws of nature were definite in nature and hence very deterministic. Born strongly believed that the ultimate outcome of any state of affairs was in the range of chance, governed only by greater or lesser probabilities. He corresponded with Einstein who was disturbed by this view and made his famous statement "God does not play dice."

An English physicist, Paul Dirac (1902-1984) picked up on Heisenberg's ideas. Dirac united Heisenberg's matrix mechanics and Schrödinger's wave mechanics into a better, unified theory, and also paved the way towards an eventual quantum field theory. In 1930 Dirac published *The Principles of Quantum Mechanics* which is still in use as a textbook and inspired the young Richard Feynman.[3]

In seeing particles as dynamic relationships, the photon became a very primal type of relationship. United States scientist Richard Feynman (1918-1988) was a physicist who specialized in bosons. Bosons are the medium of interactions between particles.

The photon is a boson, as is a meson. The difference is that the photon is involved with the interactions known as the electromagnetic force, while the meson is involved with the interactions in the nuclear force, and in that sense is like a lower-level photon. It was Feynman who worked out 'quantum electrodynamics' (QED), the theory of quantum mechanics used today to describe the electromagnetic force in terms of its mediation by photons.

Feynman continued to use the interactive approach that Heisenberg had begun so that the very nature of a particle is presented as a nexus of dynamic interactions. His famous 'Feynman diagrams' depict not only how two separated particles interact in terms of a photon, but also how one particle is a composite of such interactions. This is when matter began to really be looked upon as composed of relationships.

All these men—Born, Heisenberg, Dirac, Einstein, Schrödinger and Bohr—were publishing papers concurrently and were in contact with one another. Each built their ideas on the shoulders of the others. The standard interpretation of quantum behavior taught in schools today was actually put together by Bohr. He was working in Copenhagen and at a conference with his colleagues in Italy in 1927 Bohr presented a packet of ideas that was a composite of the ideas of these contemporaries. This package was later referred to as the Copenhagen theory of quantum reality. Professional physicists now have access to a number of other possible interpretations of quantum behavior, including the 'sum over histories,' the 'hidden variables,' and the 'many worlds,' approaches, but it is the Copenhagen theory of quantum reality that is still taught in universities, and it is this theory that asserts the unreality of light in wave-form.

As with the wave-versus-particle problem and the real-versus-unreal problem, the united-versus-divided question has been an underlying issue throughout the historical development of the concept of light. Quantum theory, as the name suggests is still a type of atomism. It entails quantization, which is the division of experience into tiny packets of energy. Quantum theory divides energy, spatial length, and even time into small but measurable units. The smallest possible spatial length according to quantum theory is a Planck length (10^{-33} cm.). The smallest possible time is Planck's time (10^{-43} sec.).

Because the Copenhagen interpretation of the (unreality of the) wave-form blocked physicists from explaining continuity in that way, one way that quantum physicists have attempted to reconcile the idea of the discrete 'packet' of the quantum with the idea of overall world unity is, as noted in the previous section, through seeing particle themselves as processes, and as a group of dynamic relationships that involve other particles (e.g., Heisenberg, Dirac, Feynman). This view in a practical sense came about through physicists working their way down through matter. Thanks to the electron microscope, macrocosmic systems like human beings began to be seen, not as individuals with individual extents, but merely as the sum of multiple atoms plus their interactions. Later atoms began to be viewed, not as objects with an extent, but as the sum of multiple particles plus their interactions. When one got down to the base of life, even an atomic particle did not appear to be a thing with an extent, but rather a centered nexus of ever changing dynamic inter-relationships. Warner Heisenberg, one of the founding fathers of quantum theory, claimed that we must now see the world as composed, not of objects or bodies with extent, but of constantly interacting processes.[4]

This interpretation of matter correlates to the AC in one very important way. As was noted in Chapter 2, each basic type of being when materially expressed is a relationship. Spatial extent is clearly shown (via the principles called Sagittarius and Cancer) to be the relationship of 'without' as it relates to and is relative to 'within.' And process is shown to be (via the principles called Aries and Scorpio) the relationship of 'going forth' and 'leaving behind' (relative to some point in space-time).

The philosophy or cosmology of the AC differs from the cosmology that emerges via quantum field theory in two ways. The first, more minor way, is that it does not agree with the 'process' emphasis. The domains in astrology indicate that every system has two basic qualities—motion and extent. What the historical view of the development of the concept of light reveals is that the emphasis on process came about partly when extent began to be seen as due to forces between particles (since a force entails a process), and partly due to 'energy' being seen as a photon and as motion (when in reality motion is a *quality* a photon possesses). But the deciding factor was when there was a shift in paradigms from atomism to holism and unified field theories. Seeing particles as processes helps deal with the constant, ever-resurfacing issue in quantum theory relative to how to connect particles to one another.

The major way in which a spiritual paradigm differs from quantum field theory is that it holds to the reality of the wave-form, and even provides a way to show how a discrete (disconnected) particle is connected to its wave-form. From a spiritual perspective the particle (as a point) is also a field (the substrate). From a soul perspective, there is part-whole duality and subsequent hierarchy in nature. What is a particle (part) from one perspective is a wave (soul) from another. It depends upon the level of experience and observation. It is by correlating wave-particle duality to part-whole duality that this issue of connectivity is resolved. What the AC reveals is that, even under the old atomism, if an atom had been seen as both a discrete atom *and* a part in higher level systems, that paradigm too could have been holistic. The division between atoms (the ends as perceived) is merely a matter of hierarchical level of observation, in other words, whether the atom is perceived as a distinct and separate individual (a whole in its own right), or whether it is perceived as a part together with other atoms in a still higher level whole.

Ultimately everything is a part of the universe because the hierarchy of the tree of life extends up to the universe as a whole, of which everything is a part.

Historically, as the paradigm shifted from atomism to field concepts, the effort to account for wholeness in the world led to a total disavowal of individuality. Hence an atom came to be seen as just an intensified bit of energy in the universal field. When, on the other hand, a concept of hierarchy in nature is introduced, an atom can be both a true individual and a part of a higher level whole.

Scientific understanding is a process that is never complete. Explanations even cycle, much like an electromagnetic wave itself. Atomism divided the world, and scientists were moving into the era of field theories. The extent of an individual had been correlated in the collective mind to limitation and division; but it does not have to be—not when a paradigm can utilize part-whole duality to account for both division and unity in the world.

It is possible to grow in knowledge without constantly bouncing back and forth between extremes. Truth reconciles such paradoxes, providing a way to move the two sides of a seeming opposition up to a higher level of understanding. The visible-versus-invisible, physical-versus-metaphysical, universal-versus-individual, united-versus-divided problems, are not separate subjects but rather closely interrelated concepts. And a major key to understanding the relationship between all these ideas resides in the now proven and accepted fact that energy-matter is both a particle and a wave. The double-slit experiment made that knowledge possible. (It actually was a proof of the existence of soul.) Seeing the wave-form as real, relating wave-particle duality to part-whole (matter-soul) duality, and understanding even this duality as due to (and explainable through) the two poles of spirit is a simple, coherent way to reconcile many paradoxes in both science and philosophy.

Quantum theory claims it is a field theory, but actually

remains a discrete particle theory. Quantum theory cannot adequately account for the connectivity between the discrete particles (quanta) that are the real entities in that theory. In order to account for real connections in the world, it is necessary to have a (real) zero point—which is what spirit consists of—and a real field (the wave-form). Whether the field is seen as a piece of the substrate (soul) or as the quantum potential, it needs to be seen as an actual existent. This problem in quantum theory is at base of the millenniums old relative-absolute controversy in philosophy and is obviously very important to a spiritual paradigm. We will return to this subject in the next two chapters.

Relativity Theory

When the new paradigm arises, it will have to account for connectivity, not just via a field concept (soul), but also in terms of spirit. That is because soul itself is capable of (and needs) an explanation. In both electromagnetic theory and quantum theory it is not philosophically sound to begin with the concept of a continuous field (whether of extent or process)—in other words taking this as a given—when it is possible through mathematics to explain how an extent (or process) becomes a *continuous* extent, and (since the concept of extent entails continuity) how it becomes an extent in the first place.

Most of what we now know about light has come through developments in the world of quantum physics. However, the special theory of relativity is also about the nature of light—and actually the general theory is as well. The latter deals with the space-time continuum of the world and is a theory of gravitation, which is described as a curvature in space. However, since there is nothing in our world that is not composed of light, the space-time continuum *is* the world's light field.

Albert Einstein (1879-1955) formulated the special and general theories of relativity. These are not only (ultimately) both about light, they are also both field theories. Unlike Bohr,

Einstein apparently believed in the reality of the wave-form aspect of the world electromagnetic field, which fits well with concepts in his general theory of relativity. In his studies with the photoelectric effect, Einstein advanced the theory that light is particle-like in nature, yet his view of the world wound up being a very holistic one.

Relativity theory is actually a particle theory as well as a field theory, but its fundamental particle is quite different from that of quantum theory. Relativity theory is especially helpful towards understanding the relationship between spirit and soul because it can help deal with the concept of 'singularities.' A singularity in physics is where a quantity becomes infinite. It can be infinitely 'tiny' (zero) or infinitely large. In relativity theory the ideal particle collapses to a dimensionless point, and the ideal field is infinite in extent (in the second sense of infinite).

Although quantum theory is primarily into processes, and relegates extent to a secondary place, it is believed by physicists that it can at least account for extent (by beginning with the concept as a given). In relativity theory on the other hand, physicists have been unable to explain how extent can emerge from a particle that begins as a dimensionless point. Physicist David Bohm believed that relativity theory was basically correct, but himself complained that there was no way to introduce extended structure into the theory.[5] And yet relativity theory is such a powerful hypothesis it cannot be discounted. So also is quantum mechanics. Theoretically, the two are only as far apart as the distance between zero and a Planck's length (10^{-33} cm). But what a distance! Somewhere in that tiny space is the means to reconcile quantum theory and relativity theory, which Einstein worked on until his death but was not able to accomplish. [In that tiny space also lays the mystery of creation, since it is the answer to how spirit becomes matter. What the astrological category system reveals is that it is both types of infinities or singularities together (the two poles of spirit) that produce an extent. This is what the mathematics of the continuum indicates, which deals with wholeness.]

It is because particles have the capacity to be singularities that all the interest in 'black holes' has surfaced. A black hole is a region where all the characteristics of matter become infinite—including time, extent, spatial curvature, and energy-mass. Very tiny (sub-atomic) black holes are of interest to physicists because it is thought they might entail centers where it might be possible to enter other dimensions in other worlds—perhaps even into other senses of time.

In addition to particles presenting opportunities to understand black holes, it is believed that there are black holes at the center of every galaxy and possibly also at the center of suns.

Quantum theory cannot deal with either type of infinity—the infinitely small or the infinitely large. As with the idea of zero point relative to spatial connections, the thought of time having a zero point is common sense. In quantum mechanics however, there is no time less than a Planck's time (10^{-43} sec.). This is the time it takes light to cross a distance equivalent to Planck's length. In quantum theory the world had to be 10^{-43} sec. old at its beginning in time, which is an irrational concept. (If there is no zero point in time, not only can there be no beginnings, there is no 'now,' which is a zero point.)

A major problem in quantum field theory is that the calculations used to describe particles lead to infinite results. These are not permissible so they are dealt with by a process called 'renormalization.' With renormalization, physicists know the finite answer they want to get, so they subtract one infinity from another, with the latter made up of infinity plus a small *desired* finite amount to leave behind a small finite amount. Many physicists, including Dirac and Feynman were bothered by renormalization, because it is cheating to make math work. Quantum electrodynamics (QED) is dependent upon this process.

These disparities and problems indicate how badly we need a

cosmology that includes discrete particles, plus both types of infinites—the infinitely small and the infinitely large. It is not that one theory or the other is wrong. Matter always has an extent, but it derives from spirit. Quantum theory has its strengths and is particularly helpful towards understanding individual particles and their individual soul. Relativity theory however may be a stronger first cosmology since, not merely matter, but soul itself (wholeness) requires spirit to explain its nature. In other words, even if the real existence of the field were acknowledged, continuity itself (whether of extent or process) would still need to be explained.

It was a turning point in history when the double-slit experiment revealed that a material system was both a discrete particle and a spread out wave, but the ontological status of the wave aspect is still being neglected, creating multiple theoretical problems. In quantum theory continuity and discontinuity are *said* to both exist. But the problem with QED is that it has not reconciled the paradox, keeps trying to quantize the field, and does not recognize the reality of the wave-form. It is not possible to adequately account for continuity if there is (ultimately) a persistent emphasis on discontinuity, and at the same time, via quantum field theory, a claim that there is not.

Consider an electron for instance. Qualitatively speaking, it is the electron that gives atoms their character. When an electron jumps from one orbital to another the atom manifests a different characteristic. Spatially speaking, the electron makes an instantaneous 'quantum leap.' (In terms of doing power, the different level of energy is what is meant by a 'quantum.') Matter is discontinuous and does entail chunks, whether the chunks are perceived in quantitative terms or qualitative terms. But what connects them? In the *process* of going from one quality (and quantity) to another, there is a zero time interval. It is known that the leap is instantaneous. But in quantum theory there is no zero time or zero space. The only way to have true continuity or zero point would be to set Planck's constant to zero. But Planck's constant does not equal zero.

To be fair, relativity theory does not allow interactions between particles to be instantaneous either. (Mechanical interactions entail motion-change, and all material motion occurs at less than the speed of light.) Relativity theory only allows for the existence of ordinary 'mechanical' causes or interactions. These occur through photon (doing power) exchanges. A photon is a unit of light. For this reason, in relativity theory nothing can go faster than the speed of light. [All ordinary mechanical interactions, not only involve some degree of time in order to occur, they also involve two things being in spatial contact ('local' contact) with one another.]

In 1905 Einstein wrote a paper on Brownian motion which, as with his work on the photoelectric effect, furthered the particle view of quantum mechanics. It also contradicted relativity theory in that Brownian motion has infinite velocities at base of its nature.[6]

In 1935 Einstein and two colleagues (Boris Podolsky and Nathan Rosen) wrote a famous paper suggesting that it might be possible to test the idea that there was non-local (and faster-than-light) causation.[7] This was later called the 'EPR' (Einstein-Podolsky-Rosen) experiment after the three men who proposed it. In the early 1980s a physicist by the name of Alain Aspect found a way to actually conduct this experiment. He and his colleagues utilized an electron that had decomposed into two photons, and separated the two spatially from one another. As rays, photons are 'polarized.' They have a specific spatial orientation. As paired photons (within a common 'whole'—the electron) the two had opposite polarities. What the experiment showed was that (no matter how far apart the two were after being separated) as soon as one was observed and manifested its specific polarity, the other one *instantaneously* manifested the opposite polarity. The interaction between the two was non-local and atemporal. Although Einstein thought this experiment would prove he was right, it proved that he was wrong. There are apparently non-local causes and interactions that are either faster than the speed of light or atemporal altogether. David Bohm, stated that he did not think that the Aspect experiment proved that the theory of relativity is inherently wrong—just that we need to realize that there is another type of causal influence in our world—formal or formative causation.[8]

A concept of soul as light in wave-form makes it so much easier to visualize what is going on. Formal causation is a soul-to-parts cause. There can be 'non-local' connections because there is a real substance that is not the parts and yet is connected to them all—is in fact their connection. This is their wholeness. It is an atemporal connection because time is a factor of the 'parts' aspect of experience. In changing a particular quality, the atom has to access potentials in its soul, first actualizing one potential or quality and then another potential or quality.

Our mathematics is only beginning to learn how to deal with the issue of zero point and continuity. Cantor was able to deal with continuity because he recognized the real being of the infinitely small and the infinitely large. (One way Cantor countered disbelief was to point out that to deny the existence of the infinitely large is to deny the real existence of the irrational numbers, since they have an infinite decimal expansion.)

The issue of how to reach zero point or how to relate the chunk-like nature of matter to the infinite character of something truly continuous is not a new one. These are the basis of paradoxes Zeno wrote about 2500 years ago (see appendix). Modern day scientists and philosophers still do not have the answers to this question.

What a spiritual scientific cosmology can reveal is that the connectivity we experience in ourselves and in our external world is due to a constant (creative) interaction with our personal and group souls, both of which entail light in wave-form. These are parts within the quantum potential. That is why there is no need to connect seemingly separated things. (From a spiritual perspective it is why there is no need to connect spirit as the point with spirit as the substrate. And why in terms of space an infinite number of points can become a line.) Recognizing the reality of spiritual substances is the way to reconcile these paradoxes.

It is difficult to imagine what scientists might do when they get past their prejudices in regards to spiritual ideas. Considering what $E=mc^2$ did for the world (including as a prelude to the building of an atomic bomb) it is actually a little frightening to think how powerful science may become when it finally puts all these concepts together, since a spiritual paradigm can even explain the *meaning* of $E=mc^2$. This is one of the reasons why it is so important to call the point-instant 'spirit,' and wholeness 'soul.' And why it is also important to link centers to conscience, wholeness to 'the good,' and so forth. When this new paradigm arrives, scientists must manifest their own spiritual nature to the utmost—and that means being ethically grounded and using this knowledge only to further the evolution and subsequent wholeness of our nations and our universe.

Bohm & the Quantum Potential

Physicist David Bohm (1917-1992) was a seminal figure in terms of trying to promote the reality of the wave-form, and hence his ideas are important relative to the development of a concept of soul and of spirit as the substrate. (He was influenced by DeBroglie, who also believed the wave-form described in quantum theory was real.) Bohm called the wave-form the quantum potential.

Bohm not only believed in the reality of the wave-form of light, he saw it as prior to light in the form of a quantum. Bohm looked upon the quantum potential similar to the way the AC sees soul or the substrate aspect of spirit as prior to material parts. He spoke specifically of the effects of the wave-form in terms of it being a force as well as a substance and hence as behind formal causation—the cause behind the characteristics that particles take on. Bohm began his career by studying plasmas. These are extremely hot gases where the electrons have come loose from their atoms. Bohm found that plasmas formed colonies. (This type of society in the AC, can be spoken of as a whole-in-the-making.) Bohm was fascinated with plasmas because they behaved like a living creature. They had the capacity to regenerate themselves and to enclose foreign objects in a wall, just as living systems did when penetrated by a splinter or some other such thing.[9] He became very interested in wholeness and formative causation. It is one thing to see formal causation within the parts of living things. In living systems, formal causation can be explained (although not adequately) as highly evolved part-to-part interactions that (somehow) produce the intense coherence in a living whole. It is quite another thing to see, as Bohm did, what looks to be formal causation occurring in the parts of groups that are not even classified yet as a whole, much less living or highly evolved.

Bohm also was involved with proving that there was non-local causation, and understandably so as this relates to the effects of a whole on its parts (and of the wave-form on the particle form). It was Bohm who inspired Alain Aspect to devise a physical experiment to test the Einstein-Podolsky-Rosen idea. The result of this convinced Bohm, not only that relativity theory was wrong in postulating only mechanical causes, but also that the Bohr-Copenhagen interpretation of quantum reality was wrong about the non-existence of the wave-form.

The effects of a whole on its parts can be used to explain both the EPR phenomenon and the instantaneous quantum leaps of the electron within an atom. Just as the photons in the EPR experiment were parts of an electron, an electron is a part of an atom.

Electron levels express what it means to be a 'type' of being. When an electron jumps to a different level it is a different kind of part. When an electron changes into a different part, there has to be a zero point at the moment of change, and the wave-form is the best explanation for this. In a similar manner, the unification factor between separated photons which previously had been parts in an electron is not in terms of material interactions as we know them. It is not involved with a material substance or a mechanical cause, but rather (per the AC) with the wholeness of the two and what is better referred to, per Bohm, as a 'formal' cause.[10]

When the wave-form is seen as real, and particularly when it is seen as a higher level or higher dimension of light, many of the contradictions in quantum theory and relativity theory disappear. In addition, the common attributes of matter become capable of being more clearly and simply visualized.

The two most fundamental attributes are motion and extent. The AC actually describes the creation of these in two ways. The first is in terms of the two poles of spirit, as was done in Chapter 3 on '**Spirit**.' The second is in terms of different levels and dimensions. These only present on the surface as two different descriptive schemata. Motion for instance can be seen as the point instant accessing the substrate (the quantum potential) one point at a time. Or motion can be seen as a three-dimensional system accessing a fourth-dimensional potential in its wave-form (i.e., a system accessing the quantum potential or substrate).

I enjoy Bohm, because he articulates so well the relationships between spatial dimensions and the qualities of motion and extent. According to Bohm three-dimensional particles are projections of a higher dimension.[11] In the Minkowski (geometrized) version of the special theory of relativity, a three-dimensional object actually has a fourth-dimensional length, and that is what 'extent' is. In the three-dimensional reality its length changes depending upon how fast it is moving. The speed of light is a limiting factor between dimensions. At the speed of light three-dimensional extent becomes zero. [12] (This does not mean particles cannot go faster in higher dimensions—just that they cannot be three-dimensional particles and go faster.)

The difference between the special theory seen in this way and the AC is that, in the Minkowski version, time is seen as some sort of fourth dimensional reality. In other words, time is seen as too closely related to space (i.e., as space-time) instead of as a *function* between three-dimensional reality and a fourth dimension-in-the-making as it is drawn out of the field potential. The AC is closer to Bohm in this respect. Bohm believed that it was the wave-form of the world that made motion serial in nature.

In the AC, the existence of a split-level reality is involved with the qualities in all three of the domains. By being both parts and wholes concurrently, systems straddle dimensions. In the AC, all three of the domains interpenetrate, and the qualitative domain is prior to the other two (motion and extent). That is because it rules the part-whole relationship per se, and hence provides for different levels of being. [Different actualized dimensions fit into one another in a hierarchical manner, and the next higher potential dimension (which is in the quantum potential) provides an infinite number of potentials so as to make it possible for an actualized extent to be whole.] Both motion and extent can be perceived as involving the *relationship* between the individual and its field potential, which itself is a higher level dimension-in-the-making. As noted, Bohm believed that three-dimensional matter entailed 'facets' of a higher fourth-dimensional reality.[13] (In relativity theory also, our world is visualized as doubling back upon itself into a higher dimension.) It was Claude Bragdon, the mathematician I mentioned in Chapter 1, who first helped me see how the wholeness of both extent and motion are dependent upon our being split-level beings. Relative to motion, Bragdon explains that we *move* (and move through time) because we are bi-level beings, straddling the third dimension and a higher-level dimension-in-the-making.[14] The wholeness of a true whole is a higher hierarchical and (for true levels) a higher spatial dimension. The wave-form of a field potential is everywhere at once and is an infinity of possible lower, three dimensional placements. Its character of being everywhere makes it possible for both motion and space to be continuous.

Where the philosophy or cosmology that emerges from the AC differs from the cosmology that Bohm pictures is in terms of the degree of the actualization of the fourth dimension. Bohm visualized this as some sort of actualized (fourth) spatial dimension and one that was actualized qualitatively as well as spatially speaking—in other words, as a higher level manifestation of law. (In the AC, each dimension, when actualized, does bring more laws of nature into play; but in the AC, *for a three-dimensional system*, the fourth dimension is not materially actualized.) There are certain aspects of Bohm's thought which indicate that he wanted, like Einstein, to see the world as more deterministic than quantum theory (or the AC) does. Bohm spoke of the wave form as the quantum *potential*, but he also seemed to see this as an actualized form of lawfulness (using terms like 'implicate order' and referring to it as a deeper form of law). In that respect he echoes Einstein when he was referring to potentials and probabilities and said "God does not play dice." In the AC, the fourth dimension may be real matter for higher level systems, but not for us. Its reality is only as a potential. Because the quantum potential is a potential, there is freedom in experience. Between spirit as our will and the freedom in this field, we are free to draw out from this potential any number of things. Despite the current limitation in archetypes (twelve) there are many ways these can be expressed; and there are more archetypes that can eventually be actualized when we scale a higher dimension—things we have yet to even dream about. At the same time, types of being are just what is meant by law. Despite unlimited variety, the ways of putting them together in matter (as matter) must be lawful. In that sense, the quantum potential is a deeper form of law if by law is meant the fact that what is actualized must be lawful—or even that spirit has everything that can ever exist in potential form as archetypes within itself. But what is in potential form is not actualized. That is what matter is.

Another way in which the cosmology of the AC is different from the one Bohm visualizes is that, although he believed in the reality of the wave-form, he did not take the Leibniz-Einstein route of seeing particles as involved with singularities and beginning in extension-less points. Instead, like Feynman and others, he visualized them as tiny bits of process. Perhaps this was because he was unable to see how a non-extended point could produce a material extent.

A third way in which the AC differs from Bohm is in regards to individualism. Bohm and Einstein were both born at a time when science was emerging from an atomistic paradigm, and they were both interested in promoting a holistic view of the universe. They did not have a cosmology to work with in which it was possible for there to be both division and unity in the world, and both felt that unity was prior—which it is. Bohm saw the quantum potential as real, and as the means for connectivity between particles, but there is much about his description of it that reminds me of spirit as the substrate rather than of soul (precisely because he had no thesis of individuality). Spirit as the substrate is the potential for anything that can ever be. Soul is a limited set of current potentials being actualized at a particular time in the evolution of a particular system. Soul is a means to account for true individuality in the universe.

As helpful as Bohm and his ideas are towards conceptualizing the reality of the soul, the archetypes are even more so. They can be utilized to overcome the individuality-versus-universality paradox, and also the relative-versus-absolute or finite-versus-infinite paradox, the latter of which keeps us from seeing matter as spiritual in nature. A soul is, not only the means to individuality, it reconciles the relative-absolute, infinite-finite paradox that has plagued philosophy for millenniums. If there is such a thing as an individual soul, there needs to be an explanation as to how a soul can be both individual and hence only a 'piece' of the substrate, and yet somehow paradoxically not be finite. The next chapter is on '**Soul**' and the chapter after entails the subject of '**Matter**.' The question is relevant for both these concepts. In a spiritual paradigm, matter and soul derive from, and are composed of, spirit. So to say that a soul is even currently limited in number of archetypes is to pose a major philosophical question about how 'limit' can even exist when everything is composed of a spiritual substance.

The thing that impresses me most about the astrological category system is its ability to bring to light the ingenious way in which the world is created. No human being could have come up with these ideas. Only the truth (and the beautiful reality it reflects) has the power to reconcile so many seemingly irreconcilable paradoxes. As you will see when we get into the nature of soul, every whole actually is unlimited in the potentials it is manifesting, and that sounds like a contradiction in terms, but as it turns out it is not.

Chapter 5

Soul

Each of us has an everlasting personal soul that establishes our identity and preserves it even after death. It is however one thing to say this is so, another to establish the truth of this. It is now possible via science and philosophy (including all the criteria for truth mentioned) to support the validity of this statement to a very high degree. This is why we are on the brink of a new way of looking at and viewing the world and ourselves.

The archetypes at base of the astrological category system can be used as a touchstone to put the available data together into a coherent whole, and in that way come to see precisely what the soul is and what it does. Astrology speaks of the existence of two metaphysical realities—spirit and soul. People have often interchanged these two terms as though they are the same thing, but in astrology the spirit of a person is said to be ruled by the sun (Leo) while their soul is ruled by Uranus, the planet that correlates to the Aquarian principle. And just as Leo and the things it governs help explain what an individual spirit is, Aquarius and the things it rules help explain what soul is. When the physician I mentioned earlier, William Davidson, was trying to come up with a key word that would best describe all the things Aquarius rules, he decided that 'cooperation-of-many-units-as-one' was a good way to express it.[1] The 'many units' can be perceived as parts within a whole. It was Aquarius, in ruling not merely the soul, but also various types of part-whole connections in humans, that helped me realize that the wholeness of a system is its soul while its parts are matter.

Just as the Leo principle does not actually rule spirit per se but rather the results of final causation arising in matter due to spirit, so Aquarius does not rule soul per se, but rather the results of formal causation which lead to material processes and states that link a whole to its parts and the parts to their whole. Just as Leo rules physical parts in the human body such as the heart that have evolved in order to express spirit immediately and directly, so Aquarius is said to rule things like the central nervous system because of its key role in making all the parts act together like a coordinated whole. The central nervous system is a perfect physical medium for handling the formal causal efficacy that soul manifests.

[Astrology has always focused on human beings, but the cosmology that can be extrapolated from astrology goes beyond the human-level world. In Chapter 3 it was mentioned that atoms must be looked upon as systems with centralized parts, and hence as having a spirit. If the laws in the astrological category system (including Leo and Aquarius) apply to atoms as well as to us—and, as universal laws of nature, they do—then these systems must be seen as having a soul as well. An atom is a whole with parts. A so-called 'living' system, such as a human being, manifests spirit and soul at an astronomically higher level of development and evolution, but the basics of an individual spirit and soul are present in any independent whole. The AC is not an anthropocentric (or emergent) philosophy. It does not exclude dogs, cats and worms from having a spirit and soul—not even atoms.]

The part-to-whole interconnectedness that Aquarius makes possible, not only has to do with the internal parts of a system working together as one, but also with the 'parts' of external wholes. Multi-faceted group 'bodies' consisting of individuals that must work together as one have traditionally been said to be ruled by Aquarius, such as congress, clubs, and other associations.

The part-whole duality that Aquarius makes possible can be seen in the personality traits of people strongly influenced by this sign. Astrologers are aware that Aquarians want to be a distinct individual and a part of some group both at the same time. The hippie movement at the beginning of the Aquarian Age epitomized this concept. These were young people intent upon demonstrating their individuality. Even their clothing was a statement to this effect. But they also wanted to be together—to belong—to be parts within a group whole—especially within the 'peace' movement, where all might be united as one.

It was Aquarius, in ruling group interactions, which helped me see that there are two types of wholes. An atom is referred to as an 'element,' not just because it is a fundamental unit of the matter of our world, but also because it cannot be split apart without producing a nuclear reaction. A molecule, on the other hand, can be taken apart and the atoms within it arranged in different ways. This is what chemical changes are all about. Yet a molecule is enough of a whole to speak of it as a whole-in-the-making, and also as a soul-in-the-making, in other words, as a group soul. These apparently come into existence when individuals relate to one another in a coherent way that is mutually beneficial to them all.

In addition to helping me understand the relationship between parts as these correlate to matter, and soul as it correlates to wholeness, it was the mythology and symbolism attached to the sign Aquarius and its planet Uranus that helped me relate these ideas to light, which also has a dual nature. Everything is composed of light and the concept of colors is linked to the ability of light to separate into different types of being. Whereas most of the signs have a particular color associated with them, Aquarius has always been said to rule multicolored prints like plaids and such. Although astrology captures a very ancient type of wisdom from a time when light itself was not well understood—much less the wave-particle duality of light—in ancient Rome the sign now called Aquarius was referred to as the sign of the 'Peacock' in reference to that bird's colorful tail. It has also been correlated to the rainbow. In astrology Uranus is said to rule the soul. In mythology Aquarius was said to be the rainbow bridge leading to heaven. Our soul does reside in 'heaven' since it is, not only perfect, but is also a higher dimension of light. Aquarius rules the connection between our earthly body and its soul. It was these symbolic and mythological connections to color and light that made me realize that the wholeness of a whole is uncolored (immaterial) light—light in wave-form—the foundation for the 'colors' that are the parts of our body. The wave-particle duality of light corresponds to the soul-matter duality within systems. By closely examining the nature of Aquarius, it is possible to bring together these three concepts—the part-whole duality of systems, their wave-particle duality, and their soul-matter duality—and that will be done in this chapter on '**Soul**.'

A deep understanding of wholeness is absolutely necessary if we are to understand what our soul is, and even what light in wave-form is. As enlightening as the growing academic understanding of spatial and kinematical wholeness is, it is qualitative wholeness that really helps zero in on the nature of wholeness, including its all-important connection to spirit as well as to matter. Qualitative wholeness is the key to understanding wholeness per se because Aquarius is a principle in that domain.

Aquarius is a child of its two parent principles Leo (+) and Pisces (-). Aquarius is primarily positive (+/-). This means that individual wholeness is produced by a process of spirit personalizing (Leo). Nevertheless, the wholeness is also a set of potentials. Thus the substrate (Pisces) is involved as well. Spirit as the substrate is an infinite set of potentials. Wholeness involves a specific, limited set of potentials being actualized by a particular material body.

In order to better understand qualitative wholeness it is helpful to first consider some different kinds of objects to see when they cross over from being an object to being a true qualitative whole or system. The first type of basic object we experience is actually just an aggregate. This is what a rock is—

a collection of molecules plus some external cohesion. Whatever 'wholeness' is there does not involve centralization and the type of unification that true wholeness entails. The parts of a rock are very similar and are merely stuck together by *external*, mechanical electromagnetic forces at the atomic level of being. When a rock is broken apart, what exists then are two rocks.

The next level of object is a whole-in-the-making. Actually a whole-in-the-making is any field of external but potentially viable interactions between systems. A molecule, as was noted, is a whole-in-the-making that is beginning to take on the role of a true whole. What is arising as the parts work well together is a source of *internal* adherence via centralization. As with molecules, the cells in the human body are wholes-in-the-making. Although they are far more complex than molecules, they are similar in that they are still not independent wholes in the way that an atom or a human as a whole is. They have no independent 'life' of their own. They die when their human host dies. It is the same with the organs in a human being, which have a degree of wholeness to them, but are not true wholes. It is not by accident that cells and organs have a dependent life. Wholeness and spirit produce independence. A whole-in-the-making becomes more and more of an independent system as the interactions of its parts become more and more coherent and hence more and more adherent.

When many different systems get together as a group and interact with one another in a cooperative manner, they have the capacity to unite and thereby to evolve into a single, higher level whole. When external interactions are orderly, coherent, and mutually helpful, the internal adherence aspect of coherence becomes stronger and stronger. Eventually what was an external field of interactions between multiple systems at comparable levels of evolution can become an internal field. In other words, what was a group begins to become a body. This process of external fields becoming internal fields is what evolution is all about. It does not involve things merely taking on better and more extensive qualitative being at their current level of existence. It entails systems jumping to higher levels of being. This creates 'hierarchy' in nature. Hierarchy is a condition wherein there are levels of being (involving levels of beings) nested within one another. The process through which systems produce higher and higher *levels* of wholeness is the means whereby atoms have evolved into complex carbon molecules, carbon molecules into one-celled organisms, and single celled organisms into very complex systems.

It is the centralizing power of spirit (as ruled by Leo) that produces the order and coherence of parts (as ruled by Aquarius) that is experienced as qualitative wholeness. Order/coherence and centralization are interactive and inter-dependent—so much so that it is difficult to say which comes first. Until the parts of a system are coherent it cannot become a higher level unit. Nevertheless, as was discussed previously, the order of (different) parts presupposes a power of centralization. As for how a whole-in-the-making can be orderly if it has no independent spirit, it apparently needs to depend upon the spirit of some higher source on the hierarchy of being to assist it while it is still a whole-in-the-making. Thus we are 'hosts' for our organs and cells.

Teilhard de Chardin was a Jesuit priest as well as a scientist, and he spoke of the sacred *heart* or the *spirit* of the Christ as a presence in the world of human interactions helping people to move towards what he called 'Christo-genesis' a point wherein all humanity would be truly one. In addition to the spirit of the Christ, as a panentheist, he considered the Christ to be a cosmic being and people to be like parts within this larger being. Hence the Christ was a 'host' in a universal or cosmic sense, and in *Hymn of the Universe* he speaks of the bread or host of the Eucharist as expressing that relationship symbolically (with wine as the life blood or spirit).[2]

The AC tends to corroborate the view that humanity is moving towards a higher level of wholeness. However, it might be better to broaden the perspective beyond an anthropocentric context. Because they are all closely related and are all true *independent* wholes, all living systems constitute the next highest hierarchical layer of systems after atoms. Our external world as a whole-in-the-making apparently involves the totality of all living creatures in process of searching for coherent interrelationships with one another.

The third type of object that we experience in the world is that of a true whole. Atoms are 'elements' not only because they are the substances from which our world is composed, but also because they cannot be taken apart and reformed into different atoms in the way that molecules can be. It is the same with a 'living' system. If an atom or person is broken in two, it is destroyed. The wholeness of a system is more than the sum of the parts, and is due to its having integral, integrated, coherent parts that are internally centralized by its own *independent* spirit. When looking at these different kinds of wholes it becomes evident that the difference between them is a matter of the degree of independence they possess. It is the degree of independence that makes the difference between a whole and a whole-in-the-making.

Everything in the world is created from and resides within a universal electromagnetic field. Every system is a 'part' of the universe. At the same time, part-whole duality allows for wholeness to also exist, and wholeness creates boundaries that allow for a 'within' and a 'without' in this field, so that there is wholeness or independence as well as part-ness and dependence.

In addition to independence, another quality that a true whole possesses is that of completeness. The very term 'to be whole' infers completeness. In other words, when a system is truly whole it is lacking none of its parts. Completeness has a qualitative as well as a quantitative aspect to it. If one of the parts of a whole is incoherent with the other parts, it cannot be a true 'part,' and hence something is missing.

The two qualities of completeness and independence are themselves closely interrelated. When the parts of a whole-in-the-making are complete quantitatively, as well as being complete qualitatively (with a complete degree of internal coherence in regards to each and every part), the system becomes a true, independent whole.

At that time the 'without' becomes the 'within.' A new higher level individual is born. At the time a new within is created, a new without also begins. Other systems that have themselves evolved to this degree of awareness and being now interact with one another at this new, higher level, producing a new external world and hence a new whole-in-the-making.

Because of the interpenetration between the domains, the spatial dimensions correspond to levels of true wholeness, and actually the qualitative domain is prior to the spatial domain. The spatial domain occurs in stages because dimensions correspond to true levels of wholeness. A whole involves a certain set of potentials or archetypes being actualized from out of the substrate. There are twelve basic archetypes in our three-dimensional world. When we ascend to a higher dimension there will be four more archetypes. That will make it possible for us to move in a new direction (actually two new 'parent' directions, or a new direction with positive and negative poles).

Each dimensional level is complete and is the substance of a world having that many dimensions. We are three dimensional creatures. The reason our three dimensional bodies and our three dimensional world is 'complete' is because, even if there were a fourth *actualized* material dimension above us, as three-dimensional beings we could not currently experience that type of body or world. Claude Bragdon discusses some of the possible characteristics of the fourth dimension in his book *A Primer of Higher Spaces*. These are an expansion upon what we do know about our own three dimensions. We can only make an educated guess at this time because what is known of the material world must be in the knower. The reason physicists speak of other dimensions in terms of other worlds is because of this. A world is what the people in it can and do experience. It is limited to their level of awareness—and 'level' is an appropriate term. The dimensions involve stages, with the highest level encompassing those below. Hence, when we refer to a higher fourth dimension, the term 'higher' is an appropriate word to use as well.

Individuals are prior to their external worlds. An external world is simply a whole-in-the-making having to do with their interactions with other like individuals. For this reason, the dimensions and types of being actually arise via individuals. At the end of Chapter 4 it was mentioned that a soul appeared to be both infinite and not infinite. When a soul is created, only a certain set of potentials within the infinite substrate are singled out. Thus it would seem as though an individual would have to be limited in extent. But the creation of something limited would take away from the infinite nature of spirit, and the material world would not be spiritual in nature.

The beauty of creation is that it occurs in stages. These correspond to true levels of wholeness and to dimensions. Because higher level realties do not exist for a whole, it is infinite in character. It is expressing all-that-is at any one moment in time. Anything else is just a potential to and for it. Soul reconciles the finite-versus-infinite paradox because creation occurs in stages and each is complete-in-time.

These, in summary, are the conclusions that the archetypes lead to. They deserve to be expanded upon philosophically and scientifically as well. All three tools are invaluable, each in their own way. The next couple of pages (sections **A** through **K**) involve a number of logical deductions that begin with simple premises and expand upon them. That is the philosophy part. After that, the concept of soul is considered in terms of physics. This entails looking first at the wholeness of a photon, and then at the wholeness of the cosmos. Finally those two systems are compared to the soul of a person. The means to developing understanding in this chapter include philosophy and science, but the end or goal entails (hopefully) a very personal recognition that we do have a soul and (hopefully also) some ideas as to what that soul is like and what it does.

Wholeness (Soul)

In order to really understand what the soul is, it is vital to explore the very meaning of wholeness. In doing so it is important to expand upon the concepts begun above of how there can be true independent wholes without negating the absolute, infinite and eternal nature of spirit as pure Unity-Being. It is important also to strengthen the aforementioned link between the three interrelated subjects of light in wave-form, wholeness and soul.

When trying to understand what 'wholeness' really means, it is helpful to look at the simplest characteristics first—things that would not normally be thought of because they are so self-evident. That is where the base of conceptualization begins. Then, from these very simple qualities, others logically follow.

A. A Whole is Metaphysical in Nature: No one has seen the wholeness of a whole. The parts of a system are just what is meant by matter. If the whole is real, but is not the material parts, then it is metaphysical in nature. There are currently two contradictory beliefs as to the reality of the wholeness of a whole. One is that what we call a 'whole' or 'wholeness' is merely the sum of a system's parts. The other is that the wholeness is something real over and above the parts. The AC concurs with this latter view.

B. A Whole is Pure Being: Once the wholeness of a whole is seen as existing, then the question of the nature of the 'being' of that substance arises. Parts are divided into this and that type of being. If a whole unifies the material parts of something, and the parts are types of being, what unifies them is not a type of being. Yet if it is real and exists it possesses Being. As was noted in Chapter 1 in regards to spirit, the word 'possesses' is not the appropriate term. Wholeness would just *be* Being.

B1. A Whole is Absolute: From the characteristic of pure Being, the quality of being absolute naturally follows. The parts of a whole are just what is meant by matter. Matter in turn is just what is meant by relationships. (In physics an atom is a conglomerate of dynamic interactions). Parts derive their character in terms of their relationship to other parts; hence their character is 'relative' (cf. the special theory of relativity). A whole on the other hand, *in its wholeness*, is not what it is based on its relationship to other parts (precisely because it is not a part), and so is non-relative. [This relates wholeness to spirit, which is absolute.]

B2. A Whole has a Stable Identity: A major quality of the types of being is motion and change. In the material world motion and change are relative and are even themselves relationships. The characteristic of non-change in the wholeness of a whole logically follows from the characteristic of absolute Being. Wholeness, in presiding over a group of moving, changing parts, apparently does not change and so provides for a stable identity.

C. A Whole is Pure Unity: If there is something that exists over and above the parts of a whole, first of all it is one. Wholeness by its very meaning is a form of unity. Furthermore, it is a form of pure absolute Unity. That logically follows because it is not the kind of (relative) unity that parts possess. It is not a *type* of unity. (The twelve types of being are concurrently types of unity, as was discussed in Chapter 2.) Nor is wholeness the type of unity that spirit possesses. Spirit is a form of pure, absolute unity; but wholeness is similar in some ways and different in other ways from the unity of spirit.

The unity of spirit and wholeness are the same in that both spirit and wholeness must be looked upon as singularities, and hence as types of unities. The personal spirit that we possess may be symbolized by zero (0) while our wholeness can be symbolized by the numeral one (1). The two are different in that spirit is undifferentiated unity while soul, although itself a potential (and not materially differentiated), unifies a specific set of material parts and hence permits the material universe to experience both differentiation and unity via part-whole duality.

[Unlike spirit or the wholeness involved with soul, the material unity of 'parts' involves different and separated things being *connected* rather than existing as a *singularity*. However, just as the wholeness of a whole express the nature of spirit, the parts of a whole express the nature of wholeness when they are correctly connected both spatially and qualitatively (expressing coherence).]

C1. A Whole is a Spatial Singularity: If something exists that is the wholeness of a whole, then it is spatially singular. It is not itself the sum of (connected) material parts, but rather the expression of their unity and the means to their connectivity.

Parts are multiple by nature. As noted, the sum of the parts of a system is not spatially singular. Each part may be singular in and of itself, but each is spatially divided off from each of the other parts, and their unity is that of connectivity rather than of singularity.

[Seeing wholeness as a spatial singularity will lead in the section on science, which is on probability fields, to defining the wholeness of a whole as consisting of a single ray of light in wave-form.]

C2. A Whole is Qualitatively Singular: Parts are multiple in their qualitative nature. In any system there are many different parts connected together and acting together as one. (Qualitative connectivity entails order and coherence.) If the wholeness of a system were just the sum of the parts it would not be qualitatively *singular* even though its parts would be qualitatively *connected*. What is qualitatively singular could be so by virtue of being one of the basic types of being. But then it would be a type of being (i.e., matter/a part). Furthermore (as was noted with spirit in Chapter 3), one particular type of being could not then unite all of the types of being. It is because it is pure Being that the wholeness of a whole is qualitatively singular and capable of uniting the parts.

[The wholeness of a whole can do so (and can be pure Being) because it has the substance of its material parts in potential form. The difference between spirit and wholeness is that spirit as the substrate is the potential for all that can ever be while soul is a specific set of material potentials.]

D. A Whole is Both Pure Being and Pure Unity. As with spirit these are the same thing. [Wholeness presents as a specific, individual representation of the absolute Being and Unity of spirit. Like spirit, wholeness is the source of the Being of the types of being, and also the end (unification point) for the types of being it is in dominion over.]

It is easy to think of wholeness as giving unity since that is what 'to be whole' means. It helps, when trying to visualize wholeness as giving being on the other hand, to think in terms of levels of being. An atom is a being. When atoms unite into a higher level whole, such as us, the system is given higher level being. If the wholeness is more than just a summation of parts, then something real is added whenever there is a hierarchical concrescence. A 'being' comes into being and is added to the universe.

[The wholeness of a whole also gives being in another closely related sense. The wave form of a photon is a repository (in potential form) for energy.]

E. A Whole is a Monad: Philosophers speak of a 'monad' as the simplest possible unit out of which everything else is made. Since a whole is singular, it is simple and cannot degenerate into parts.

And yet, paradoxically, what is a whole is a part, depending upon the level of being. A whole, when seen as matter, is an 'atom' so to speak, although 'photon' is a clearer way to visualize this.

A photon occurs in both wave and particle form. A single ray of light in *wave-form* is the wholeness of a system; in particle form it makes up the parts of a system. A photon is hence a monad, being both the ultimate whole and the ultimate part.

[Spirit too is both the ultimate whole and the ultimate particle. As a particle or point-instant it is an 'atom.' Everything derives from it. As a whole (the substrate) spirit encompasses everything. It is this duality that leads to the part-whole duality expressed by a photon.]

F. A Whole is Complete: A world expresses a particular set of types of being. These correspond to the spatial dimensions, as was discussed in Chapter 2 relative to the archetypes themselves. Each higher level dimension is a higher level world. The systems in a world are not only expressing the dimensions in their external world, they are themselves an expression of a particular set of spatial dimensions. (In a cosmology where individuals are prior to external worlds, individuals are the real source of dimensions.) As dimensions relate to completeness, the beauty of a hierarchically constructed universe is that each dimension entails a world (with a specific number of dimensions peculiar to that world). Because any particular world (or individual) is not expressing other dimensions, it is complete-in-time at any one period in time. Spirit does not realize all the potentials in the substrate when it creates, but rather does it one world at a time.

The terms 'to be complete' and 'to be perfect' are closely related. Hierarchy in nature corresponds to levels of wholeness. Levels of true wholeness, in turn, correspond spatially to spatial dimensions. Because of the step-like nature of creation, in a hierarchically constructed universe it is possible to be perfect, and yet evolve everlastingly towards higher states, and hence greater perfection.

[The soul is always complete and perfect, but material

perfection exists only *when parts reflect and express the whole*. If one of the parts is missing, they are not a complete set. Parts must be qualitatively whole as well as spatially so. In other words, they must be totally coherent with one another. When something is wrong in a body, a material 'part' (e.g., a bunch of molecules) may be present in one sense, but something will be 'missing' (e.g., the formation of molecules into a cell). The incoherent parts are not true 'parts' of the system—but rather foreign objects within the nexus of true material parts for that level of being. Only when material completeness is present do systems know the satisfaction of perfection in all three of the components of their trifold nature. The astrological category system shows clearly that this is the state they are naturally (by natural law) intended to manifest.]

G. Wholeness is a Power: The wholeness of a system is not only a *substance* that is the unity of the parts, but also a *power* or force that presides over them and actively unifies them. Parts are different and multiple. Whatever unifies them gives them their form. When thought of in terms of causes, the type of influence that wholeness manifests has been referred to as formal causation.

In addition to giving unity, wholeness is also a power in terms of giving being as was mentioned in **D**.

H. Wholeness is a Higher State and Value: There is a simple, natural distinction between wholes and parts. Wholes are over parts. Parts are not over wholes. In philosophy 'value' means that which in some way is prior to or higher than something else. Higher value can mean that which comes first in time or before something else, that which is greater in power (relative to something else), that which is prior in dependence, and so forth. In the AC spirit is prior to wholeness, and wholeness is prior to matter—prior in creation, prior in dependency, and prior in power.

Relative to which comes first in time, in the AC, the priority entails first spirit (the quantum potential), then wholeness or soul (a specific set of potentials), then matter. Relative to dependence, the only thing wholeness depends upon is spirit. Wholeness has its own form of absolute being, and is not dependent upon parts. (Our soul would not survive the disintegration of our parts if it did not have this absolute, independent characteristic.) Parts are the last in the series of dependences, and are dependent upon the wholeness which sustains them and on spirit which sustains wholeness.

[Various senses and examples of 'value' were discussed in Chapter 3; however, one that has not been discussed yet is potentiality versus actuality. It would seem as though spirit and soul would lag behind matter in this respect, but such is not the case. Pure Being is not a potential. It is a reality that must exist and be real before any type of being comes into being. The term 'potential' is only as seen from the material side of experience. It is a term used in reference to matter. Both spirit and soul are real substances but only potential matter. Relative to soul, in a hierarchically constructed universe, what is actual matter when seen from one level is potential matter (i.e., soul) when seen from another.]

I. A Whole is Self-Sufficient & Independent: To be 'absolute' (**B1**) is to be non-relative or independent. What follows from or is produced by something else depends upon it for its being (and unity). What is prior in creative order is independent of what it creates (in the sense of depending upon it). Parts depend upon the wholeness of a whole for their creation and for the Unity-Being that sustains them in existence. Wholeness is dependent only upon its source—spirit.

The quality of independence is true of the nature of wholeness as well as of its creative character and capacity to stay in being. What is absolute does not depend upon, and hence is not relative to, anything else. The material parts of a system, in being relationships and being relative to one another, depend upon one another for their nature (and ultimately upon the whole). A whole is independent in this respect. (This quality of independence that is a characteristic of both spirit and wholeness can be seen in their shared character as pure, *absolute* Being. In material systems, both spirit and wholeness produce the very essence of what it means to be an independent individual because what is truly independent is absolute.)

J. A Whole is an Individual: All the qualities and powers of wholeness converge to produce individuals. Individuality arises because a whole (in its role as a whole) is not a part of anything, and because a whole is non-relative, is self-sufficient and is independent. It can stand alone. It does not derive its nature from its relationship to other parts as parts do.

Wholeness is a reflection of spirit in all of its powers, qualities, and capacities. Spirit has the capacity within itself to individuate. It is naturally creative and hence naturally material as well as naturally spiritual. Spirit expresses itself creatively through producing individuals (perhaps even at its highest level of expression by being the highest, most all-encompassing individual, i.e., the 'Most High' being in the universe—which is to say, the universe itself).

K. A Whole is Infinite: Wholeness reflects the character of spirit by being infinite in its own way. It does this through being complete. This attribute leads to seeing a whole as possessing the attributes of a world. A world has nothing outside itself—precisely because it is a world. It is unbounded and unending, producing an individual expression of the substrate.

K1. A Whole is Infinite Qualitatively: Because there are unlimited potentials in spirit, a particular whole would seem to be finite rather than infinite, but such is not the case. As was noted in **F**, a whole contains the complete set of the archetypes (in potential form) for a particular system manifesting in a particular dimension. Higher dimensions have more archetypes, but these have no reality for three-dimensional systems. Because of this a whole is infinite.

K2. A Whole is Infinite Spatially: To be 'bounded' is to come to an 'end' by virtue of having something else material that is encountered. Parts are bounded by other parts. Where one comes to an end another different part begins. A whole is not bounded by other parts, precisely because it is a 'whole' rather than a part. This makes any whole infinite in extent.

Creation occurs in hierarchical layers and these layers of wholeness correspond to spatial dimensions, which are also step-like in character. Each soul actualizes the spatial dimensions appropriate for its evolution. (Each dimension is a singularity that encompasses those below.) At each stage, there are no higher dimensions that are real. Hence each is complete-in-time.

[The fact that it takes an infinite number of points to make a whole or extent be what it is, means a whole must be unbounded externally (i.e., must be infinitely large). This is necessary in order for there to be an infinite number of points within it. In other words, it is not just that an extent requires an infinite number of point instant to make it what it is, but that an infinite number of point-instants has to have the kind of 'space' that corresponds to that idea. Only a world possesses the needed characteristic. Physicists have begun to speak of different dimensions as different 'worlds,' and they are correct. Each higher dimension achieved is a new world, and a 'world' is always unbounded and unending in extent. (This characteristic of being infinite is true of every whole, and individuals are prior to worlds. The world possesses this characteristic because it is a whole, not just because it is a world. This will be discussed more in depth in the section on '**The Cosmos.**')]

L. A Whole is Everlasting: Spirit is absolute, infinite and eternal. The wholeness of a whole reflects and expresses spirit by also being these things. What is infinite in extent is infinite temporarily as well (reference the general theory of relativity in regards to the unity between space and time, e.g., the space-*time* continuum). The wholeness of a whole cannot come to an end in space or time because there is nothing to bound it.

External continuity is related to internal continuity. The wholeness of a whole is everlasting, not just because it is unending in extent, but also because it is a spatial singularity internally, and (unlike parts) cannot come apart.

The wholeness of a whole not only gives stable being to a system; it also makes it possible for the system to endure (as an individual) forever. This is what the soul is.

[The parts of a system are everlasting when they reflect the whole. They continue through time when they are each and every one properly connected, but they can only be so when they are coherent with one another.]

[*These qualities of wholeness are not separate characteristics. Each is logically interconnected with the others (just as is the case with the characteristics of spirit).*]

Soul, Wholeness & Wave-Forms

The two goals being pursued in this chapter continue in this section. Those include furthering the case that true independent wholes do exist (and without compromising the nature of spirit), while concurrently working to establish an identity factor between soul, wholeness, and the wave form of particles.

Thus far the concepts of soul and wholeness have been correlated; but the concept of the wave-form also needs to be brought in as well.

The Photon

The photon is a monad. It is at base of every system. By examining its nature it is possible to understand what soul and matter consists of.

The photon comes in both particle and wave form. As a particle it is a fundamental 'material' unit. Even though still in the form of energy rather than mass, a photon is the basic particle underlying the composition of the world. It is the basis of our atoms, and hence of our matter. [Within the atom there are lower-level 'photons' (e.g., 'mesons') but that is getting into the world of the atom's 'matter.' The systems we use to construct our matter are atoms, and the photon is our ultimate particle.]

The photon is an ultimate whole as well, since the wave-form of an atom or person is just light in wave-form. If the wave-particle duality of light is part-whole duality, then the characteristics of wholeness as given in the previous section should correspond to those of the wave-form of a photon—and that wave-form does indeed seem to fulfill the criteria for wholeness outlined in the previous section.

Everything that is true of a whole is true of the probability field of a photon, including the important attribute of formal causal efficacy (**G**). Light in wave-form is a probability field, having to do with where a photon, atom or molecule will manifest when it materializes, and what it will be.

As with the wholeness of a whole, the probability field of a photon is not visible or material. It is not energy because it is not a photon. Because it is neither energy nor matter, it is *meta*physical in nature (**A**). Light in wave-form cannot be seen and will never be seen. Seeing immediately turns light into a photon, the particle-form of light. The wholeness of a whole cannot be seen either. Only parts are visible. (Likewise, no one has ever taken a picture of soul.)

At one point in time scientists claimed that we should leave metaphysical subjects alone since, like religious subjects (to which they are closely related), they could not be proven to exist. Now quantum physicists are avidly studying something that is non-material and that will never be capable of being examined materially except through observable effects on matter. The rules are changing because it is necessary for them to change. We *can* learn about probability fields through their effects on matter (and about souls as well).

What cannot be seen is not a type of being. Matter is visible and is experienced as the types of being. A probability field possesses the types of being in (real) potential form, but not in a physically actualized form. This is what makes it meta-physical. If a photon in wave form cannot be accounted for in terms of actualized types of being, but at the same time is real and exists (expressing the words 'to be'), then it needs to be conceptualized as pure Being (**B**).

A major characteristic of light in wave-form that relates it to wholeness is its absolute, non-relative nature (**B1**). The various qualities that material parts possess depend upon other bits of matter to make them what they are. For instance, the characteristic of motion we call speed is dependent upon how fast one material object is moving relative to another. When a person is sitting on a train next to another train moving in the same direction, and at the same speed, they get an eerie feeling of not moving at all. According to Einstein's special theory of relativity, this is not just a perception; they really are not moving *relative to the other train*. The same pertains to spatial extent. According to the special theory of relativity how long our extent is depends upon how fast we are moving relative to some other object. Both of our most powerful theories as to the nature of matter have this relative-relationship perspective. In accordance with relativity theory material qualities are relative—and, in accordance with quantum field theory, matter is *composed* of relationships.

The one exception to the relativity rule is light. According to the special theory of relativity, so long as it is not hitting something material, light manifests a speed of 186,300 miles per second no matter what it is relating to. What is not relative is absolute. Of course, this is not an infinite speed. Rather, it is a limiting factor for our three-dimensional world at its level of evolution. It is complete-in-time (**F**). No three dimensional system can go faster than the speed of light without (presumably) entering into a higher dimension. The speed of light is infinite and absolute (**K**) in the sense of having nothing beyond it (again, for *our* world).

In addition to being infinite and absolute, another characteristic of the wave-form of light that makes it very similar to a concept of wholeness is its everlasting nature (**L**). According to the law of the conservation of energy, no matter what material shapes light takes on, and no matter how many particular photons come and go, the amount of energy in the world remains constant. Nothing is lost. The wave-form gives stability, constancy and an everlasting character to the world.

Because it is so close in nature to spirit, light is very fundamental. It is the most fundamental part or particle, and the most fundamental whole as well. As the ideal particle (a point) it is infinitely small; as the ideal field (in wave-form) it is infinitely large. As the infinitely large, it is possible to compare it to the quantum potential which provides for the wholeness of the entire universe. Spatially speaking, as a field potential, it is everywhere—but in potential form. It is no actualized where. Qualitatively speaking, the wave-form is every type of quality that can exist, but again in potential form. In wave-form it is no actualized thing. Light is the substance of which everything is composed and from which everything is born. When the particle-form of light is perceived as beginning in a singularity it becomes possible to see just how close light is in character to spirit. In addition to being a point-instant (Leo), its wave-form is like the substrate (Pisces)—and very well may be the substrate. If so, what we call 'light' may actually be spirit in the process of creating the world.

Despite all these correlations between wholeness and the wave-form of a photon, the most important characteristic of wholeness has to do with its ability to provide for individuality (**J**) within the universal space-time continuum. The current scientific paradigm does not permit true individuals to exist. And yet there are concepts within that same paradigm that are inconsistent in this respect. To begin with, the terms scientists currently use to express the nature of light are contradictory, and that very fact is a clue that the paradigm is excluding something that is difficult to exclude. As was noted in Chapter 4, when waves are described, whether they are said to be individuals or not depends upon how the physicist wants to describe them for a specific purpose. Light is an electromagnetic wave and, in electromagnetic theory, physicists speak of individual rays of light going this direction or that direction, indicating that there is a measure of spatial independence. Also as concerns spatial independence, different rays have different wave lengths. Qualitatively speaking, the fact that a ray of red light has a different frequency compared to a ray of green light is an indication that there is some capacity within light for there to be qualitative as well as spatial independence. (Internally, a ray of light is centered and has parts, indicating that it is a whole.)

In addition to the inconsistent manner in which physicists describe light, another thing that provides evidence for the ability of light to manifest as individual rays is the double-slit experiment. The wave-forms studied in this experiment are specific to individuals—to the particles they guide. They recognize individuality. Different types of wholes and wholes-in-the-making have a wave-form—from photons, to atoms, to electrons, to molecules. But these must be a whole or a whole-in-the-making like a molecule. Huge molecules called bucky-balls consisting of sixty to seventy atoms were recently utilized in the double slit experiment. These have their own specific wave-form that guides the placement of the bucky-balls. Sixty to seventy atoms that are not occurring together within a single molecule (e.g., atoms in a drop of mercury) do not have their own specific wave-form. It is this thing called 'wholeness' that makes the difference and produces a probability field. The fact that an atom or molecule has its own specific correlated probability field while an aggregate of atoms does not is one piece of evidence that a probability field is just what is meant by wholeness. It is also another piece of evidence that the idea of individuality within light is a valid concept, and that there are individual souls.

Another piece of evidence may be the existence of non-local causation (the EPR experiment mentioned earlier). Anything that exists or happens in our world is composed of and due to light. Therefore, it is difficult to believe that there is any causal influence that is not due to light. The fact that there is non-local causation indicates that light as a causal agent may manifest in more than one way—and the other way may be in terms of a higher level of light—the wholeness of a whole acting upon lower level parts. Light from the parts level of being cannot break the light barrier (186,300 mps). Three-dimensional matter cannot go faster than the speed of light. However, non-local causation does break the light barrier and this may be evidence, not only of the existence of the wholeness of a whole, but also of its character as a higher dimension of light. This would explain the atemporal character. In terms of connecting this particular causal influence to the wholeness of a whole, the experiment with non-local causation involves two systems that have previously been together as parts of a whole, i.e., two photons that have been together within an electron. Apparently only parts of wholes demonstrate this non-local connection.

Another piece of evidence involves that which was mentioned earlier—the nature of the speed of light. The speed of light (186,300 mps) is not an infinite speed. And yet it is for our world. Nothing in our three-dimensional world can go faster than the speed of light (without presumably entering into a higher dimension). Thus for our world this is an infinite speed. This finite-but-not-finite character is to be expected if light is occurring in stages (and dimensions), while at the same time *each stage is a whole and is complete*. Presumably higher dimensions (higher true levels of wholeness) have higher limiting speeds. In addition to being infinite, the speed of light is also non-relative (absolute). Both the infinite and absolute characteristics of light attest to what it means to be 'complete,' and this is what occurs when there are dimensions of being—in this case expressed in the kinematical domain in terms of the speed of light.

If everything is composed of light, then our spatial dimensions are what they are based upon the nature of light. It is in looking at how each higher level spatial dimension unifies the 'parts' of the dimension below, that the truth of the overall premise really comes home; in other words, that light is not only occurring in stages, but also that it occurs as an individual ray of light at the highest level. Each higher dimension gives wholeness to the one below. Dimensions occur in layers, and geometry teaches that higher dimensions encompass lower ones in a hierarchical manner. Each overarching dimension is a continuum or singularity. It is a whole, and unifies those below.

Physicists are intent upon seeing the universe as one unified electromagnetic field, and it is. But light also occurs in terms of different dimensions—and concurrent different levels. And because the highest level is a world as a whole (or individual) this makes it possible to explain the existence of individual rays of light.

It is in a paradigm where individuals are prior to external worlds that dimensions can be so satisfactorily perceived in terms of levels of wholeness as well. At the end of the previous chapter we talked about how physicists after the nature of Bohm and Einstein have suggested that the motion a complex system like an atom enjoys (not to mention its spatial extent) can best be explained when systems are seen as straddling dimensions. What better way to perceive this 'straddling' than in terms of the part-whole duality of systems, which in turn involves levels of light?

Once the paradigm shifts and creation is perceived in terms of levels of wholes, each of which express part-whole duality, then the wave-particle duality in the double-slit experiment begins to make sense—even the fact that we make the difference between an atom being a particle or a wave. We are the next true level of wholeness after the level of the atom. The atom itself is a whole and has a wave-form. But it also serves in the capacity of our matter and hence is a particle when we perceive it and relate to it.

In addition to the thoughts in these pages, there is other evidence as well of the existence of levels of light, and of singular rays of light serving as the wholeness of systems. But for this evidence it is necessary to get into the subject of wholes at a higher level such as the solar system and the cosmos as a whole.

The Cosmos

The astrological cosmology entails a monadology in that the simplest part or particle is a simple ray of light. This is also what a whole is.

The AC is not only a monadology; it is a macrocosm-microcosm metaphysical system. In the AC, every system is essentially alike. Simple rays of light are behind the parts and the wholeness of every system. Complexity is simply due to layering. Because of this it is possible to understand experience (and in this case wholeness) by looking at either a photon or the cosmos.

If our world is a whole (or whole-in-the-making) it should be unified in terms of a single, simple ray of light. And in fact, the world is very clearly a whole. It presents as a singularity—and in both a qualitative and spatial sense. Relative to the former, the fact that the laws of nature are coherent with one another, and are integrated integrals is evidence that the cosmos is (**C2**) a qualitative whole with the power to unify different expressions of the types of being.

Relative to being a spatial whole, it is via the general theory of relativity that the spatial attributes of the world have been studied and articulated in the fullest manner. (The general theory attempts to explain gravity in terms of a curvature of space.) The general theory refers to the space of our world as the space-time 'continuum,' in other words as a spatial singularity (**C1**).

Quantum theory and the general theory are not currently correlated, so it is not possible to say whether the so-called space-time continuum is light in particle form or light in wave-form, but it must be one or the other. That is because everything is composed of light. The wave-form turns into particle form whenever there is an interaction with matter. Outer space is dark. We only see something when a photon materializes due to the wave-form of light interacting with a material object, such as occurs when light reflects off a planet and ultimately interacts with our eyes. Since outer space is dark, the space-time continuum needs to be seen as the wave-form of light.

In addition to its essentially dark nature, there is another reason to connect the space-time continuum to the wave-form of light, and that is because it is a continuum. Even in quantum theory it is the wave-form rather than the particle-form that is said to give continuity to the world.

[If this space-time continuum is the wave-form of light, it is not a physical type of actualized physical being but rather is just pure (**B**), absolute (**B1**) being and is metaphysical in nature (**A**).]

In addition to being a singularity or continuum with no breaks or ends internally, another thing the general theory has revealed about the world space-time continuum is that it has no ends externally either. It is unending (**K**). Physicists do not speak of the world as infinite in extent because it has a limited extent (that can theoretically be measured). In physics, the fact that the world is finite but unending is seen as a paradox, but if there are stages and levels of wholeness (and hence wholes that are less than the substrate as a whole), this begins to make sense. Most people can accept the fact that the world is not bounded by anything external to itself, and think that this is logical due to its being a world. But that thought may need to be turned around. Instead it may be due to its being a whole. The wholeness of a system would have to have something to 'bound' it in order to come to an 'end.' It is material parts that bound other material parts and are bounded by them. A whole is not a part and hence cannot be bounded by other parts. This is an essential characteristic of wholeness.

Since the space-time continuum does not end, where it is that it goes when it gets to the 'end' of our three-dimensional world is currently seen as a paradox. Bohm and Einstein conceived of it as circling back and up into a higher dimension, possibly an actualized dimension. From the perspective of the AC, until we ascend to a higher level of being, and our soul becomes our matter, we will not be fourth dimensional, and what is above us is a material potential, not an actualized material fourth dimension. This is true for our external world as well.

The characteristic of being without ends or breaks internally (continuous) is related to that of having no ends or bounds externally—i.e., being unending in extent. What the mathematics of the continuum reveals is that it takes an infinite number of point-instants to make a continuum continuous. In order for something to be whole, it must be infinite in extent and conversely, for something to be infinite in extent, it cannot have any internal breaks and must be whole.

The rest of the characteristics of the world begin to be understandable once a world is seen as a whole. The cosmos is self-sufficient and independent (**I**), and this can be thought of as being due to its being a world, since we have commonly come to define a world in this way, or it can be seen as due to its being a whole. (Being independent is very closely related to being non-relative or absolute. A world stands alone. 'Independent' is probably the best term to describe the world rather than absolute, so long as it is remembered that to be truly independent is to be absolute.)

Because the world is unbounded, it is complete (**F**). The finite but unending character of the space-time continuum is one piece of evidence that it is a whole. A whole is a finite set of potentials—but paradoxically at the same time an infinite (complete) set. As was pointed out relative to wholeness in the previous section, if there is no space external to our world (and for now no laws of nature external to it) it must be seen as complete. There are other potential laws (potential qualities) and other potential spatial dimensions in its field potential, but these are not in actualized form, so there is nothing 'missing' in our world. Being finite but unending seems like a paradox, but the reason that the world is unending but not infinite (as being infinite is usually perceived) is because it is a particular whole. A particular whole is less than the universe as a whole and even less than spirit as the substrate. (The next section discusses other, higher worlds.) The very character of the world as unending but 'finite' is evidence that the world is complete but may have the potential to become something more.

In addition to being absolute and infinite, a whole is also necessarily everlasting. In addition to the fact that the world is a spatial singularity internally (a continuum), and unending in extent externally, another thing we know about the world is that something is preserving the energy so that this is everlasting (**L**). Out of the substance of the world field, individual photons and gross material systems form, and these are the 'parts' of the world. These parts come and go and change shapes, but *as a whole* the energy in the field is everlasting. (In this way the light field of the world provides for stability (**B2**) in a constantly changing material world, just as spirit does. And that makes sense since spirit and wholeness are both expressions of pure Being. (As this relates to individual souls, this is what gives people their stable identity. Even as various parts change, they are still themselves.)

It is the law of the conservation of energy that quantum physicists refer to when speaking of the everlasting nature of light, but there is another way to know that the world light field is everlasting—a way that relates more to the physics of the cosmos than to quantum physics. The spatial continuum of the world—which is its light field—is spoken of as a space-*time* continuum. Since the spatial continuum of the world goes on and on in space unendingly, and this is a space-time continuum, it is logical to presume that the temporal aspect of this field goes on and on in time unendingly. (Visualizing this field as a space-time continuum helps point out the relationships between being infinite in time and infinite in space, and actually it is common to use the term for both states.)

All of these qualities that the world possesses are interrelated and interdependent. It cannot be a continuum if it is not infinite in extent. It cannot be infinite (unending and unbounded) if it is not absolute (having no external relations). It cannot be everlasting if it is not infinite, etc., etc. It is not possible to exclude any of these qualities without them all failing to exist—and that includes internal spatial singularity.

Because wholeness entails singularity, although the light field of the world may, as described in electromagnetic theory, have multiple rays *within* it (at a lower dimension of being), the light field of the world *as a whole* (at its highest level of being) needs to be perceived as an overarching, single ray of light in wave-form. Despite all the small waves that are factors of quantum theory (e.g., the electromagnetic field and the electromagnetic-like field within the nucleus of the atom), in the field of the cosmos as a whole the general theory of relativity postulates a simple, constant curvature. According to the general theory, the space-time continuum of the world does not just curve at the 'end' of the world. It is constantly curving. If our world *as a whole* is united by a single, simple ray of light in wave-form, that would explain the simple, constant curvature.

This explanation for why the space-time continuum of a whole (any whole) is curved is different from—and yet parallel to—the general theory of relativity, which is a thesis used to explain gravity. In the general theory the curvature of space is said to be just what is meant by gravity. Gravity in the AC on the other hand becomes a force once more, and is seen as expressed through the Leo principle, which is an immediate mode of spirit manifesting at the center of each and every whole at whatever level of being.

These two explanations for gravity do not need to vie with one another. The paradigm simply needs to be shifted. We need to see gravity as a force, i.e., as a spiritual power at the center of every system, but also involved with the curvature of that system's space. This is no more than saying that a single ray of light is a force as well as a substance, and that the force at its center gives the ray of light (as a field) its curving nature. It makes it a whole, with a center and a circumference. The 'extent' of anything, even of an electromagnetic ray, is its 'space.' The space around any system is curved because its wholeness *is* a single electromagnetic ray.

The constantly curving nature of the space-time continuum of the world is evidence that it is a spatial singularity. All these concepts put together give evidence, via the macrocosm that we experience as the cosmos, that there are true, individual rays of light in wave form, and that they are the wholeness of a whole.

In addition to the space of our world being curved, there is experimental evidence that the space around our sun is also curved, as would be the case if there are different levels of wholes and each is unified by a single ray of light in wave form. The sun's space-time curvature is currently explained as due to gravity. If the solar system is a system, as its name implies, then, as a whole or whole-in-the-making, it is rational to assume that (as with the world as a whole) the substance that makes the solar system a whole would have a singularity to its nature, and (as a single ray of light in wave-form) would have the type of curvature to its space that it has.

One way to verify that a higher level whole such as the solar system is a *quantitatively* single ray of light may be to observe one and see if it fulfills the criteria of *qualitative* singularity. A ray of light as a qualitative singularity is a probability field. It is a single, simple ray of light manifesting a specific set of potentials—twelve in a three-dimensional world. It is possible to confirm the qualitative singularity of our solar system (in its role as a probability field) due to the atoms that have collected at certain spots in that field. Reading the qualitative character of the archetypes manifested within our group probability field (as this can be seen thanks to the planets) is what astrology is all about, and astrology shows clearly that the field of our solar system is a single ray of light in wave-form. We will come back to this subject later.

There is much evidence available that relates worlds to wholes, and wholes to worlds, and some of these things were discussed in this and the previous section. Another piece of evidence has to do with the nature of centers. If we are a centralized system, as we seem to be, it is rational to suppose that something is centralizing the parts. It is also rational to see this something as existing in the center of our bodies. As was mentioned in the chapter on '**Spirit**,' and illustrated through the graph of a spatial coordinate system, if a center was within something in a *material* sense of 'within' it would not be a true center. Only the center of what we normally think of as a world is a true center. Only the center of a world really has no extension into any of the spatial dimensions. If we are centralized—if we have a center, and a true independent spirit—then we are a world or, more to the point, we are a true, individual whole.

The idea of comparing individuals to worlds has broad philosophical implications. It even has practical humanitarian implications. What would happen if, when we looked at our neighbors or our children, we began to see them in a new and different way—in terms of each as a unique, evolving regent of a world? As a young person, Alexander the Great was given as a student into the care of the philosopher Aristotle. Perhaps we all need to see the same type of future potential in our children—and actually an even greater future potential than that of any earthly king that has ever existed.

Worlds within Worlds

As a whole our world is a singularity that is complete and stands alone. Although the world is independent, this does not mean other worlds do not exist. A number of physicists are currently suggesting that there are many other worlds in addition to our own. This is the thesis of the AC as well.

Because physicists have not yet addressed the fact of (or meaning of) hierarchy in nature, most many-world hypotheses speak of parallel universes. Per the AC, there are parallel universes, just as there are parallel atoms within our body. But atoms also exist together within us as our parts. We have already discussed how part-whole duality leads to hierarchy in nature. If there are systems nested within systems (and there are) and, if a whole is a 'world' so to speak (with an accompanying outer world in addition), then there are bound to be worlds within worlds. (In order to keep concepts and definitions as clear as possible, it may be best to speak of *a* world as a 'world' and the composite of all worlds as the 'universe.')

Each higher *level* world entails a higher *dimensional* world. When quantum physicists speak of other worlds, they usually link this notion to the idea of other spatial dimensions. The AC makes a distinction between the domains of space and that of part-whole hierarchy in nature, but also recognizes the parallelism and interpenetration of the two domains. When a system unites with other systems at the same level of development, and together they rise to a new level of qualitative being, if a new, higher-level *world* is created, a new *dimension* is also created. A higher hierarchical level of qualitative wholeness allows the new system to be qualitatively enriched. It can produce more types of being. The spatial dimensions are one manifestation of the types of being.

Just looking at the hierarchy in nature around us provides empirical evidence of how pervasive this impetus within matter really is. A look at the nature of the spatial dimensions provides further proof. The nested manner in which the dimensions fit within one another (i.e., points within lines, lines within surfaces, surfaces within cubes) is an excellent piece of evidence that hierarchy is the manner in which the world is created.

A worlds-within-worlds hypothesis has a chance for empirical verification, whereas a general many-worlds hypothesis may not. People cannot verify the existence of something they have nothing to do with. In the AC, atoms are independent wholes, but we can experience (and hence 'prove') the existence of atoms because they are a part of us. We affect our atoms and they affect us. Furthermore, if we are parts, not only in our immediate external world, but also in higher level worlds, it is theoretically possible that we could interact with the denizens of those worlds and they with us. If angels, saints, people in heaven, or beings in higher realms can watch over us and influence us, we must be a part of their world (and if they are substantially higher, perhaps even a part of them). In a hierarchically constructed universe, the 'part' aspect does not end one level up. There is a vast tree of life above us, and we are a part in each level that encompasses us, including a part of the universe as a whole.

Higher Worlds and the Highest World

Spirit is a principle of creativity, and it would not be itself if it was not creating. Creation occurs through the creation of individuals. Even external worlds are wholes-in-the-making. Every bit of 'matter' is a soul, depending upon the level it is experienced from.

Because the world is created via individuals, and these

possess part-whole duality, that state of affairs leads to hierarchy in nature. The hierarchy in the universe is self-evident.

Unless all creation was made just for us human beings, we are not the highest creature on the hierarchy of being. Nevertheless, the levels above present as different from the levels below. When looking down I see solid things. I am a solid body with solid systems like organs and cells. When I look up, there are solar systems within galaxies within meta-galaxies, and on and on. But if there are material systems above me, they do not have a solid extent as I do.

When I go back to thinking about how solid I am, I realize that I would not appear solid to an atom. The only thing that presents as solid, and that really can be referred to as 'matter,' is the *relationships* between the systems immediately below me. There is no matter for me except for the systems I am using for matter. (The 'matter' of an atom is nuclear in nature—its own 'parts.') There is no way I can see the matter of systems above me—although I myself *as a whole* might be being utilized for the matter of the next highest system.

When I look up, I do see a hierarchy above me as well. But the matter I see is only in terms of the molecules I am familiar with. What is different is that these molecules within the stars and galaxies are collected into huge concentric shapes that could correspond to centers within macrocosmic systems if they were truly there. Astrology indicates that, within my own solar system, the planets are transducers for energy coming from our group probability field. The AC also indicates that these planetary centers are unifying the types of being below which are manifesting that particular archetype. Living systems are the real 'parts' of that group whole; and yet atomic-level molecules have collected at those centers. This is due to gravity (the Leo principle). Apparently the atoms and molecules that I use for my matter are attracted to centers at higher levels of being. Are there systems above me that I can 'see' to some extent because of this—

systems that are both higher dimensionally and qualitatively?

If creation involves creating individuals, and these evolve in terms of stages, why would I think that I am the epitome of creation? Is this not anthropocentrism, and a logical fallacy? I think it is. It stands to reason that there are systems above me as well as below me.

But, if there are systems above me, what kind of systems are they? The Milky Way Galaxy is above me. I feel certain that I am much more evolved than that galaxy. And I am, if the 'galaxy' is the atoms that have collected around the centers in the sky. On the other hand, if those collections of atoms are just secondary to centering forces, and there are indeed higher level beings, then the next question I need to ask myself is, do higher level systems present as more highly evolved? And the answer is that, from what we know of evolution, the more highly evolved a system is the more centralized and complex it is. That complexity is due to layering. If I am not the epitome of creation, then those systems above me are most likely higher dimensional systems with, not only more layers of matter to them, and hence greater complexity, but also more archetypes being utilized as well, precisely because dimensions increase the number of types of being.

In addition to greater complexity, a question also arises relative to manifestations of pure, absolute spirit. Again, looking at the world, it would appear that the higher and more qualitatively complex a system is, the more it is possible for pure spirit to express itself. I am much more evolved than a one-celled organism. I am capable of centered, ordered thinking; I have the capacity to love; I have a well-developed conscience, and so on. The pure expressions of spirit at the center of a being increase as the body becomes more complex. Even at one level coherent parts amplify the effects of pure spirit. The more kinds and layers of parts, and the more coherent they are, the more varied and strong are these pure manifestations. If there are systems above me, it stands to reason that they possess any expressions of pure spirit that I do, but even more, and to a greater degree than I do.

Our solar system is merely a whole-in-the-making. That I know from astrology, which looks to that probability field for the information it can impart. It is the probability field for the cell-of-living creatures in our solar system. At the same time, there are many levels of true individuals, such as atoms and people. Even if our solar system is merely a group soul, might the Milky Way Galaxy entail a personal soul? Our solar system is like a cell in that galaxy. There are at least two hundred billion stars in the Milky Way. That is how high that system is compared to our little sun, which sun in turn is astronomical in size compared to us. If that is a being up there, and if his level of awareness is astronomical compared to ours (perhaps even great enough to be cognizant of us), then I suppose he would be like an invisible king having dominion over us—but with a degree of love, awareness and righteousness we cannot even begin to imagine.

These thoughts in turn bring up the question of the universe as a whole. If the nature of spirit is to create—and specifically to create via individuals—then the universe as a whole may be an individual. If there is a God, might the universe be God (Allah, Yahweh, Brahma)? Or would God just be pure spirit, unadulterated by matter? Matter would certainly be an adulterating influence on spirit if it were not spiritual in nature. But, what the astrological category system indicates is that matter is spiritual, and meant to be perfect—just like spirit. Actually it *is* spirit. The photon can very plainly be seen to be spirit, and everything is composed of light.

When looking at the pure expressions of spirit like awareness and love, it is difficult to think that God would not be personal. Pure spirit without mental, emotional and physical vehicles would lack something that spirit with vehicles would possess. What is Love until it loves? Loving is a process of expressing love, of doing something—and processes are material things.

It would seem as though God would be missing something that we possess if he did not possess material vehicles. He would also be missing something if he did not encompass all material bodies. In a hierarchically constructed universe, it makes sense to consider the possibility that there is a most high being. Perhaps the entire universe is the body of God, and all creatures below are parts within his body.

The average modern concept of soul is not currently identified with the world (or universe) as a whole, but rather is perceived as peculiar only to humans—something that preserves our personal nature after death. This was not always the case. Plato believed that both the universe as a whole and individuals had a soul (*Timaeus*). It is virtually impossible to think about the existence of an infinite, absolute and eternal substance (especially one with awareness as a principle attribute) without thoughts about the possible existence of God (Brahma, Allah, Yahweh). It is a valid subject, and one that deserves attention, but let us leave it for now and come back to it in Chapter 8, which is devoted specifically to this subject.

The Human Soul

In a microcosm-macrocosm metaphysical system like the AC it is possible to learn about the wholeness of one system by looking at the wholeness of any system, from the highest to the lowest, and we have done that by looking at a photon and the cosmos. According to the paradigm that emerges from an examination of the archetypes, there is a single ray of light in wave-form that is a regent over many layers of dependent parts, and that preserves the place of the material parts of our body even when they are not in material form—in other words, something that is a stable, individual, everlasting part of us, even when (in a gray world) our parts temporarily disintegrate. This is our soul.

Parts that are not coherent can come apart. We experience this as disease and eventually as death. A single ray of light, on the other hand, cannot come apart. Furthermore, light has a tremendous capacity to register, retain, store (and restore) information. We know this via the computer and other electronic technology. What the AC indicates is that the soul records the personality, producing an everlasting personal identity for individuals.

If the universal light field did not permit true individuals to form within itself—and specifically individuals that could be everlasting as individuals—then the 'soul' of the universe would be the only soul there was. And, although they would not call the universal light field 'soul,' that is the conclusion that many materialists have come to. They believe that when they die and their bodies disintegrate, their matter returns to the atoms from which everything is formed, and consequently that their individual identity ceases to exist. The only everlasting nature they believe they have is in terms of their atoms being used some day to form other things in the universe—grasses, flowers, and so forth. The materialists are right in their view if only the world has a soul and they have no individual soul.

The Buddhist slant, as has been noted, actually has some correspondences to the materialist viewpoint in regards to personal being. I disagree with the Buddhists about matter or our personal self being negative or an illusion, but that is not the point. What they are right about is that anything personal or individual is material in nature. We talked a little about that when discussing the attributes even of mind. It is not possible to have qualitative differentiation without becoming 'material.' If we reached the Buddhist goal, and returned to a state of pure Awareness (or pure Bliss), then we would not be individuals. And we would have nothing differentiated as a vehicle to use—no bodies, whether physical, mental or emotional. At that point we would have nothing to be aware of or enjoy inside ourselves or in the external world either. If we became pure spirit, then we would have to stop, not just all physical doing, but also thinking and feeling as well. There would be no 'us.' There would be no external world either. There would be no creation, since individuals and their external world are the wholes and wholes-in-the-making that creation consists of. Not only is anything personal material in nature, anything material is personal. And it is not just that there are no qualities that are not the parts of some whole. There can be no qualitative differentiation at all without wholeness and the specific borders wholeness produces for material parts (as can be gleaned through studying the archetype called Virgo). The 'subject-predicate' distinction is a matter of the level of perception and experience. Individuality is the means to create a material world.

In order to know the truth about the existence of our soul and of its individual nature we need to take a hard look at the world. There are individuals everywhere. The centralization present in those individuals is a proof that spirit is a principle of individuality. Biologists are aware that the parts of a living system are centralized upon a particular system. They are not centralized by or upon a neighboring rock or tree. Astrophysicists are also now highly interested in the centers they call singularities, as are particle physicists. If there are galaxies that are singularities, solar systems that are singularities, and particles that are singularities—not to mention evidence that living systems are centralized—then there is converging evidence in the various sciences that there are individual centers.

In addition to the visible centralized physical bodies that can be seen everywhere, there is evidence that these have unseen individual souls. In addition to the individual probability fields present at the atomic level of being, evidence is arising in the field of biology that there are individual human fields of a similar nature. This data will be discussed more in depth in Chapter 6. If we are a true whole, we may need to consider the possibility that each of us has an individual ray of light in wave-form that surrounds us, provides for our wholeness, and hence also for our form. A number of fringe or 'organicist' biologists have come to believe that there is some type of field—possibly electromagnetic in nature—that is guiding the character, placement, and processes that the parts of a living body manifest. Eastern biologists are much more open to these ideas; and those who work with so called 'energy medicine' in our own country, like acupuncture, Reiki or Healing Touch, have been telling us that there is such a field of light.

In addition to empirical evidence of individuals everywhere, there is also evidence that the same power of centralization that individualizes systems also superiorizes them, causing them to constantly join together in hierarchical concrescences. One of the strongest pieces of evidence both that there are true individual wholes in the universe and that there is a force of individuation manifesting is the existence of hierarchy in nature—the state in which there are individuals within individuals nested within one another. There is hierarchy everywhere. We are a whole in our own right, but we are also composed of parts, such as our cardiovascular system and our nervous system. These are called 'systems' because they themselves possess part-whole duality. Relative to the cardiovascular system, in and of itself, this is a whole. At the same time it is a whole with parts, and the parts in this system include veins, arteries and the heart itself. Each of these is a whole with parts as well. Our heart is a whole with various parts, including heart cells. Each cell is a whole made up of carbon molecules. Molecules are wholes made up of atoms, and atoms are wholes made up of atomic particles. To be sure some of these are not independent wholes as we are (and as atoms are) but these various layers all appear to be following laws of individuation and subsequent hierarchy in nature—even to reaching up for ever higher levels of individuality. Molecules and the very cells in our body are individuals-in-the-making.

People are like atoms, and they too unite into molecules. That is what a family is. These loose wholes-in-the-making may have developing probability fields. The human race as a whole may have a species probability field that records the progress of the human race and unites all people. The same may be true of other species of animals and plants. Some biologists (e.g., Rupert Sheldrake, *The New Science of Life: A Hypothesis of Morphic Resonance*) believe that there are real species fields, but apparently hesitate to speak of them as probability fields because, according to quantum theory, nothing beyond the level of molecules has a wave-form nature.[3]

The hierarchy of being does not end with us as the epitome of creation. As we look above us there are solar systems within galaxies, galaxies within meta-galaxies, meta-galaxies within meta-meta-galaxies, and so on. It is time for science to stop ignoring something that is obviously such a lawful part of the universe, and instead seek to understand what is behind this phenomenon. The AC shows clearly that it is the power of individuation present in spirit, soul and matter that produces hierarchy in nature.

Each of these three substances is involved with states of affairs and processes that preserve and promote individual wholeness (and ever greater levels of wholeness). The soul or wholeness of a system is what makes it an individual, and its spirit produces that wholeness, centralizing the parts. Matter supports the process from below. Each part is a perspective on the whole. Material parts are not only types of being; they are types of unity intended to promote wholeness—and specifically wholeness *in the systems they belong to*. As with spirit, parts are forces. They are center-seeking and actively seek to be whole. In doing so they are promoting individuality.

Gravity is one of the most powerful forces in the world, and the purpose of gravity is to create individuals. It is a manifestation of the Leo principle of centralization. Although it works to centralize the parts of individuals, it is currently studied in its role as a force that manifests between individuals, drawing them together. Since it works in accordance with inverse-square laws, the strongest pull is in terms of systems that are close to one another. It encourages matter to remain in any interactions they are involved in, especially if they are coherent interactions, since these magnify the effects of spirit as the source of qualitative Unity as well as of spatial Unity. (Coherence expresses qualitative 'closeness.' Inverse square laws are currently measured in quantitative-spatial terms—how spatially close something is. But coherent parts also amplify the effects of spirit.)

When two or more systems interact in ways that are coherent, they are mutually beneficial to one another. This saves overall energy for them personally. The law called the 'principle of the least expenditure of energy' kicks in where interactions are viable. The more coherent that the internal relationships in a group are, the less external energy is needed, and the more independent the collective can be. At a certain critical level of increasing wholeness, the collective becomes capable of self-sustenance, self-maintenance, and hence of independence. At that point, the whole-in-the-making becomes an individual, and a new higher level whole is born.

Everything is constantly moving even groups of systems towards independence, and independence is about individuality. The move towards independence occurs because spirit itself is self-sufficient and absolute, and hence is the very essence of independence. So also is wholeness. Everything in nature is intended to promote true individuality, and all of these things collectively provide proofs that *we* are true individuals with individual souls and individual spirits.

Personal Proofs of the Existence of Soul

One of the best ways to know the truth about something—whether or not it really exists—is through personal experience. If we have a true, individual soul, then we can learn about this through our own nature—and our personal experience of that nature. This was true of spirit in Chapter 3 (when discussing our life, our awareness, etc.) and it is true of our soul. Below are just a few of the things that attest to the existence of a personal soul.

First of all, there is our individuality. Science currently does not recognize the existence of any true whole other than the world as a whole. But we *are* individuals—true wholes—each of us with a distinct, individual, stable identity that sets us apart from the rest of the world. We are a set of characteristics, and we know very well the difference between those that belong to us, and those that belong to the external world. There is a valid distinction between the within and the without. Furthermore, because our parts are constantly changing, we would not have any stability, permanence or identity if we did not have a soul. We know deep within ourselves that, even if we accidently lost many of our parts, we would still be ourselves.

In addition to our individuality, there is the existence of our internal bodily order. What is orderly is centralized. And it is not just our physical bodies that are orderly—so also are our minds. Our life, our awareness, our judgment seat, our will—these are all evidence that we have a center and that we *as individuals* are centralized. Where there is individual order and individual centralization, there are individual wholes. [It is the harmony and cooperation expressed between the 'parts' of families and nations that also make it possible to know that there are true 'group' souls or wholes-in-the-making.]

Soul is a formal cause. A cause produces effects in matter. These effects give a proof of the existence and nature of their causes and, in the case of the soul, proof of its existence as a substance and as a causal agent. All the patterns within us, and that are centered upon us, provide a proof of the existence of our individual soul.

David Bohm could not understand why physicists were ignoring signs of formal causation. He was fascinated by signs of formal causation occurring in plasmas, and these are very loose wholes-in-the-making. Atoms speak much more strongly of the existence of formal causation in the world. True individuated systems higher on the hierarchy of being speak even louder. Some of the most valid proofs of the reality of individual souls have to do with the patterns visible in living bodies, where spirit and wholeness manifest to a stronger degree than in atoms, and the degree of order and complexity is truly amazing. There are many biologists who have asserted for some time now that the patterns in living tissues cannot be explained simply in terms of DNA. It is their belief that the order of living parts and their incredible coordination is due to more than biochemistry. Rather than discuss their ideas here, I will save them for the next chapter which is on the subject of causation and includes a section on formal causation.

Another proof of the existence of our soul and its individual nature is ordinary, common healing. The soul is a power as well as a substance and we experience that power every time we have a cut that heals. Our bodies are constantly trying to be whole and to heal when they are not. Every time that wholeness is reinstated after being compromised, that is a proof of the existence of our soul. It is not just DNA that does this, but rather both DNA and the soul. As with ordinary wholeness, it is very difficult through a concept of DNA alone to explain healing. Like the central nervous system, DNA has apparently evolved as a medium to express the power of the soul as a probability field. As for healing being, not just a proof of the existence of soul, but of the existence of a personal soul, if the power exerted in healing was just for the sake of the universe trying to be whole, there would be no need for personal healing. The universe is a whole and would continue to be so even if there were no individuals within it.

Another proof of the existence of soul may be found in energy healing, which deals with personal centers and personal wholeness. Energy healing entails the field of an outside source encouraging our own personal field towards greater efficiency and coherence. Through the spirit at the center of a system energy enters into the whole and (via the soul) is disbursed to the body. This is coherent light—the stuff of true wholeness and of healing. There are many people currently involved with energy medicine—trying to understand it and use it. I have two friends, both with their doctorates, who became involved in energy healing. One is retired now but she was a research pharmacologist with a specialized interest in neurological processes. The other is an attorney. The latter learned his method of energy healing through an oriental man whose family had been a line of physicians serving the emperors of China. The former studied whatever she could that is available on this subject in the U.S. and after retirement became a certified practitioner. She had to hide her interest while working. She said it would have negatively affected her career. My other friend continues not to speak of his interest overtly at the work place. It is a shame for science not to look into these things, and it is a shame for people to have to hide something they know works but which has no current explanation. It is the materialistic paradigm that is responsible for this type of thing (and of course the pressures that people exert upon one another to conform).

My friend who studied with the Chinese man actually healed a knee injury I had sustained—one that traditional therapy had been unsuccessful with. I went from three years of dealing with slowly contracting tendons and (when in a store) leaning on a shopping cart for support, to total freedom and a straight leg after three treatments. That was eight years ago and I have had no trouble with my knee since. This is the type of personal experience that makes it difficult to deny the existence of the field that surrounds us. Energy medicine is an uncommon experience of the power of an outside influence on the soul rather than a common, internal one as occurs in normal healing, and is perhaps all the more impressive because of that. We little appreciate the common things.

Next to our own individual wholeness and the healing it can facilitate, one of the most common things we experience—and that attests to the existence of our soul—is our experience of time. Currently scientists do not know what time is. By using a concept of wholeness (and part-whole duality), it can be explained in a simple way. According to the AC, the material bodies of systems exist in the present, but they move through time because of their connection to their future potential. Soul actually provides for past, present and future. In terms of the future it links to spirit as a field potential—what Bohm called the 'quantum potential.' All particles (including people) have a future thanks to this field potential. Since this is where light in wave-form connects to spirit as the substrate, it is everything that can ever be. It is a 'free' future precisely because it is unformed in a material sense. We also have a past thanks to our individual probability field. This is an *individual* record of everything that we have done in the past as this has produced our 'present.' We have a 'now' not just because of our spirit, but also because of our soul. Time is not a fourth spatial dimension as some physicists have supposed. Nor is time 'out there' independent of individual systems. We are not 'in' time; we are time, and time is a proof of our individual being. We are the ones that have a present that leads to the future. Time is the unification of the material aspect of a system with its spirit and soul. We are bi-level beings that progress through time because we are time. (We share time with others when we form a collective. But even that is a whole-in-the-making.)

In addition to our experience of time, another common experience that attests to the existence of our personal soul is our ability to conceptualize. Our soul is not just the wholeness of our bodies; it is also the wholeness of our mind. This was discussed in Chapter 2 in regards to Plato's theory of the Forms. The question that Plato asked himself was how it is possible to identify similar things as the 'same' if they are different, even though the variation between (for instance) two blue things is slight? Plato reasoned that, in order for two blue objects to both be blue and be identified as blue, there had to be a higher state or species of blueness that linked the two. He spoke of this higher unifying species as a 'Form' of true blueness. He reasoned that it had to be something real, or else there would be no true qualitative connections between the varieties of blue things (and no real way to identify these qualitative interrelationships). Because no two material things are the same, he believed the Form had to exist somewhere in a higher, non-material 'world of the Forms.'

It is now possible to update Plato's theory of the Forms via a modern theory of hierarchical layers of light in wave-form. In the wave-form there is a potential (an archetype) for blueness.

The beauty of this newer version is that the Forms do not have to be in some realm of their own. They exist in our soul, which is light in wave-form. There is no need to posit a world of the Forms 'out there' somewhere. We are able to think and conceptualize because of soul—and not just because of 'soul' per se (which could conceivably be the soul of the universe) but because of our individual soul. It is our own personal minds that relate things (and seek to be whole). These experiences are a proof of individual wholeness.

There are so many things that our soul does for us, and we little appreciate these things. Spirit for instance is the judgment seat relative to knowing that which is good as well as a touchstone for knowledge in general. But soul plays a part as well, since soul is the material expression of the good. Soul is that which is whole. It is good to be whole—both internally (in our own body), and externally (as parts in the 'bodies' of the groups we belong to). Our soul lets us know when we are not whole, and makes efforts to heal any rifts. (And again, it is not just soul per se that lets us know this, but our soul. Our personal sense of our personal need for wholeness is a proof of the individual nature of soul.)

Everything that was mentioned in Chapter 3 in regards to our spirit and the experiences and powers it makes possible for us, is also true of our soul. That includes happiness and love. These two experiences may be spiritual in essence as absolute Bliss and absolute Love, but materially speaking we feel happy and good when we are whole, and love is present when we experience wholeness as unity with others. These common experiences are proofs of the existence of wholeness—which is to say, of soul.

Many people do believe that they have a soul, and that it preserves their individual personality after death. This is a tenet of most religions. There is such a thing as the perennial wisdom. Some ideas will never die because they resonate so strongly with the spirit of truth at the center of our being. It is perfectly all right to believe in the existence of soul—and in its ability to preserve our individual personality after our death. It is even better to have a basis for this belief. And the evidence is available.

Group Souls, Astrology and Solar Systems

Next to a photon, a solar system is one of the most fundamental systems in the world, and (if a probability field is a single ray of light) there may be an identity factor. It is actually in doing astrology that it becomes evident, not only that what is involved is a probability field, but also that it is a single ray of light in wave-form.

According to quantum theory atoms and molecules have probability fields that determine what they will be. These also determine where the system will materialize in its external world. Those who are familiar with astrology, on the other hand, know that the planets supposedly determine both what a person will be like, as well as their external life circumstances. Since people are the 'parts' of a group of living creatures, looking upon something that expresses the wholeness of their union as a probability field is the most logical thesis possible. Since everything in our world is composed of light in one form or another, it is also the simplest and most comprehensive.

There are several similarities between a ray of light and the solar system. (We will talk about the differences later.) First of all, a ray of light travels through space-time with its alternating undulations of cycling waves at right angles to its forward progression. This is true of the solar system as well. In addition to the planets circling around the sun, the entire solar system is moving forward in space-time, just as a ray of light does. Furthermore, in a single ray of (uncolored 'white') light, there are potentials for coloration. When actually doing astrology, it is possible to look upon the different cycles of the planetary placements as corresponding to frequencies that involve parameters for 'coloration,' in other words, as related to the way a probability field produces a specific set of relationship potentials and by means of these unifies the types of being actualized by the material systems it is linked to.

Our solar system entails a group probability field. Due to both wholes and wholes-in-the-making having probability fields, it seems likely that not all solar systems are the probability field of a group. There was a philosopher who helped form the doctrine of the early Christian church. Because of some of his unorthodox ideas (including a belief in reincarnation), many things he wrote about later were branded as heresy. His name was Origen (185-254 BC). As a philosopher, Origen was apparently intuitive as well as intelligent, and perhaps a student of esoteric knowledge as well. One of the things he claimed was that each human being possessed a field of light that was a miniature solar system the size of their own body. This caught the imagination of later medieval mystics, who thought perhaps the heart in a person (and their 'solar plexus') might be related to a personal 'sun' center, and also somehow to the sun we see above us.

This idea actually predates Origen. In the *Timaeus* Plato sets forth a creation epic which may have derived from the old pantheistic, nature-astral religions, since these were the types of religions in existence when he lived. What makes it difficult to determine where Plato's creation epic derived from is that it has so much philosophy in it. In explaining how the world is created, Plato speaks of the world as a whole having a soul. He says that the world-soul (which he presents as a being) is itself divided into parts, and these consist of the individual souls of human beings. Each soul, from the world-soul to that of a human being, according to Plato, is assigned to a star (*Timaeus* 41d). He even speaks of the capacity of each person to dwell in their native star (*Timaeus*, 42).

The new paradigm may not be so new after all, or it may just be that intuition precedes knowledge. If matter is composed of spirit, and if we have a spirit at the center of us, then we know the truth intuitively as to how matter is made. Our minds are actually 'material' in nature, although not necessarily physical. Our intelligence—our science, mathematics and philosophy—is like a garment for our spirit. As our understanding (and the words to express it) evolves, we grow in the capacity to clothe a wider form of knowing. But it is possible to intuit things above and beyond the level of current scientific knowledge.

If any probability field is a single ray of light in wave-form, and if this is what a solar system is, then we *are* a solar system. Furthermore, that which we actually call our solar system is a probability field for a group soul. Because we are a part of that higher level whole-in-the-making, we are a perspective on *that* whole. By looking at it (from our perspective) we can tell something about our own nature.

We cannot see either our own probability field or that of the group we belong to, but the probability field of our group soul is paradoxically visible to us through the molecules that have collected at unification points in that higher level field. We call these planets.

Astrology claims that, when you and I are born—when we *materialize*—there is something about the solar system that makes us encounter the types of circumstances that we do. This sounds fantastic and unexplainable scientifically, which is why astrology has been classified as a pseudoscience. But, if what is involved entails two closely interconnected probability fields, then the astrological phenomena begins to make some sense. From antiquity Aquarius has been said to rule astrology as well as part-to-whole relationships. This is because astrology entails reading a probability field—the group soul of the 'cell' of living creatures that we call our solar system as it reflects and affects lower-level parts like you and me. We are affected by our external world. We are a part of it. According to relativity theory we are a part of the entire universe. It affects us and we affect it. What better way to explain this than through quantum field theory and the wave-form of light, plus of course the added perspective of hierarchy in nature.

There are differences between what we think of as our solar system and its existence as a group probability field. The sun and planets are material objects. A probability field on the other hand is a non-material entity. Astrologically rather than astronomically speaking, our solar system is a ray of light in wave form with coloration potentials. There is a certain set of potentials that are being realized in our three-dimensional world and these are singled out from a continuum of potential colors in that ray of light. The types of being manifest dynamically and spatially as frequencies, and the particular frequencies singled out correspond to the orbital positions of the planets.

The parts of a whole or whole-in-the-making are unified by a ray of light in wave-form. It is 'above' them because it is their wholeness. Relative to the solar system, which unifies our personal set of living creatures, what the AC indicates is that each planet resides at a unification point for a particular quality or color; in other words, at a center that unites all those of the same type manifesting below in the parts that this ray serves as the wholeness of. As with all centers, there are centripetal (center-seeking) attractive forces generated. These have a drawing (gravitational) effect. Their primary role is to unite the types of being immediately below them; however, the attractive forces generated have an effect on still lower level particles as well. Atoms are attracted to these spots, and this has produced the physical planets. The center of this ray is a center of centers and (as ruled by the Leo-sun principle) is an immediate mode of spirit with very strong centripetal forces. That is why the lower-level matter that has collected there is so massive. We call that center the sun.

When looking at the twelve basic forms of differentiation that astrology deals with it is possible to see that these arise in the same way that a single, simple photon in wave-form produces colors in our world as was discussed in Chapter 2 on the category system. Although the solar system is a huge, macrocosmic-sized ray of light, astrology shows plainly that it is indeed a singular source for the simple coloration potentials in our world. Astrology reveals that the wholeness of our group soul is a qualitative singularity. (The spatial curvature around our physical sun reveals that area to be a spatial singularity as well—a single, simple electromagnetic ray in wave-form.) That probability field is a form of pure Being with the potential to express the twelve archetypes in our world (and also to unify them—precisely because it is pure Being-Unity).

What to Call Soul

This chapter has been an effort to understand wholeness, and to relate it to the wave-form of light as well as to a concept of soul. In the process evidence was presented for the existence of true independent wholes that are less than the universe as a whole. These are proofs of the existence of our own soul.

Our science is growing by leaps and bounds. We stand on the

brink of a major shift in consciousness. A new scientific paradigm is already beginning to arise that will have the potential to account for spirit and soul. But why call whatever exists at the center of a system 'spirit' or the wholeness of a whole 'soul?' Why offend materialists, and potentially create greater resistance to an emerging new paradigm still in its infancy?

One reason has to do with related characteristics. What exists at the center of a system corresponds to the characteristics of 'spirit,' i.e., being infinite and absolute. Likewise soul has historically been seen as everlasting, as well as the means to preserving the individual being of a person, even when they are not in 'particle' form, i.e., not matter in this dimension. This is what a probability field does. When something has the same characteristics, it is the same thing. When scientists begin to understand that quantum level wave-particle duality entails part-whole duality, it is to be hoped that the term 'soul' will then be utilized for the wholeness in wave-form. I agree with the concept of the 'identity of indiscernibles.' If our soul is everlasting, if it is the basis for the holistic nature of our individual personality, and retains that personality even after death—then what is the same is the same. There are too many things being revealed about the wave-form of light and about the world as a whole to continue to deny the existence of soul. We must not balk at where this new knowledge may be leading. This is a chance to really know that we have a soul, and what it is like. And as with the term 'spirit' it is irresponsible not to use the term 'soul' if that best fits.

The second reason for using the term soul is a personal one rather than a conceptual or philosophical reason. We need to realize our own divine, eternal nature. (We need to also understand what the consequences are for us when our material parts do not reflect the potential in our souls. Our life and the continuing existence of our external collective depends on this.) The subject of spirit and soul needs to be openly discussed, and explored with the best logic, science and math we have available to us.

Chapter 6

Soul & Spirit as Powers

Spirit and soul are not just substances; they are powers. In matter they produce certain types of states and processes, and hence may be referred to as causes. The theory of causes that science now utilizes posits only the existence of material causes in the world. These are what are referred to as mechanical causes. Mechanical causes are also called 'local' causes because they involve the agent and the recipient being in material contact with one another.

The cosmology that is supported by the archetypes indicates that there are three types of causes in the world—mechanical causes, plus what philosophers have referred to in the past as formal and first-final causes. Spirit and soul are behind these latter two causes. A formal cause can be looked upon as a whole-to-parts (soul-to-matter) type of cause wherein the whole has the ability to give the parts their internal design. A first-final cause is a spirit-to-parts cause—that which was spoken of earlier in reference to 'alpha' and 'omega.' A mechanical cause is a part-to-part (atom-to-atom) influence. If the world shows evidence of having first-final and formal causes as well as mechanical causes, these provide proofs of the existence of spirit and soul.

There are many people who have written on the subject of first-final and formal causation. For the most part, these are authors whose life work has led them to look at certain conditions in systems that necessitate the postulation of spirit and soul, although they do not normally use these terms, and usually not even the terms first-final and formal causation. I would like to reference some of these authors here—and with gratitude and appreciation, since their observations have helped me with my thoughts. It is important to access the knowledge available through these men and women. Their work provides added proof that spirit and soul are real substances and powers.

This chapter begins with the concept of first-final causation, and then moves on to a section on material causation as it is explained today through science. Even in the most current theory of causation—quantum electrodynamics—mechanical causes cannot be adequately explained without recourse to a concept of the infinite, and hence of spirit. The chapter ends with a section on formal causation, which then becomes a prelude into the last couple of chapters in this book, which have to do with astrology itself. Astrology is the science of interpreting the results of formal causation. It entails 'reading' a probability field.

(A) First-final Causation: This type of causal interaction involves spirit affecting the parts of a system. Spirit is a first cause when it gives Being to the world or to a system. It is a final cause when it gives Unity. A *first* cause entails creativity, which is the bringing of new energy-matter into the world. While a first cause is a beginning, a final cause is an end towards which something moves. Unity is one of the most generic and inclusive final causes that philosophers have supposed might exist. We have talked about how the spirit in a system makes it whole. It also moves it to unite with other systems, thus producing the attractive effects of electromagnetism and gravitation. As an impetus towards ever greater unity and being, spirit impels individual systems to move towards ever higher and more extensive forms of wholeness by uniting with other systems at the same level of being. In this sense, spirit (as a power of Unity as well as of Being) is an evolutionary impetus.

Although the production of greater Being is a (creative) first cause and greater Unity is an end cause, the two are actually one since, for instance, when several atoms unite into a molecule, the wholeness achieved produces a higher level system or being, and hence increases both Being and Unity in the world.

When things arise from a source or first cause, and move

towards an end or final cause, there is an 'arrow' to the process. They move in a particular direction. One way it is possible to know that first-final causation exists is if there are directed movements in the world. And in fact, there are directed processes in all three of the basic material domains.

In the domain of kinematics, first-final causation makes it possible for things to move in the first place. Motion is the result of absolute being producing the characteristics of 'now' from an ever changing new perspective. Spirit, as a director, is selecting a sequence of changing potentials. Any change in position involves a new set of spatial and qualitative potentials being actualized from the field potential.

Motion has a direction to it. It is a 'going forth' in time and space. The coming from somewhere, leaving some state of affairs behind, and moving towards some other state of affairs is what 'process' is. Time is also involved. We move 'through' time, into the future, leaving the past behind. This gives direction to the world processes.

It is not just as individuals that we move. We move other systems. This is what is referred to as a 'cause.' An 'effect' is a reciprocal or receptor in the relationship. As with ordinary motion, causes have value and direction. There is an arrow involved. A cause has priority over an effect and it has 'priority' in two different senses. First of all a cause comes first in time, preceding the effect. Secondly, a cause has power over its effect.

Just as the kinematical domain gives evidence of the existence of first-final causation, so also does the domain of space. The extent of a system is a certain set of potentials being actualized (in the 'now') by the spirit of a particular system. In order for a system, not merely to move, but even to have a spatial extent, it may need to sustain its extent in the present via constant recreation due to the ever-changing nature of the spatial coordinates for a system in the material universe.

In addition to this re-creation, in the AC there is truly new creation. As was noted in the chapter on '**Spirit**,' the AC is a 'constant creation' cosmology. Added energy makes a system (and any world as a whole) increase in content and hence in extent. Our scientific observation of the expansion of space, when interpreted in terms of a constant creation cosmology, is one piece of evidence that first-final causation exists. We cannot see matter above us but, thanks to the stars in the skies, we can see the expansion of the patterns up there and thereby know that, just as we ourselves and our world grows, so do higher level systems and worlds.

Although there are directed movements in the kinematical and spatial domains, because it has so much to do with spirit per se, the qualitative domain provides the most telling evidence of the existence of first-final causation, and does so via the processes of centralization and superiorization to be found in systems. Relative to centralization, the life spirit in a system actively keeps all of the parts together in an orderly, coherent manner. In terms of superiorization, there is qualitative progress in the world. It is in the nature of systems to constantly become more centro-complex and to evolve. Because of this, complexity and coherent differentiation in the universe is increasing. There is some loss of useful energy in our gray world, but on the whole there is progress. In a paradigm where 'life' takes on an expanded meaning to include all true (centralized) systems, then biological evolution is everywhere. Evolution is a qualitative proof that first-final causation exists.

What is important about studying first causation is that it is a creative power. As such, it has something to say about the character of creation, and hence about the character of spirit. Whether as a first or final cause, what becomes obvious through studying the archetypes is that the power of spirit is directed towards creating individuals, and that the power of creation continues through individuals. First-final causes are not 'out there' (somewhere). They exist in the center of systems. New photons for the use of a system internally are produced via their own center. [This is also true of any part of a system when acting as a unit itself (e.g., an electron).] New photons for use in the collective, external world also come about through the center of the systems involved. It is the same with the energy needed for superiorization. In addition to entirely new photons coming into existence new systems are created through hierarchical concrescences. If it were just diffuse energy coming into the world, Being would be increased, but not Unity. It is through individuality that both are furthered, and specifically through personal energy.

When systems are seen as the source of their own ongoing creation and superiorization, then their energy needs for increasing growth can be seen as coming from within. The chapter on '**Soul**' discussed how the soul is a source for independence. So is an individual spirit. The energy a system needs is provided in a way that can (or should) entail independence. When first-final causes are visualized as arising within a system then, when a system becomes a true, independent whole, the system is independent in regards to its being and its life; in other words, it is self-sustaining and self-creating. Per the AC, once a system is independent, the spirit within it keeps it in being (providing for its motion and extent), and continually providing more substance so that it can grow both quantitatively and qualitatively.

Our cells are dependent upon us because they are still dependent systems; but we are supposed to be independent in regards to our energy. In an independent (and truly whole) system, the power of spirit is sufficient for both maintenance and growth. Living systems are the next level of true independent systems after the level of atoms. We are supposed to be independent creators. When we lack wholeness, our bodies are not as they should be, and spirit cannot shine through in its fullness.

In living systems, eating appears to have developed as a necessary compensation mechanism due to the depletion of energy that occurs in systems that are not whole. This depletion occurs due to the second law of thermodynamics being brought into play in a negative way in a world where wholeness is compromised. Living systems are naturally anti-entropic unless their parts experience disease, abuse or misuse. And of course any lack of wholeness carries through from lifetime to lifetime—with species in a gray world producing ingenious defenses, such as eating and digestion, to counter the destructive effects of energy depletion. (I mentioned earlier that my favorite author in regards to this issue is Mae-Wan Ho. Ho is a biophysicist who makes a case that living systems experience entropy only when their parts, which are naturally coherent, become incoherent.)[1]

Most main-line scientists do not believe that first or final causes exist. Relative to first causes, physicists would have to believe that the universe was in a state of constant creation with more energy coming into the world. The primary reason scientists are reticent to consider the existence of a first cause is because the world is material, and something at base of matter creating matter would not itself be material. This brings up a problem relative to the existence of something metaphysical (something being created from no-thing).

Final causes are equally problematical to materialists. They suggest that the world process is going somewhere, in other words, that it has direction—an end—perhaps even that it has meaning and purpose. In the AC, the teleological end matter moves towards is unity, being, and ever more unity and being. In a spiritual paradigm there *is* a purpose. The purpose of creation is to express and further Being and Unity, and specifically through creating new individuals as well as furthering the individuality of those in creation.

[In addition to the overall purpose of life, there are also individual purposes. We have a spirit, and by postulating a spirit within us, it is possible to explain individual purposeful behavior—something the current paradigm cannot do. (Of all the multiple purposes and ends we pursue, they pretty much dissolve into the desire to Be and to be United—we want to live, to be whole, and to find joy and love—the latter of which are themselves expressions of Being and Unity.)]

Some philosophers have disliked seeing the world as possessing an overriding purpose or final cause because they assume that, when matter achieves whatever end it is reaching for, the process may end (and possibly also the systems moving in that direction). That is not a concern when the end is spiritual. Spirit knows no 'end' and is infinite in being. As the matter in a system is drawn to its spiritual source, the parts involved not only achieve unity, they also are replenished by their source. Spirit is both a first and final cause, and as a first cause more being is constantly created. [Nor is there a problem with an unreachable end being a tease and an ongoing disappointment. We *are* able to achieve wholeness (completeness), the material equivalent of a spiritual 'end,' although even there we have the capacity to achieve ever higher levels of wholeness—unendingly.]

Matter is dynamic and moves, but not just randomly anywhere. The outcome of its processes is weighted, producing the 'arrows' discussed here and in chapter three on '**Spirit**.' The ultimate 'weight' is the question of why there is something rather than nothing. To sense the real reality and power of spirit, it is only necessary to contemplate the nature of Being. Being (by definition) exists, and non-being (by definition) does not exist. Incoherence between parts can make a part of us temporarily cease to be a part of us; but it remains part of the atomic-level world. (A wound has dead tissue in it but the atoms in that tissue are real. There is no non-being, even in a wound. What is missing is the correct level of being.) Buddhists claim that only pure Being is real and that the world of process is an illusion or dreamlike state, but (in our gray world) it is a fall in level of being that creates disunity and the dark missing places in *that* level of wholeness. Since Being is Awareness, perhaps the Buddhists are right in that we are like sleep-walkers currently, existing in somewhat of a dream-state (and sometimes in a nightmare). Thank goodness non-being does not exist, and thank Goodness that what we do experience of it *is* a temporary state involving incoherence in parts and a drop in level. And we can thank Goodness too that all real power resides in spirit—that which really exists and is Being. The 'power' aspect of the Real is why it is possible to know that the grayness in our world will inevitably be overcome, and why we have the power to do so. Being does not create non-being. Our future is to Be, and to be ever more. We have a creative and directive point at the center of us that is our will, and is the source of all power. It is also the source of our matter. Every aspect of our triune nature—body, soul and spirit—is working to heal our wounds and that of our collectives.

Spirit is primary in this task, and those who work with energy medicine know just how important first causation is. They also know precisely where the energy produced via first causation comes from. They are knowledgeable about centers. In order to understand first-final causation it is helpful to access people who have themselves spent their lives trying to understand the creative centers within our own bodies. The resource material is vast. It ranges from 'New Age' practices, to discussions of the 'ancient wisdom,' to the work of biologists, philosophers and theologians. The reader can look into Eastern or Western discussions of the Buddhist-Hindu concept of chakras as sources of life energy (e.g., *The Chakras* by C.W. Leadbeater).[2] Another source of interest in the chakras resides in all the schools of energy medicine that currently exist, from Reiki to Healing Touch to acupuncture (e.g., *Hands of Light* by Barbara Brennan).[3] These authors help illustrate the presence of a hierarchy of centers reaching from a primary center in the human body, down to centers within individual organs, even to those within individual cells. They also discuss the presence of centers in different kinds of bodies—e.g., within the physical, mental and emotional bodies. The goal of healing for energy medicine practitioners is to facilitate the flow and alignment of potential energy as it arises through the chakras and at these points is transformed into energy-matter as we know it.

Any data on yoga, qigong and other similar Indian and oriental practices is very helpful. Most forms of yoga entail an effort to open the chakras as centers of awareness as well as of energy. (Magic practices that were prevalent in Persia entailed an effort to open the chakras as centers of will.)

For a good scientific, philosophical, and religious discussion of centers and centralization, there is none better than Pierre Teilhard de Chardin. By being a biologist interested in evolution, he relates the concept of centralization to that of superiorization, elaborating upon how spirit makes systems evolve, producing ever greater and more extensive qualitative centro-complexity in living organisms. He relates spirit to the unity as well as the being aspects of first-final causation, and in doing so points out that love is a form of unity—one that he believes is the primary driving force behind evolution. Chardin is also interesting in that he speaks of humanity as a whole being drawn to a higher level of wholeness through and by means of the sacred heart of the Christ spirit (but also by means of the Holy Spirit within the center of each person).

Mythology, especially (East) Indian mythology, is an excellent way to access the ancient wisdom in regards to chakras as centers—even their connection to the solar archetype. For an astrological discussion of centralization and superiorization (as these are ruled by the sun-Leo principle), see Dr. William M. Davidson, "New Light on the Signs of the Zodiac."[4]

While Chardin is best for understanding the relationship of centers to *qualitative* unity, Gottfried Leibniz (*Monadology and*

Other Philosophical Essays) is an interesting author relative to an understanding of centers or points-instants and *spatial* unity. Reference also Georg Cantor (*Contributions to the Founding of the Theory of Transfinite Numbers*),[6] and Claude Bragdon, (*A Primer of Higher Space: Explorations into the Fourth dimension*).[7]

(B) Material (Mechanical) Causation: Most scientists today recognize only material (atom-to-atom) causes. This kind of cause is called a physical or mechanical cause. Our atoms are electromagnetic in essence, and their causal interactions are described in terms of photons. When one atom affects another, it is said to give photons to the other system as packets or 'quanta' of energy. This gives the other atom, not only more substance, but also more 'doing capacity' and hence it moves and changes. The giver is the source—the power or force. The recipient is the effect. What happens when one atom affects another atom is that the electron in the causal atom emits a photon which then affects the recipient atom via its electron, which absorbs the photon. How this happens is currently explained through the quantum field theory called quantum electrodynamics (QED). Richard Feynman is responsible for formulating QED as a hypothesis to explain mechanical causation.

QED utilizes the two theories called 'electromagnetism' (which sees light in terms of a waving ray alternating between electric and magnetic components) and quantum theory (which sees light as a localized particle). It combines these two into a field theory but ultimately quantizes the electromagnetic field. (Feynman needed a real wave in order to have a field and explain mechanical causal interactions, but he also needed to quantize his ideas in accordance with quantum theory, so he combined the two.)

The electron is negatively charged electrically and by interacting with the electromagnetic field, per QED it has a 'magnetic moment' of interaction. During these moments virtual photons are said to pop into existence from out of the field and also to return to the field, thus the field becomes the medium for the interaction between the two electrons. The problem is that the field or quantum potential is not generally seen as real in quantum field theory; plus QED quantizes the so-called field into the discrete particles called photons. And the interaction is not so simple—not in terms of a single photon disappearing in one place and appearing in another place. The intermediate distance has to be accounted for, and then the intermediate distance of that distance, and on and on (reference Zeno's paradoxes in the appendix). What happens is that one virtual photon is said to have the capacity to produce other virtual particles ('internally' so to speak relative to *that* interaction). Then these have the capacity to produce other virtual particles (again 'internally'). Instead of the term 'internal' physicists use the term 'self-interaction.' This creates a descending hierarchy of self-interactions, and altogether these lead to an infinite regress. This is why infinities are a problem in QED and why renormalization is needed. A small, precisely known number must be added to the calculations. Because it is small, it is thought that it does not matter. (In addition to the infinitely long number of potential exchanges required, zero point itself can never be reached because there is no zero point in quantum descriptions, either of time or space.)

[These 'self-interaction' processes as described are actually a very apt way to illustrate how spirit produces a causal interaction or even simple motion, the latter of which presents as a constant recreation of itself as the point-instant. And the 'infinite regress' aspect is reminiscent of, not only how the point produces wholeness, but also of how wholeness entails the other pole of spirit—its infinite potential as the 'field' that is the substrate. In other words the AC utilizes the same description but has the substrate to explain connectivity. It begins with the reality of both types of infinities instead of reaching towards them but denying their existence.]

It is not that QED is wrong. The priorities just need to be reversed, and of course infinities added as true realities. All three concepts—particles, soul and spirit—are needed to understand how atoms affect one another—and ultimately how we affect one another. A 'force' does not just entail a 'packet' of energy (a photon being exchanged). That is only one-third of the equation. There can be no continuity to parts or particles without the wave-form or soul, and there can be no wholeness or zero point without spirit.

In addition to atoms affecting other atoms, we believe that we ourselves as individuals have the capacity to affect atoms, and through these all sorts of things. If we are really doing this as a true causal agent in and of ourselves, then we need some explanation as to how this is done. Atoms are our parts, and they are the parts of our external world. QED and the materialistic, mechanistic paradigm in general, tend to visualize atoms as the systems behind any cause. From this perspective, human beings are not individual systems in and of themselves, and they are not true agents.

In the AC, on the other hand, there are true individual systems, and they exist at different levels of being. We are an astronomically higher level of being than the atoms that make up our matter. When we use atoms to do something or make something we become a force or causal agent. Will was discussed in the chapter on spirit and hence was not reintroduced in the previous section on first-final causation but, if we have a will, it is one of the strongest pieces of evidence that first-final causation exists. Furthermore, spirit or will is actually behind mechanical causation.

It is important for us to understand *if* we are true causal agents, and if so *how* we do this. (A theory of how we are a cause is not the same as a theory of how, for instance, the atoms of our fingers cause something to move.) Per the AC, a concept of spirit is needed, just as it is for ordinary atom-to-atom interactions. The will is an expression of pure spirit, and pure spirit acts through centers. A force center *is* a center. When a higher level system affects a lower level one, there is a hierarchy of centers involved, just as there is a hierarchy of parts. The mechanistic paradigm only takes the hierarchy of causal effects from the atomic level down (atomic interactions via electrons, electron interactions via photons, photons via the endless hierarchy of regresses just mentioned). When as a living creature, we are doing something, the ultimate overarching whole (and hence the center of that whole as a force center) is the primary cause, not the center between the material atoms utilized to produce the mechanical equivalent. Per the AC, we are true wholes and, when we are initiating a cause, we affect atoms (and ultimately the center of the photons involved) through our individual center. By directing our will to a particular material circumstance we can affect the atoms involved. We can do so because spirit is a power, because spirit acts through centers, and because centers are not ultimately separated, no matter what level of system they occur in.

Our spirit is creative when exercising will power; and that creativity can be present in everything we do. The use of our will power does not just entail ponderous, momentous decisions. Every time we do something intentionally we are being a creator. But the movement, thought or feeling must be our conscious decision and not just a situation in which we are moved by the external world. The difference between a cause and an effect is the difference that exists between a creator and that which is created.

Whether a cause is initiated by something outside ourselves, or we ourselves are the initiator, a cause is a creative first cause. It produces something—sometimes something entirely new, sometimes just a change in what already exists. Under a materialistic paradigm nothing comes from nothing and hence nothing can really be created. A 'cause' therefore only changes something already in existence. If the change is striking and dramatic then materialists speak of the production of something new; but the cause has merely produced different arrangements of the same old energy-matter.

That view becomes a problem under QED, which itself is a product of a materialistic paradigm. In QED particles do materialize out of nothing. Materialists seek to neutralize these contradictions by claiming the amount of energy remains the same.

If you believe that matter is limited in amount and finite, the configurations it can manifest must also be seen as finite. The relativist philosopher Friedrich Nietzsche (*Thus Spoke Zarathustra* and *Beyond Good and Evil*) maintained that whatever has happened in our world before is doomed to happen over and over again in unending cycles exactly the same way each time.[8] (The *same way* aspect was due to the lawfulness of matter.)

This is somewhat related to the reason materialistic biologists cannot believe in natural evolution. If every system is becoming more complex, creating new and more extensive configurations, those systems must be taking away from other systems. In the future other systems will take away from those systems—and so on. There is a definite upper limit to new configurations, and so either the process comes to an end or the same things just keep recurring. (Nothing really has a purpose under this type of paradigm—nor any beauty or majesty. Creativity seems ultimately pathetic and useless—even sad.)

To go even deeper into these thoughts is to realize that, not only would there be no new configurations without constant creation, no motion would exist, and neither would a single interaction. Motion, change and interaction entail a relationship between a particle and the quantum potential. There would be no motion, no change, and no cause-effect situations at all without constant recreation, as everything accesses the quantum potential and moves into the future.

The idea that ordinary causation implies the existence of first-final causation has bothered more than one philosopher. The relativist philosopher David Hume (1711-1776) was well aware that the notion of causes and effects entailed an 'arrow.' Thus Hume claimed that different things just happened together. He maintained that there was no such thing as a causal connection. From this perspective, even the laws of nature can be presented as having no causal efficacy. They can be explained as merely involving correspondences between things that are invariably connected.

Relativists with this type of philosophical bent have suggested that the concept of causation has a human origin. The argument is that the mistaken idea of power begins with our own false sense of power. We believe that we can move, change ourselves, and change our world. Hence we believe that this power exists in nature as mechanical causation—that a system really can move itself and can move external things as well.

Per the AC, we need to trust our own judgment that there are causes and effects in the world. Furthermore, the idea that ordinary mechanical causation begins with self-causation or will power is also a correct one. All power begins in spirit.

We are powerful spiritual beings. We have the power to create ourselves and (collectively) our world; and we are doing so whether we are fully aware of this or not. Whatever the level of belief, in order to manifest will in its fullest sense, we need to be more aware that we are creators—because the will is the awareness. Science can help us with this knowledge, but first we need a better paradigm—one that shows the role of spirit even in atom-to-atom mechanical causation. A spiritual cosmology that can assist us in visualizing how causation occurs, including how we affect atoms and create our world is tremendously empowering.

(C) Formal Causation: Highly evolved systems are very complex, and to understand formal causation it is necessary to understand complexity. We have talked quite a bit about all the things that the presence of soul makes possible in the world, from wholeness to hierarchy in nature. But we have not yet discussed the 'pattern' and 'order' capacity of soul. A whole does and does not unite its parts in terms of one over-riding ray of light. Our soul is something that is in dominion over an incredibly complex hierarchy of dependent parts. Those wholes-in-the-making have rays of light involved with them. It is the layering that produces the complexity. Despite the complexity, there is an underlying simplicity. Each layer involves a simple, single ray of light with twelve simple 'coloration' potentials.

At the end of this chapter, a complete theory of how complexity arises is presented based on what the astrological category system indicates. Then, beginning in Chapter 9, it is possible, via astrology itself, to see how the practice of this ancient science verifies the thesis.

In trying to understand how formal causation works, it is helpful to look at living systems because of the high degree of patterning, order and coherence their forms possess. Although the answers are not scientifically available yet (or perhaps because of the lack) the complex form of living systems helps highlight the need for a good concept of formal causation. Trying to understand formal causation in a living system includes trying to understand the manner in which a probability field produces form in the first place, how it maintains it, even how it reproduces that form when a part of the system is damaged.

In searching for understanding, the place to begin is with the question of whether or not formal causation even exists in a higher-level system. As has been noted, there is currently a controversy going on in physics as to whether the probability fields of even quantum particles exist, much less whether living systems have probability fields. This is where biology can be extremely helpful. As with physics, it too is in a state of evolution.

There are a number of authors writing on the subject of

biological form and its development, and some of these will be referenced here. Their work is vital to the new paradigm that is arising.

Currently, it is simply not known how a multi-faceted living organism can develop from a single, simple egg or seed. DNA is not an adequate explanation. Rather DNA appears to be a correlative—a means *through* which parts can develop. Bruce Lipton was mentioned in the preface as a modern biologist working on this issue. Classics on the subject of biological form production include: D.W. Thompson, *On Growth and Form*[9], Joseph Needham, *Biochemistry and Morphogenesis*[10], Paul Weiss, *Principles of Development*[11], and Brian Goodwin, "A Phase-shift model for the spatial and temporal organization of developing systems"[12]. A number of these authors are what have come to be called organicist biologists. They tend to believe there is some sort of field guiding the development of a system from the time it is an egg or seed through processes that bring it to the peak of its individual complexity.

In addition to individual fields, some authors have written on the existence of group fields—examining such things as the subject of how there can be correlated, interactive behaviors between different members of the same species, even when they are at long distances from one another (e.g., Rupert Sheldrake, *Presence of the Past: Morphic Resonance and the Habits of Nature).*[13] These explorations into group fields provide evidence of the existence of the group souls referred to as wholes-in-the-making in this book. They also provide an important type of proof relative to the existence of formal (versus mechanical) causal interconnections. The reason evidence of group fields is so helpful is because, although the form of a single individual can be explained (albeit inadequately) as due to biochemical interactions, it is very difficult to account for group interactions in this way because the 'parts' (in this case individuals) are not materially contiguous as is the case with the parts of a living body. In many instances the various members of a group or species are miles apart. A classic example of a group causal effect is the 'hundredth monkey' story. A biologist was studying primates on a Pacific island—several islands in fact. A female monkey named 'Imo' learned how to wash sand off her food to make it more digestible. Her skill spread inexplicably to all the monkeys on all of the islands. They went from none of them having this skill to many having the skill in a very short period of time after Imo learned how to do this, even though those on other islands were not in contact with Imo.[14]

Other, less specific changes in behaviors are shared by members of the same species all over the globe—including the human species. We are subject to regular, world-wide cycles in monetary inflation and recession. There are cycles of birth and marriage. In species (like trees) where it is possible to monitor growth (via tree rings) there are growth cycle correlatives that cannot be simply linked to rainfall, which is regional. I have been particularly interested in the manner in which tree growth appears to follow a period of a little less than twelve years, similar to the period of the planet Jupiter, which in astrology rules growth. In cases where it is possible to measure growth through fossil records or whatever, this particular period manifests in terms of growth cycles for many species, not just trees. A classic on this subject is Dewey and Mandino, *Cycles: The Mysterious Forces That Trigger Events*.[15]

A good book on the subject of individual, internal (biological) cycles is *Biological Rhythms in Human and Animal Physiology* by Gay Gaer Luce.[16] As was just mentioned though, group behaviors are much more telling, because the existence of cyclic behaviors in a single system can be partially explained through mechanical causal efficacy because the 'parts' are contiguous. The more interesting books on biological cycles pursue the subject of group links. Bio-rhythms in living creatures in different areas of the world have been noted to be quite similar. They have even been known to be temporally correlated relative to the timing of their peaks and troughs, which is very difficult to explain. These correlations occur not only in different members of a species, but even in different species altogether. Because of the temporal correlations, a number of authors have attributed some biorhythms to non-biological causes, such as lunar and planetary influences (e.g., F.A. Brown in "Living Clocks".)[17]

There are many, many books and articles written on the subject of formal causation, from how it is that snowflakes get their patterns (J. Maddox) [18] to how we think and remember. Relative to the latter, one that I particularly enjoyed was Ninian Marshall's "ESP and Memory: A Physical Theory."[19] The number of synapses needed to account for all of our memories is so great that the physical brain cannot be the source of long term memory. (From the perspective of the AC, individual probability fields make it possible for us to have memories of past lifetimes, as well as memories from this lifetime.) Marshall also discusses the subject of ESP. [It is through a concept of group probability fields (and spirit) that ESP finds a logical explanation.]

I suspect that someday even healing will be explained through formal causation and the postulation of soul. When cells in a body are damaged or die, they are able to reincarnate. Scientists do not understand how healing takes place. To see the place of the cells as retained and regenerated through patterning within the body's probability field is the simplest explanation. And as those who deal with energy medicine know, it is not merely the body as a whole that has a field, but also each individual cell as well.

I like to think of reincarnation itself as a form of healing. It is not normal for cells or for us to die. These things occur because of the current state of our gray world. If we are reincarnated into different lives, in different places and different times, then there is a higher probability field than just the one that goes into the formation of this specific body and its characteristics. And if that is so, then earth lives may be as cells to this higher self, and reincarnation may be occurring because of the need to heal a part of that self. We have been told through our religions that we have a higher self—a soul in this higher sense. I have memories of a number of past lifetimes. So do many people. The simplest explanation for the retention of character after death (via soul), as well as how rebirth or reincarnation might be enabled to occur, is the postulation of a probability field.

Thus far we have been talking mostly about how form develops from an egg or seed, and how it re-forms via healing; however, even when a living body is not forming or re-forming—when it is in a steady state so to speak—there is tremendous activity going on. There is a constant *dynamic coordination* of all the parts in a normal, healthy, physical body. This coordination cannot be explained in terms of ordinary physiological interactions. The time period required for these virtually instantaneous interactions is too difficult to explain in terms of chemical responses. Electromagnetic connections are a much more likely source; but these would surely have to be centralized and (if occurring via formal causation) produced by a wave-form with singularity. Our bodies are incredibly complex. Yet all the cells, organs and systems work together in harmony—like a whole. The simplest way to explain this is in terms of the unification potential in whole-to-part connections.

[The soul, as a higher dimensional reality, has atemporal causal efficacy relative to the three dimensional matter it unifies, which makes even instantaneous interactions possible.) Instantaneous interactions have been a subject of much interest at the atomic level of being. Because the parts of an atom are electromagnetic in essence, the interactions between the parts should follow the nature of electromagnetic interactions, which includes the attribute of occurring at less than the speed of light. What quantum mechanics has discovered is that, instead, even parts currently inside an atom do not obey laws of local (mechanical) causation. Electrons, when responding to an impetus to change levels, do so instantaneously—not in terms of an infinitesimal second—but truly atemporally. For another example, as was discussed in Chapter 4, the parts of an electron (photon pairs) have shown themselves to be unified even when they are artificially torn apart and transported long distances away from one another (cf. the Alain Aspect confirmation of the Einstein-Podolsky-Rosen experiment).[20] Once a theory of the reality (and causal efficacy) of the wholeness of a whole is posited, then non-local causation becomes understandable.]

In conclusion, there are many specialists, in many fields, writing currently on the subject of formal causation. None yet are claiming that the fields they believe exist in living systems are probability fields. That is because quantum physics currently claims individual fields stop at the molecular level of being.

Quantum physicists themselves are not unanimous about the existence of even atomic-level probability fields. But every instance we see in the world of formal causation is a proof of the existence of probability fields, and hence of soul. Of course *a* proof is not *the* proof; but no proofs of any cosmology are final, and when the magnitude of these things (and the magnitude of books on the subject by serious enquirers) is so great, after a while the preponderance leans to the side of soul—and that is what the criterion of comprehensiveness is all about.

For those who do see formal causation as real and as involved with light, only a few authors have attempted to specifically explain how light might give form to systems. Two of these include David Bohm, the physicist mentioned earlier, and Karl Pribram, a neurophysiologist at Stanford University who became interested in formal causation when he discovered that memories are not localized in any one part of the brain. (Somehow any part of the brain has access to the whole field of memories.) Both Bohm (*Wholeness and the Implicate Order*)[21] and Pribram (*Languages of the Brain*)[22] independently came to the conclusion that a hologram might be a way to at least metaphorically explain formal causation, and so began a dialogue among a number of scientists in regards to the 'holographic metaphor' (reference Ken Wilbur, ed., *The Holographic Paradigm*).[23] Holography is a method of photography in which light reflected from an object is recorded on a plate as an interference pattern. In 1947 Nobel Prize winner Dennis Gabor developed a formula for the construction of a hologram. The mathematics he used was invented by Jean B. J. Fourier and entails a way to change things seen in the external world into a language of simple waves, and then back again to typical figures as we see them (and even to sounds). The math steps are known as Fourier transforms, and are the basis of our ability to have pictorial and auditory patterns transferred via electromagnetic waves from a television studio to our individual television sets.

Two very interesting things about the holographic metaphor include the fact that, first of all, it suggests that form can be developed through frequency and interference patterns in light waves. This has relevance for a concept of formal causation in general and for astrology specifically since, per the AC, formal causation occurs in terms of twelve basic types of being, which in turn are mediated through frequencies in light. These involve 'houses,' 'signs' and 'aspects' as well as planets, and what we will be doing in Chapter 9 is relating that astrological terminology to wave-lengths and interference patterns within light—patterns which carry the language of the laws of nature in this way. As noted, astrology is the best way to see how formal causation actually works.

A second way in which the holographic metaphor is relevant to the AC has to do with the part-whole concept which is also important in the astrological paradigm. In a hologram, each part has access to the whole. Any small part of the hologram can be used to reproduce the whole. This is hauntingly reminiscent of the (two-pole) part-whole character of spirit, and ultimately of light with its part-whole duality. A part is a perspective on the whole. At a higher level of being it is the whole. Because a part is a perspective it infers in its own character the 'place' of other parts, and knows the character of the whole implicitly if not explicitly.

The metaphor helps explain some of the procedures used in the practice of astrology. An astrological chart is a group probability field, but it is person-centered. The chart of a person maps everything in the group field in terms of how it relates to that person. It maps each individual in terms of where on the earth they exist (the 'houses'), and draws the planets, not as they relate to the sun (as in astronomy) but as they relate to the person. (For example, from your perspective on earth, Mercury will seem to go backwards at different times each year—and it really does—*from your perspective*. Astrologers call this 'retrograde' motion.) Astrology is very relativistic. According to relativity theory each thing is what it is based on its relationship to the entire universe. In that sense, the whole is in the part as well as the part being in the whole. Astrology looks at the whole (a group probability field for the world-of-living-creatures), but looks at it from a precise perspective—because that is what a 'part' is—a perspective on the whole. In this way it has much to say about a particular part such as you or me.

The way in which the AC is different from the holographic metaphor has to do with the fact that it has a hierarchical orientation and utilizes, not just waves and interference patterns to explain form production, but also layering. Every probability field—even the main overarching one for a very complex system is simple—but within its depths are smaller layers or octaves, and astrology indicates that this too is a factor in electromagnetic recording.

In explaining how formal causation works to produce a complex form the AC utilizes four concepts. One (**1**) entails seeing the qualities that material objects possess as produced via twelve differing wave-lengths in electromagnetic vibrations. Another (**2**) is seeing these twelve as interacting with one another in terms of differing interference patterns at different times. A third (**3**) entails seeing these patterns as present in hierarchical layers. And a fourth (**4**) entails seeing these layers as themselves connected and interactive. In other words, electromagnetic vibrations do not just interact at their own particular level of being, but are also capable of acting upon one another, via resonance, down through many layers of being.

Formal Causation and the Astrological Cosmology

(1) The Types of Being as Differing Frequencies: An electromagnetic vibration is a wave. Waves manifest different frequencies. The concept of frequency encompasses the two primary attributes of matter—motion and extent (wave-length). Per the AC, the different types of the laws of nature are created via different types of frequencies.

In light that is visible to our eyes, we interpret different frequencies as different colors. Color, as a simple expression of the types of being, is a way to visualize the laws of nature as frequency variables within light.

Color corresponds to qualities, not just in terms of how color is reflected externally from light striking objects, but also in regards to internal parts. Scientists can identify any atom in the world through the types of colors it emits and absorbs. Each atom has a very personal spectrum that is a colorful, material expression of its character, which is to say, of the qualities it possesses. Biophysicist Mae-Wan Ho (*The Rainbow and the Worm*) also discusses this concept in regards to living creatures.][24]

It is easy to visualize how the quality of coloration is represented spatially through certain frequency variables in light—less easy to visualize how other types of being might be represented, perhaps in this same way. And yet quantum physicists believe that the fundamental qualities of experience (the laws of nature) will one day be found in a few basic particle regularities such as the spin of a photon or electron. (In the AC these complement one another. Particle regularities are the 'parts' way to explain differentiation. Differing frequencies on the other hand are the way differentiation occurs in the wave-form.)

(2) Differing Interactions: A second way in which rays of light differentiate is via interference patterns. The basic frequencies are produced via the connection between the types of being as actualized and their potential in the wave-form. Material systems are moving 'through' time (actually creating time). Differing frequencies have differing tempos. Two or more waves will relate to one another in different ways at different times in terms of their differing internal phases. Physicists speak of this as the phenomenon of interference patterns.

In our group probability field, as manifest above us, the points of unification (planets) for the differing frequencies that are born and unified via this field can be seen to change in their relationship to one another. This gives variety to the twelve basic types of being. It is a means for them to combine in differing ways. These interactions (and their qualitative equivalents) produce a slightly different spirit-of-the-times at any one moment of time.

(3) Layers of Colors: Per the AC, the third way in which rays of light produce matter is through layering affects—that which has been spoken of as hierarchy in nature. This is how complexity, not only arises from essential simplicity, but also remains at heart simple.

Seeing material complexity as produced through a hierarchy of singular layers of light is one of the most fruitful (and simple) hypotheses available to explain complexity, and it does not require light to change into something else in order to produce a material system.

A human being is one of the most complex systems in existence. But in looking through its layers, it becomes simpler. A living body is composed of systems, which are composed of organs, which are composed of cells, which are composed of molecules, which are composed of atoms, which are composed of particles (like electrons), which are composed of light. When looking down through a hierarchy of layers, it is possible to see, not just that each lower layer gets simpler, but that each layer (in its soul aspect) *is* simple.

If each higher level of complexity that a material system achieves is unified in terms of a simple ray of light with a few coloration potentials, and if a whole is a part, depending upon the level of experience, material complexity can be seen as due to the multiplication of layers of simple photons within photons, or rays of light within rays of light—in other words, as layered probability fields when seen as wholes within wholes—as layers of matter when seen as parts within parts. In this way matter can be looked upon as composed of many, many layers of simple expressions of a few laws of nature, which satisfies the puzzle of how matter can be so complex when there are only a few basic laws. (Fractals in mathematics give a visual representation of this.)

No matter how many layers there are in a system, they are all simple in nature. Hence, in order to understand how formal causation works, it is most productive to study just one layer. When trying to understand that layer, in a hierarchically constructed universe, it is possible to look up or down. We cannot see the probability field of an atom or molecule. But we can see the group soul that has jurisdiction over our world of living creatures. There is such a probability field. We call the involved area in the heavens above us our solar system. We can see it because of the planets that have formed there. It is simple, it involves only one layer, and it is working the same way that any probability field does—precisely because every probability field is a single, simple ray of light in wave-form. If we want to understand how a probability field works, if we want to become much more informed experts in terms of information theory, we need to look up (and to study astrology).

Not merely wholes-in-the-making, but also true, individuated wholes have a probability field. Many of the old religions believed that a particular star in the heavens was assigned to each person. Afghanistan is a land that sits between (and has been torn between) larger dominating nations for eons. It was once a part of Persia and once a part of India. The Aryan tribes that descended into India and Persia 2,000 years before Christ, apparently crossed the Oxus River in northern Afghanistan, and poured south, east and west via the Wakhan Corridor created by the Oxus. For centuries the Zoroastrian (Avestan) religion predominated in the Wakhan even after the Muslim incursion, and the remoteness of the area preserved many of the ancient beliefs. As late as 1890 one visitor to the Wakhan (an Olaf Olofsen) reported that the inhabitants told him that each person had their own star, and that when they died a shooting star was seen.[25] We think of the stories in the old religions as myths (in the pejorative sense of 'myth'), but there is more truth than poetry in the myths. A star is a solar system, and a solar system is a probability field. Thus not every star is above us. If the thesis that has been presented is a true one, then we are indeed given a star when we individuate, at our first birth—and not merely given a star—we become a star. As for the stars above us, they could be the probability fields of true individuated wholes or of wholes-in-the-making.

As with matter, the stars in the heavens are arranged in hierarchical layers. There are galaxies within meta-galaxies, within meta-meta-galaxies. Physicists do not know how far up this hierarchy extends, since after a while it becomes beyond our ability to explore. If hierarchical layers correspond to dimensional layers (e.g., to worlds within worlds), the stars in those galaxies, meta-galaxies and meta-meta-galaxies above us could encompass the probability fields of four, five or six dimensional beings, and/or their interactive worlds. It boggles the mind to think of it. In fact, when I do think of it I am put in mind of the story of the Christmas star, and wonder if that story too has more truth than poetry to it.

Matter is very complex but it may not be as complex as we think. If something (in this case complexity itself) can be explained in a simple, coherent and comprehensive manner, these attributes are an indication that the explanation is a good one.

Per the AC, the world has not evolved from light. The world is light, in an unchanged but constantly growing form. Every system is literally a being of pure light—layer upon layer. Creation is both incredibly simple and incredibly complex.

(4) Connections between Levels of Being: Hierarchy is a primary means whereby individuals can separate out from the universal field of light into separate rays of light. But it is the means also for things to stay ever unified. That is because anything is always both a part and a whole. We are individuals, but we are also parts in a group soul. That developing 'cell' is connected to a higher whole, on up the Tree of Life. We are connected to all that is because we are a part of the Tree.

In addition to being a part, we have parts. These levels are all interconnected. We are connected to our various systems, organs and cells. [When we become an independent (complex) whole, we become a regent over the (dependent) layers we encompass.]

All the levels below that we experience as 'material' levels are connected as well to the solar system above. The psychotherapist Carl Jung was a student of Sigmund Freud. He believed in the veracity of astrology, and even suggested ways it could be tested empirically; but he believed that the connections occurred because of synchronicity (*Synchronicity: An Acausal Connecting Principle*).[26] This is reminiscent of the idea that perhaps there are no causes at all—just correspondences between things. I disagree. Everything we know and understand works through causal principles. Everything is composed of light, and light is a force as well as a substance. All that is required in the case of astrology is an understanding of how light itself occurs in layers (and dimensions) and, via this thought, how it is that causal forces work down through layers of being. Our cells are not separated from the molecules that uphold them, despite the huge difference in size. They are causally interconnected. Likewise, there is an astronomical difference in size between the nuclear electromagnetic-like force and the electromagnetic force—the difference between the size of a baseball and the world as a whole. Despite the size differential, there are causal interconnections between the parts of an atom and the atom as a whole. Our solar system as a whole is no farther away from us than the circumference of an atom is from its nuclear parts, or than we are from our atoms.

The means to interactions between hierarchically separated layers is formal and first-final causation. Astronomically different levels of being are connected through both types of causes. Although the primal causal influence is a formal one, even formal causation has the power of spirit (centers) at its base, just as material causation does. How formal causation works can best be seen by looking at the heavens and by studying astrology. What the field above us indicates is that there are twelve basic archetypes in our three-dimensional world. These manifest in terms of frequencies and, as was mentioned, an easy way to visualize them is in terms of certain potential colors. In their probability field, they are merely potentials and remain such. It is the relationship with a lower level system that produces the actualization.(This is no more than saying that a photon changes from a wave to a particle when it interacts with matter.) Within the set of basic frequency potentials actualized in our world there are centers created by the linking of the colors below to the coloration potentials of the light in wave-form above them. Any connection of any type entails a zero point, and a zero point entails spirit and final causation. Centers produce centripetal (center-seeking) forces. Atoms and molecules are attracted to these centers and, as a result, planets have formed.

When looking at the solar system, it is possible to see that there is a primary center and secondary centers. The sun is where the primary center of the ray exists. The planets are within the frequencies of coloration potentials—and manifest those specific frequencies allowable within a three-dimensional world. The sun center expresses the unity of the whole. It is the center of the single ray of light in wave-form that actually is the whole. When seeing this in terms of Plato's theory of the Forms, the sun corresponds most closely to *absolute* Being. That center is an immediate mode of spirit. The other (planetary) centers represent points having to do with the unity of the *types* of Being. They each have their own specific potential frequencies.

These centers are focal points for *types* of being and *types* of unity. Plato believed that all red things on earth were unified by a higher-level 'Form' of 'redness' in the 'realm-of-the-Forms,' but was unclear about where that realm existed. As an astrological thesis, what must be visualized is that all 'red' photons on earth are linked to the higher level field in wave-form above us in terms of the same frequency (which produces among other things the color red). Due to the center-seeking effects of this unification point, molecules have collected at the unifying point of the 'red' principle, and we call that collection of molecules 'Mars.'

The archetype of 'redness' (the wave-form) is merely a potential. As such it is non-material, and thus fulfills the criterion that Plato (through reason) came to the conclusion must be true of Forms. The sun-center verifies the criterion he visualized in terms of the hierarchy of the Forms themselves—that there had to be a Form of pure Being, and that it stood even over the Forms that relate to the types of being (over the Form of redness, the Form of blueness, etc.). Thus the planets actually are 'parts' of the sun. The sun is higher in the hierarchy.

There is not only a sun center above us in our group field, as a true whole we also have a sun-center, where our spirit resides. This is the source of our life and of our awareness, as well as of our wholeness. We have secondary centers as well, in terms of the eleven other basic types of being. (Actually, as has been noted, we have many levels and layers of centers, since cells and molecules, as wholes in-the-making, have simple photon fields that register their own level of wholeness.)

Whether within us or above us, each center (and over-wave) has a Being as well as a Unity aspect to it. The wave-form nature of an over-wave is the source for the energy-matter of the parts when they manifest. Understanding first-final and formal causation is at base of understanding how a ray of light not only records information and thereby unifies the parts, but also how it causes parts to come into being and be what they are. Our solar system can be looked upon as a giant photon in wave-form. The center is a source for energy. Each of the secondary sources is a distribution point for energy of that type. If probability fields did not have a Being as well as a Unity aspect, then matter would have no being, since a probability field is the source that energy-matter 'appears' from when it materializes.

As for the actual mechanism of the interactions, whether considering the interactions between hierarchical levels of being or the more basic soul-to-part influence, I suspect that it entails resonance. As this relates to hierarchical levels, the process involved can best be visualized in terms of music since this, not only has to do with different wave frequencies (as in color), but also with different octaves—higher and lower sets of waves. Consideration of musical tones gives a visual picture of how different levels of being relate to one another, in other words, how electromagnetic waves that are astronomically different in wave-length, might yet affect one another.

In astrology, there are seven basic 'angles' (potentials for differentiation) in one single cycle of an electromagnetic wave as the wave makes its way from its starting point to a point 180° or half way around (0°, 30°, 60°, 90°, 120°, 150°, 180°). Then the cycle turns back to its starting point to form five more angles, essentially repeating those that came before but in reverse order (210°, 240°, 270°, 300°, 330°) arriving back at the beginning. Altogether there are twelve phases in one complete cycle, and these correspond to the twelve basic principles or potentials for differentiation in our three dimensional world.

Music is also based on a twelve-fold tonal system. What is called an 'octave' is fundamental and natural. We divide a single octave into twelve portions or 'tones.' These entail seven fundamental tones plus five sharp/flat tones (seven white and five black keys on the piano). These twelve tones have a certain 'pitch' and the pitch depends on the frequency of their vibration. The set of twelve is called an octave. Different octaves entail an entire jump from one set of twelve to a higher or lower set of twelve. It has been determined that the twelve internal parts within an octave—those that are harmonious with one another—involve a 'power' or 'root' relationship to one another. If 'c' is considered as a starting place within an octave, then the relationship between middle 'c' and high 'c' (c in the next higher octave) is 1:2. Going through just one set of twelve, if c = 1, then c# =(1.05946), then d $=(1.05946)^2$, then d# $=(1.05946)^3$, then e $=(1.05946)^4$, then f $=(1.05946)^5$, then f# $=(1.05946)^6$, then g $=(1.05946)^7$, then g# $=(1.05946)^8$, then a $=(1.05946)^9$, then a# $=(1.05946)^{10}$, then b $=(1.05946)^{11}$, and then finally the return point c $=(1.05946)^{12}$. These are not the actual vibration rates, but rather the ratios or interrelationships. (I do not know if there is an actual correspondence between the notes of an octave and the twelve principles in astrology. Although no one has yet made these correlations, I suspect they exist; and for anyone who wants to try, a good book to start with in this regard might be Sir James Jeans, *Science and Music*.[27])

What I think probably happens in terms of interconnectedness between hierarchical levels of being is that 'high c' and 'low c' resonate with one another. You would think that astronomical differences would cancel the effects but, as noted, the electron is connected to the nucleus of the atom despite the astronomical size differential. I think the reason there is interconnectedness may be because a truly higher (dimensional) level may have a wave-height (amplitude) that is far higher, just as its wave-length is far longer, but the comparable wave length to wave height of 'high c' may not be essentially different from the same ratios in 'low c.' In a same-dimension interaction involving different hierarchical layers this may work through the fact that, with resonance, very small and brief applications can have strong effects with continuing applications, increasing the amplitude of the effected wave until there is a match.

Although musical resonance is a matter-wave phenomenon, all resonance appears to have a deeper, quantum-level equivalent. Furthermore, all particles, including atoms and molecules have a dominant, over-riding resonant frequency. (This would correlate to what in astrology is a complex system's 'ruling' principle.) In addition to atoms and molecules as a whole, the parts of atoms also demonstrate resonances. When an electron absorbs a photon, it is said to be 'excited' by it. This is a resonance phenomenon.

The simple soul-to-part interaction may involve the same principle as occurs with hierarchical layers of material being. Physicists are aware that all particles in their creation from a probability field demonstrate a resonance-like phenomenon as the probability of their manifesting (as a particle) increases in accordance with their correspondence to the information in the wave-form. It may be that, the closer the 'pitch' for a particle or wave (or octave thereof), the higher the probability for an interconnected subsequent reaction. In other words, the wave-form of the potential and the material particle being actualized may be in perfect 'pitch' at the moment of materialization.

This is apparently why and how astrology is useful towards

picking a good time to initiate a new venture. It correlates the (desired) character of the effect to the character of the causal influences manifesting in a probability field that is visible.

Free Will vs. Determinism

It would not be appropriate to leave this chapter on causation without addressing the issue of the extent to which mechanical causes in particular are deterministic, since these are actualizations of the laws of nature. Is there freedom in the world? Or do the laws of nature determine exactly what will happen in any particulars circumstance? The AC has very clear answers, and as usual there is a paradox reconciled. The answer, per the AC, is that there is both determinism and freedom.

To begin with, we do have freedom, the degree of which we in no way fully imagine or appreciate. We have a free will. We are the creators of ourselves and (collectively) of our external world.

But there is law as well. Everything we do, think and feel has a specific, correlated consequence. This is the deterministic aspect of experience.

Furthermore, habitual patterns of thinking, feeling and doing translate into personality. These habits become our mental, emotional and physical bodies. These are registered in the soul. We pick up where we last left off when we reincarnate. Our soul is the pattern for our present based upon our past and what we have done and become to date. It is a probability field, and is what makes it most likely that, when we (re)materialize on earth we will be where we are and what we are. We carry our personality with us from lifetime to lifetime as it slowly evolves. The means we use to create our vehicles are the laws of nature. These are strongly binding, and we would not want it to be otherwise. The law is intended to protect and promote wholeness.

There are an unlimited number of ways to create without disobeying the laws of nature. The will can be free without doing what is wrong or unlawful. When we create circumstances that are not right and that do not promote individual and group wholeness, there are consequences. These can lead to wounds within us and within our collectives. In Hinduism, the belief that every action has a consequence is referred to as 'karma.' The word carries with it a better connotation than 'determinism,' since it also implies that we have responsibility for whatever happens to us. To understand and respect our freedom, we must understand what so-called fate really is. There is a reason for everything that happens, and 'reason' has to do with rationality and law. Everything we encounter in any lifetime, including our parents, siblings, children, mates and friends, is due to who and what we are. To understand anything that has or is happening to us, we need to look far more deeply for the personal reasons than Westerners who have been indoctrinated into gene and early conditioning theories have been habituated to do. In a new and better paradigm, the source is the soul and ultimately our own creative will power. Genes are material vehicles that have developed to convey the messages of our personal field. Likewise early parental conditioning (the parents we encounter) reflects the incredible meaningfulness of life—who we are and who they are as they match our level of soul development.

Karma is the expression of law. Law is the 'parts' aspect of experience. All parts are intended to express unity and hence goodness.

In our world, there is both good and bad karma. Given the current problems in our world, it is especially important to take responsibility for any negative conditions, and to search for personal reasons. To find the reason is the first step towards finding the cure. The archetypes show clearly that we are creators, and that spirit and soul are prior to matter. This paradigm promotes a sense of responsibility rather than of victimization—and with it the power to control our life circumstances. This is true of both our character and our external circumstances. Character is destiny. Who and what we encounter is determined by who and what we are. Only when we take responsibility for creating ourselves can we then recreate what is wrong with ourselves, so that what we are never again attracts negative external circumstances. Wounds attract flies, and there are collective dark forces in the external world. When we heal and seal the wounds, the flies go elsewhere.

We are the ones who must make ourselves whole. It is because every condition creates a related condition, and because like follows like in a very lawful way, that it is difficult to get off wheels of negativity. Ultimately getting back on track takes a miracle, since what is changed is superseding the law which mandates that a certain cause will produce a similar, related effect. But this does happen. Were this not so we would never heal from our wounds. We heal because spirit is prior to matter, and is the source of matter. Parts are just what is meant by law; but parts must constantly access the substrate for even ordinary change. That process is where (and why) we get a chance to be free and *really* change things, including anything that is wrong in ourselves and our external world. Getting back on the right track is a very common, unappreciated miracle that occurs every day. We are both free and enmeshed in determinism. However in terms of which is the boss, spirit is prior to and stronger than law.

Chapter 7

Matter

The purpose and goal of a new paradigm is, not just to recognize the existence of spirit and soul, but to better understand the nature of the material body in terms of its *relationship* to these other two substances.

From the beginning, the thesis has been that matter is composed of spirit. This hypothesis poses some real problems due to the nature of spirit as pure absolute, infinite and eternal Unity and Being. Matter has always been thought to be relative, not absolute, and in fact is just what is meant by a nexus of relationships. Furthermore, things that relate to one another are divided off from one another. Spirit on the other hand is pure Unity. Material bodies, especially if they are true individuals, would seem to be divided and set off from one another, and hence to have 'ends' or limits to their being, while spirit is infinite in character. Spirit is also eternal while our body certainly does not present as everlasting, as is brought home to us every time we go to the funeral parlor. The last characteristic is especially important, or will be some day.

The true nature of relationships, division, limitations and endings in our world must be clearly understood if we are to understand the relationship between spirit, soul and matter, and hence the true nature of our body. This chapter begins with the material quality experienced as '**relationships**' and '**relativity**,' then moves on to '**division**' and '**limitations**.'

After those sections we will talk about the unnatural division that the material parts of a body are subject to in our gray world. In the chapter on **'Soul'**, the everlasting nature of the soul was discussed. However comforting it may be to know that we have a soul and that it is everlasting, to someday have a soul without a body is to lose something—part of our nature. To have a whole without parts is to upset the order of the world to put it mildly. The purpose of the first part of this chapter is to make the case that a personal body, like a personal spirit and personal soul, is everlasting, when it is created correctly. The purpose of the second part of the chapter is to pursue the subject of sickness and death, including why it is that these have no ultimate, lasting power or being.

Relationships

Matter is now seen as just what is meant by a dynamic set of interacting processes or inter-relationships. Relationships involve two or more things being set off from one another and then interacting. That entails division, and spirit is neither extended nor divided. If matter is spiritual in composition, how can there be relationships?

The archetypes indicate that this paradox is resolved by the fact that the 'relationship' used to create our material world is one in which spirit simply perceives itself and relates to itself in different ways. This makes it possible for spirit to be the material universe and still be itself as well. When spirit actualizes its two poles it becomes a soul. Soul is the wholeness of a whole.

There are three types of wholeness—qualitative, spatial and kinematical. The qualitative domain has to do with wholeness per se. It is the other two domains that produce the actual material qualities themselves. These are the spatial domain and the kinematical domain. Of the two material attributes of spatial extent and motion, it is through spatial extent that it is easiest to visualize how spirit produces matter. Space occurs in terms of dimensions. (These correspond to levels of wholes.) Before there is even one dimension, there is spirit as an undifferentiated and hence totally non-exended substance which, in the material world, is experienced as a point. A point has no spatial extent. If a point is visualized as relating with the substrate an infinite number of times, the result is a one-dimensional extent. The spatial whole or continuum produced is what soul consists of in spatial terms in a one-dimensional being. A whole consists of two expressions of spirit (the infinitely 'small' and the infinitely 'large') relating to itself. There are quotes around 'small' and 'large' because extent is a material quality and it derives from the *relationship* between the point-instant and the substrate.

It is the same with motion as with extent. Leibniz was right when he came to the conclusion that motion involved the point instant being manifested part-at-a-time. He just did not see that the point and the substrate (the two types of infinities) are one. Thus he was not able to connect the 'dots' so to speak, and decided that motion (and causation) were not real.

For a two-dimensional system (as symbolized through a surface) it again takes an infinite number of the 'wholes' that are lines, to make a surface. Just as the point had to access the substrate, so does the one-dimensional line-being. The connectivity between the infinitely thin 'separated' lines is again made possible due to the substrate pole of spirit as well as the essential unity between these two poles. It is the same when a two-dimensional being becomes three-dimensional. The material extent of a system (or world) entails an infinite number of (spatial) perspectives on spirit. Philosophers have wondered how even an infinite amount of something with no extent could ever have an extent. But the extent is the relationship. Matter is a relationship, but the relationship is one of spirit relating to itself.

This is not a different view from modern mathematics, as has been discussed in terms of the mathematics of the continuum. It is not a different view from science either. Spirit as the substrate correlates to the field that Bohm spoke of as the quantum potential. Spirit as the substrate is everything in the material universe that can ever be, and an infinite number of potential types of being. It provides the infinite 'quantity' that is needed if an extent (or motion) is to be continuous, whether it is the extent of a one-dimensional, a two-dimensional or a three-dimensional system.

[It is actually the infinite and singular nature of the soul (rather than the entire quantum potential) that is being accessed. This was mentioned as well in Chapter 4 when Bohm's way of perceiving motion and extent (as the relationship between a particular system and the quantum potential) was discussed. It is the soul that makes motion serial and extent spatially whole.]

As enlightening as the kinematical and spatial domains are, it is in looking at the qualitative domain (the part-to-whole relationship per se) that it is possible to see that, even for three-dimensional systems, the point continues to exist at the center of the system as the source of creation for the matter of that system. And this too is not a separate idea from science. The photon is a simple system, and the basis of all systems. In accordance with relativity theory, the ideal particle is a dimensionless point and the ideal field is infinite in potential. Light consists of the relationship between these two (very real) ideals. When an individual wave-form becomes an individual photon, a certain set of potential is actualized from out of the substrate. This gives the photon a particular extent. The spirit at the center of the photon is able to create the material particle because it is the ideal field or quantum potential (and ultimately the soul or probability field). The actualized specific extent is what matter is. Quantum theory deals with the *relationship*—which is the material particle. A photon is a material relationship. But in order to understand the photon it is necessary to recognize the reality of the 'ideal' that is spirit at the center of the system, and also the reality of the 'ideal' field—the quantum potential. These two types of infinities make creation possible. Spirit does not change when it creates energy-matter. Any material actualization is an aspect of or perspective on soul, and ultimately on spirit. Different kinds of material parts are just different perspectives on the whole to which they belong. And any particular whole is a part of the substrate, to which it belongs. Different perspectives are different ways of experiencing the same thing in a different way. Matter is spirit perceiving a part of its own being.

To summarize, all of the relationships in creation are internal ones. This is spirit producing a material image of itself via its own substance. Because all relationships are internal ones, there is no overriding division, and this preserves the unity of the universe.

Relativity

The idea of something being relative is tied in with the idea of relationships, but with the added concept of dependency. Spirit depends upon nothing outside itself to exist or be what it is. Wholeness is dependent only upon spirit to exist and be what it is. Material parts are dependent upon the whole to exist and be what they are, and hence upon all the other parts as well. Spirit is absolute and matter is relative, and it would seem as though these two things were opposites. The reconciliation of this paradox resides in the fact that, the only thing parts are really dependent upon is some other version of them. Together these versions are not just parts of the whole, they are the whole. (As the awareness of a particular part grows and becomes aware that other parts are perspectives of its own self, it eventually expands into the whole, and becomes a higher dimensional individual.) What is ultimately self-contained is ultimately absolute, and really does not depend upon other (external) things to make it what it is. Whether something is relative (matter) or absolute (soul) depends upon the level of experience and perception.

Division

Division in particular would seem to take away from the spiritual nature of creation—especially if Being is Unity. The scientific paradigm known as 'atomism' divided the world into discrete atoms. According to the newer paradigm brought about through relativity and quantum field theory there is no division at all in the cosmos, only one continuous energy field, with only certain areas of more 'intense' energy. To have real division in the world would seemingly entail vacant spots of nothing, whether the nothing was seen as creating the division between two systems or creating separation anywhere. So far as we know there is no nothing in the universe.

Although the AC is a cosmology of Unity, it actually permits division. The manner in which the AC reconciles this paradox is by allowing for the real existence of no-thing. No-thing is not nothing of course. It is simply not a material thing. It is pure Being and ultimately pure Unity as well. For this reason, there is and is not division. In a spiritual paradigm, there can be real division in the world because spirit is real. At the same time the ultimate unity of the universe is preserved because zero-point is simply no-thing.

To allow for the existence of zero in the world, whether as a zero-point connection between objects or as a true zero-point cut separating objects, it is necessary to believe in the reality of the infinitely small. Under the materialistic, relativistic paradigm this position has been frowned upon. Cantor believed in the reality of the infinite—both the infinitely large and the infinitely small. The story of Cantor's life is fascinating, and in one way parallels that of Pierre Teilhard de Chardin. Chardin was a Jesuit priest as well as a scientist. Because he tried to unify science and spirituality, and because he saw evolution in matter as due to spirit, he was banished to China to work. We think of churches as often being antagonistic to science, but science has been just as entrenched in establishment ideology. Cantor's life was an example of the other side of the equation—the suppression of original mathematical and scientific ideas by predominant academic authorities. Today Cantor's theory on the nature of continuity is universally accepted, but he was blocked from expression of his ideas by such authorities as Leopold Kronecker and Karl Weierstrass who held predominant chairs and positions in the world of mathematics at the time. They strongly opposed Cantor's theories (and also successfully countered any public recognition that Cantor might have received during his lifetime). As a result he worked to establish international congresses of mathematicians and other societies that later had the ability to shake the powers that be. What led to Cantor's rejection by the establishment was his belief that the infinite was real. Weierstrass in particular was a true materialist. He was highly skeptical of any mathematics that dealt with infinities, and apparently believed that anything infinite could only exist as a potential—that it could never be actual or real. This of course depends upon what you call 'real' and is at the heart of saying a probability field, a field potential, or a point-instant itself is unreal.

A spiritual paradigm utilizes, not just spirit as a way to explain division, but also a concept of soul. Qualitative separation and spatial separation both involve different levels of being. Soul is what wholeness is and, as was mentioned in the chapter on '**Soul**,' a whole is complete and independent. Once an individual is seen in this way, it is then possible to see how the universe can have division and yet be one. In a hierarchically constructed universe a whole is a part, depending upon the level of observation. Because a system is a part as well as a whole, its dual nature preserves the unity of the universe. At the top of the hierarchy of worlds is the universe as a whole. Ultimately, everything is a 'part' of all that is. At the same time, because a system is a whole as well as a part, *in its wholeness*, it is a true individual with all the attributes of a whole, including the ability to be independent. Wholeness gives true boundaries. [It is precisely because a system is distinct that it can concurrently be a useful (distinct and different) part in a still higher level whole.]

Division becomes a matter of the level of experience—i.e., whether something is being perceived as a part or a whole. It is not that division is not real. It *is* real, just as different levels (and different perspectives) are real. Only a new paradigm that includes hierarchy in nature can solve this paradox—and it is the archetypes that reveal the secret of this truth. In astrology, the so-called 'earth' signs provide an explanation (and proof) of how qualitative differentiation and spatial division come into the world. By looking closely at their nature, as produced through their parent principles, it is possible to understand the role of soul in producing the two types of separation.

In the qualitative domain, Virgo rules qualitative differentiation. The astrological category system indicates that Virgo is a 'child' of Leo and Pisces. In other words, it is the creative process of individualization (Leo) that divides the substrate (Pisces), the latter of which is a spiritual whole. These two expressions of spirit entail a spiritual form of part-whole duality. The result of their union and interaction produces material part-whole duality. Aquarius and Virgo represent that material part-whole duality. Aquarius governs parts as they are parts of a whole. Virgo governs parts as they are apart from the whole (qualitative differentiation). Aquarius is an immediate mode of soul. Aquarius is the polarity partner to Virgo. It takes the precise boundaries that an individual (as a whole) creates in order to have differentiation.

In looking at the spatial domain, Capricorn (which rules spatial division and limits) is a child of the Cancer and Sagittarian principles. These rule the within (Cancer) and the without (Capricorn). There is a within and a without due to levels of being. Division is the result of hierarchy in nature, which in turn is due to individuation.

Limitation and Finitude

The concept of division is closely related to that of limitation, and the concept of limitation in turn is related to the idea of finitude. To divide or set something off from other things is to limit its extent. When something is set off or divided from other things, it then has a visible limit to its spatial extent, and the extent is then 'finite' in nature. It comes to an end. This would on the surface of things appear to be a quality of matter that is in diametrical opposition to spirit, which is infinite.

What a spiritual paradigm reveals is that matter is and is not finite. Creation occurs in terms of levels of wholeness and in terms of dimensional stages. Each dimension is a complete 'world.' Because of this, for spirit to perceive some of its potentials does not limit it. Once a system is seen in this way, it no longer can be looked upon as finite. A world has nothing external to it. *In its wholeness* a system is infinite because it is complete and has nothing external to serve in the role of bounding it. Due to concurrent part-whole duality and hierarchy in nature, parts are finite, but in their higher nature they are infinite.

Even at their own level of being (as parts relating to parts), the parts within a whole have no real ends. This is so for two reasons. In a world, there are no parts external to the set of parts that are present. So they cannot be bounded externally where the whole set ends. Internally, the 'end' of one part is where (as a part) it links to other parts. In a world that is whole, at a place where one part meets another, there is merely a shift in spatial perspective (and of qualitative aspect), and what occurs is not a real separation in the sense of a vacant spot separating the two parts. Each part is just a different perspective. They are not only different aspects of one another, they are all perspective on the whole, which itself is unending in extent.

The laws of nature are composed of light. The parts of system are intended to be unified, with no empty, dark spaces between them. (This is true of the 'parts' of wholes-in-the-making as well.) Processes of individuation in nature makes it possible for each part to be different and distinct; but when they are parts they are all perspectives on some higher level whole, and linked to one another without gaps when correctly expressed and created—i.e., when they relate with one another in a lawful, and coherent manner.

Unnatural Division & Subsequent Finitude

The universe is a continuum internally and externally. To be a continuum internally is to have parts that are continuous, with no places of non-being—no disconnected places. To be a continuum externally entails going on and on unendingly. The general theory of relativity described the space-time continuum of the world as having both these attribute because they go together. This is true of time as well. Systems create time. It is a function of the relationship between material parts and soul. It entails a movement of an extent from the present into the future. A space-time continuum is continuous internally and externally, and does not end temporary. These three things are correlatives. They go together.

In addition to these correspondences, there is also a correlation between spatial and qualitative unity. When a system—be it the cosmos or a person—is qualitatively whole, it is spatially whole.

When the parts of a system do not fit together coherently, they do not cohere. They not only do not work together, they do not stick together. They do not express the unity of the whole to which they belong.

The lack of internal unity produces disjointed places. When a whole is not a continuum internally, it cannot be a continuum externally in terms of the linkage between its present and the future. We are intended to be whole, and to be a space-time continuum. Our material parts are intended to be everlasting. Right now the parts of our material body are unnaturally finite. They come to an end in time because and only because there are incoherent places within us. We come to an end in time as an individual whenever a certain critical level of internal continuity ceases to exist.

To put this simply, we are a whole. When we are not whole, we fall apart and die. Unity is Being.

Our parts are naturally coherent and lawful; but the will is free, and strong enough even to oppose and break the laws of nature. It takes an act of deliberation to create things in a wrong way, since they are naturally intended to be lawful. Because the laws of nature are forms of unity, ill will produces disunity and hence darkness between parts. It is possible for systems to not relate correctly with other systems at their level of being. When they do not, they produce incoherent, disconnected places within their collective world. Systems can also form internal relationships within themselves that are incoherent (for humans, via thoughts and emotions as well as through acts). These create incoherent, disconnected places within their mental, emotional and physical vehicles. These relationships are 'wrong.' They do not manifest the laws of nature correctly and produce dark, disjointed spots in bodies and in the world. This type of 'division' entails a very different kind of separation than was discussed in the previous section. It is not natural, and it damages the wholeness of individuals and of their collective world.

We must look at our human experience and see the truth about its problems. There is darkness. The material world has many things in it that are wrong and bad. We are not whole. We suffer from mental, emotional and physical illnesses. Ultimately there is death. Our world-of-living-creatures is not whole either. There is pain and suffering all around us. It is not 'being negative' to look at these things. It is blind not to acknowledge their existence. Many people think that we must accept the bad with the good, and focus on the good—and that is fine as a practical plan of action in a world that will not be suddenly released of its darkness. However, what a new paradigm can reveal is a higher, ultimate truth—that this world is intended to be whole, and wholly good—that the darkness can be (and will eventually be) utterly and totally eliminated. In order to do so, we must accept responsibility for our world—both our personal world and our collective human experience—and work to bring both these back to what they are naturally intended to be. Rather than being negative, this is the basis of true hope because it begins with a faith in the beauty and goodness of creation—not with ignoring the bad or accepting the bad with the good.

One of the major historical philosophical difficulties wrestled with in trying to understand the relationship between matter and spirit (and even the relationship between the material world and God), is what has been called 'the problem of evil.' The material world has many things in it that are wrong, bad, and very dark. There is sickness, pain and suffering. Do we blame God (Brahma/Allah/Yahweh) for these things? Are they just the way things are? Are bad things necessary in order for the good to have a contrast?

One of the major benefits of a new paradigm entails a switch of responsibility for the circumstances of our lives from fate, God or other people to our own selves. Per the AC we are creating ourselves, and our individual being is prior to our external world. In fact, the character that we have produces our external circumstances. At birth our probability field determines what we will (most probably) be, and that probability field is a record of all our past lifetimes. It is our present based upon our past. That probability field will also determine where we will (most likely) find ourselves in our external world. The parents we have will cohere with who we are and what we have become to date; and it is up to us to look within and come to understand these correlations. Nothing happens by accident—no person or circumstance we encounter. If we take responsibility for ourselves we can evolve (and be resurrected back to everlasting, non-interrupted life) so much faster.

Because we are responsible for our own world, other worlds are not damaged by our mistakes. We are responsible for only ourselves and our group whole-in-the-making. When we do not create correctly it does not harm the rest of creation. Atoms are and are not our matter. Atoms are worlds. We use them to create our matter. But the real matter is the relationships between atoms. As creators we are using these atoms to create ourselves and our world. When we do not put atoms together in viable ways we do not harm the atoms. We only harm ourselves. (This is not true of the cells in our body because they are not independent. If we do not take care of ourselves, we can harm them.)

All matter is composed of light, and in a hylozoism, all matter is living. Living tissue entails *coherent* light. (Biophysicist Mae-Wan Ho speaks of this in *The Rainbow and the Worm*.)[1] Our world is gray because of lack of *coherent* light. When a human being has a wound, there are atoms making up the pus and dead tissue, and there is nothing wrong with those atoms. What has gone wrong is the coherence between the parts at the higher, human level of being. There is no non-being in the wound. Non-being does not exist. But, due to the stage-like nature of experience, there can be lack of being at a particular level. When we have a mental, emotional or physical wound, there is a disconnected place. Lack of coherent light at a particular level of being produces a qualitatively disconnected spot for that system. This produces a concurrent spatial disconnection but only in reference to that particular level. What has happened with the wound is a fall in the level of being. What we do wrong to ourselves and to our fellow living creatures creates darkness in ourselves and our world. It entails a Fall—our Fall, not that of other worlds.

If you have followed the diverging and converging lines of the astrological cosmology this far, you may at times have wondered what all this intense philosophy has to do with you. Philosophy is not a mind game. Wholeness in particular is a tremendously important subject. It is the cornerstone of the new paradigm, and it cuts right to the heart of what is most important to us and our loved ones. We need to understand, not just how the world is made, but also how to make it better, and the one can lead to the other. If we do not know that the world is intended to be whole, like unthinking animals we may tend to just accept the darkness along with the light. If we do not know that we have a free will and create our own internal and external environments we will not put strong efforts into change. We need a paradigm that can explain these things for us—and the time for this paradigm is now.

In the chapter on '**Spirit**' the subject was discussed relative to how the characteristics of being absolute, infinite and eternal go together. In the chapter on '**Soul**' the characteristics of being absolute, infinite and everlasting were also discussed—and specifically in terms of why the three go together and why they express the nature of spirit. Material parts in turn are formed so as to express the nature of soul. As divine beings we are intended to be infinite and everlasting. We are body, soul and spirit, and even our bodies are spiritual in essence and substance. Unfortunately the material parts, because they are multiple, are susceptible to division. They must express unity if they are to continue to be. For them not to be unified destroys their character. In having disconnected spots internally, the body is then not a space-time continuum. The result is an end in time as well as in space. The only reason we die is because of pockets of darkness within us. These have arisen because of the incorrect expression of the laws of physics.

In physics, 'work' is defined as doing something. Every time something is done that is not right from all perspectives, useful energy is lost due to the negative effects of the second law of thermodynamics. The entropy or loss that occurs is our loss, and it is due to friction. This loss occurs not just because we fight with other nations and with our marital partners, but also because of all the struggles within ourselves. And the 'doing' is not just in the physical realm. We 'do' things every time we think and feel something. Every fear, every sad thought matters. We even do negative things when we do not do the things we should. To not realize our potentials—to not be who we can be right now given our level of personal evolution—is to create vacant, incoherent places in our character.

We are powerful, spiritual beings, and we are intended to have a million facets to our nature composed of clear, coherent light, with all the parts working together in harmony, expressing the beauty of creation. I warned you at the beginning of Chapter 1 that it was possible, through the AC, to come to see who and what we are; but coming to understand this is a two-edged sword which also makes it possible to see what we are supposed to be but are not. We live in a fallen world. Every religion has legends of the Fall, so it should come as no surprise. The thing is, it is not Adam that fell, or just our collective world. *We* are a fallen world. Within ourselves we have spaces wherein our parts do not connect right. Because being is light and unity, these are dark pockets of incoherent light—of 'non-being' (at our level)—and they prevent the fullness of our unity. We are divine beings composed of the substance of spirit. Spirit is Unity-Being, and dark spots separate us from our own true self—and that includes our higher self.

In terms of the latter, a Fall entails a fall from a higher level and dimension of being. This lifetime is as a part or cell to that higher self. It is obvious that these 'cells' have problems, causing them to become sick and eventually to die. It is also obvious that the cells in our three dimensional body (those that we traditionally call 'cells') have the capacity to heal and hence to reincarnate after they have been damaged and die. The field that is the wholeness of our body makes this possible, and there is no reason to think that higher level bodies do not follow the same laws of formal causal efficacy.

Just as death is not normal, so reincarnation is not normal either. I do not know when exactly I began to believe in reincarnation. Perhaps it was when I began to get glimpses of past lifetimes that I have lived. Now it makes sense to me logically as well. The archetypes further validate this concept. In astrology the eighth house (the Pluto/Scorpio principle) rules both rebirth and healing. Reincarnation is a type of healing process.

We may have fallen from paradise; but what the AC reveals is that all worlds are heaven, at each level of being, when they are correctly made. Heaven may be our higher 'home,' but this world too can be heaven, and will be heavenly again someday when it is healed.

Speaking of religions and older forms of revelation, the Grecian god Hades (called in Rome 'Pluto') was said to rule over a lower world named after him—a world that was also called Hades. We utilize that term now for the concept of hell. As was mentioned in Chapter 5, the cosmology that develops from astrology indicates that heaven and hell are not 'out there' somewhere. Rather, they are states of being—individual and collective. When someone we love, or we ourselves, are tormented by wounds and in pain, we know exactly what and where hell is.

No doubt there are even different levels of hell—depending upon how much pain and sorrow a particular world is dealing with. One thing the philosophy behind the astrological category system attests to is that there is no everlasting hell. *By definition, the only thing everlasting is a heavenly state of wholeness.* That is because wholeness is continuity.

Because of the power of the world (and of any whole) to heal itself, it is possible to know rationally and with confidence that the unnatural points of incoherence and separation within us (and between us and other systems) are being corrected and healed. The darkness ultimately has no power because all power resides in spirit. And it is not merely spirit that has the power to promote wholeness. The power of all three substances—spirit, soul and matter—is directed towards producing wholeness. All that is necessary is to follow the recommendations of another piece of revelation that actually has found expression in several different religions—and that is to follow the Way. Taoists call this the 'Tao,' which means 'Way.' To follow the Way is perceived as being obedient to natural law. (Christians relate the 'Way' to Jesus and to his showing people how to perfectly follow spiritual laws.) It is actually easier to follow natural law than to oppose it. Our resurrection once again to a state of everlasting life will happen sooner or later. It is inevitable. (I personally would prefer it to be sooner rather than later.) It is time to stop resisting a return to the happiness that is naturally ours.

Many people in many religions have tended to see the material world as evil by nature and as something to escape from rather than redeem. We are the world, and nothing could be farther from the truth. We are good, and worthy of redemption. There is a difference between the black wounds that dark, unnatural separation makes, and the type of division that the world is naturally supposed to have. The world is intended to have spatially separated states of affairs, including separate individuals. It is also intended to have qualitatively different states of affairs.

Hierarchy in nature allows for these things and still preserves the essential unity of the universe. There is nothing wrong with the material world. Even relative to the way that it is now, there is nothing that we have not caused, and that we cannot correct. We are powerful spiritual beings. We simply need to utilize the power that we possess. A major part of the ability to do that resides in our capacity to see ourselves in a new and different way—in terms of a spiritual paradigm.

Chapter 8

God

The purpose of this book has been to use the three tools mentioned—philosophy, science, and twelve archetypes—to help people understand who and what they are. The *New Paradigm* is not a book about God. At the same time, if people are creators, this is a power that has been attributed to God. Furthermore, if people have a spirit that is infinite, absolute and eternal, these are characteristics attributed to God. For this reason, it may not be possible to fully explain the nature of people without discussing the nature of God, and any interrelationships that may exist between God and people.

First of all, I will share with you that I do believe that God exists. I also have a growing image in my mind about what he is like. This belief and general image has developed slowly over the years and is still in process. Some of it has come from philosophy, some from the archetypes, and some of it just from my heart. I have mentioned the three roads to knowledge that I use, but I have not thus far spoken of the knowledge of the heart. If we have a spark of spirit at the center of us, intuition is also a guide. Intuition is a 'just knowing' type of thing. As much as I value intuition, I do believe that this spark of the divine is meant to be embodied. In terms of the mental body, that means knowledge—which itself includes reason.

I am a Christian. I was baptized Evangelical as an infant. I was confirmed Lutheran as a teen-ager. When my son was small I taught Sunday school in the Methodist church we belonged to. My beliefs about the character of God began to emerge within these traditional churches.

As far as knowledge of the heart goes, I always knew that God was real. I have always visualized him as a being—not a person of course—just a being.

I have a very strong sense of his reality, and that he hears me, knows me, loves me, and is right 'here.' When I speak to him it is "Dear God" or "Heavenly Father," although I do not see him as male or female.

I cannot imagine what it would be like if I did not have him to confide in. My image of him is changing as I myself grow in knowledge, but in my heart he will always be my Heavenly Father. He is my rock, my guide, and my comfort. Sometimes I get a sense of how great he is, and that humbles me, because he is also so very personal.

This belief in his reality I have always had. The rest of the details about God, and concepts connected to religion, I have had to work out over time—ideas about heaven and hell for instance. I do not believe in hell any more—at least not in the traditional sense. Even as a child I knew that my Father would never sentence one of his children to everlasting pain. Strangely enough I had a problem with heaven too. The idea of being perfect did not appeal to me. I loved school and learning. I knew that learning entailed growth, and growth is about change. It seemed to me that, if I went to heaven and was perfect, I would not change; and that bothered me.

I have never feared God. I have never been afraid that he would be angry if I did something wrong, or if I did not go to church, or if I visualized his nature incorrectly. I now know this intellectually as well as intuitively. Anger entails friction, and friction is antithetical to the nature of Unity. My thoughts have led me to believe that God is pure Being, pure Unity, pure Love and pure Goodness. Although the correlations between these things entail philosophy, somehow I have always known in my heart that he is totally loving and good. This image is different from the angry or vengeful God that has been depicted in some of our religions.

I was taught that God is infinite and absolute, and I believe that—infinite and absolute in power as well as in being. I believe that he is omnipotent and omniscient—that he can do anything, and that he sees and knows everything. I believe he is the one being who really knows me (in this gray world, better than I know myself).

I was taught that God created the world. There was a time when I let him know that I was angry about the way my personal part of the world was created. I no longer blame God. I now know, both deep in my heart and in my intellect, that I am creating myself and my world, and that I am responsible for anything that is wrong with it. I know that I possess the things mentioned in Chapter 3 (expression of pure Spirit), including free will and the power to create.

In a way it is usurping God's domain to say I am a creator; but in a way it is not, because that spark of the divine within me is a piece of God. What I realize is that I am a true child of God—flesh of flesh and blood of blood. The blood I see as the 'life' that courses from out of my heart—the pure absolute aspect of spirit in action. This power of life and creativity I see as at the very essence of who and what I am. I am a piece of my Father, and a microcosm of that macrocosm.

My flesh too expresses the nature of my Father. This entire book has been an effort to reveal the true nature of the material world in its character as infinite, absolute and everlasting. In a similar manner as with my awareness and my life, my flesh is God expressed in me.

I am aware that my bodies are imperfect. But what I have come to realize is that, when the parts of my mental, emotional or physical bodies are incoherent, they lack Unity. Then God is not being expressed in me. Lack of Unity is not God! Any wounds I have are the only things about me that are not God, and thank goodness they are not.

I once wondered why, if God is omnipotent, he did not prevent me from wounding myself (or all of us collectively from wounding our world). Now I know that he does—when we listen—and that the voice is within as well as without. Because he is wise and good, he has created us to be wise and good, and to learn to listen to that spark of intuition and conscience. Most of all he has created us to be self-creators, and in that respect has truly withheld nothing of himself from us.

This image that I now have of God and of my relationship to God is called panentheism. The term was coined by a German philosopher by the name of Karl Krause (1781-1832).[1] Krause believed that it was possible to synthesize two seemingly opposite ways of visualizing God—pantheism and theism.

In pantheism the material world is seen as divine. There are slightly different versions of pantheism. All the creatures on the earth can be visualized as divine. Or, the entire world as a whole can be viewed as a deity. Or both can be seen as true. I believe that the latter is true, i.e., that every living creature is divine, but also that the world as a whole is God, and I will explain why I believe this in a minute.

As you might guess, I now am very open to pantheism. This is partially due to the archetypes. The ones I use for philosophy are the same as existed in the old pantheistic nature-astral religions. The cosmology that can be extrapolated from the archetypes is amazing. I believe the view they provide of creation is a true one.

The Christian, Jewish and Muslim religions are not pantheisms. (Their archangels and twelve 'patriarchs' are the same archetypes as existed in the old pantheisms; but I will leave that for another time and another book.) Christianity, Judaism and the Muslim religion are referred to as 'theisms.' Theism is the view that God is transcendent to the world. I was raised with theism. I was told that God was infinite and eternal, and I believe he is. The reason theists believe God is transcendent to the world is because they believe the world to be relative and finite. There is also the problem of evil. The world has things in it that can be very bad. In theism God is seen as pure Goodness—the source of all that is good and only what is good. I also believe this.

Unlike pantheism, in panentheism there is something about God that is transcendent to the world. That view corresponds to theism. We now know that the quantum potential is involved with awareness, and that it appears to be at base of the world experience—the source of all the energy-matter that exists in our world. This is a field of infinite potentials. I consider God to be pure Being and the source of all that exists. Because of all these characteristics combined, it is natural to correlate this field to God. I believe it is God. It cannot be outside God and exist, since God is being, and it really seems to entail pure infinite, absolute Being, not to mention the awareness aspect. Because I correlate this field to God, and because there is no system that expresses the substrate in its totality, I believe that God is transcendent to the world.

Nevertheless, I also believe that a piece of this substrate exists in me. In fact, the archetypes indicate that there are two kinds of pieces within me. One is my awareness, my will, and other expressions of pure spirit as discussed in Chapter 3 on '**Spirit**.' The other is the set of potentials that make me an individual. That I have spoken of as my wholeness and my soul. In Chapter 5 on '**Soul**,' it was explained why wholeness is infinite, absolute and everlasting.

Furthermore, I believe that the parts of my body, when they are coherent, reflect and express the set of potentials making up my soul. In other words, they are just as much expressions of the infinite, absolute, eternal substrate as my soul is.

Furthermore, and this too I have shared previously, I believe that every system in the material world is living, precisely because all systems are centered. In addition to possessing the spirit of life, as individuals they are realizing a specific set of certain potentials. This is their soul.

For this reason, I am a panentheist. I believe that everything created is a living, organic system and is a piece of God—is in fact a microcosm of that macrocosm.

By visualizing systems as a piece of God, I do not mean that they just exist within the substrate, although that too. In the substrate they are non-individuated and non-differentiated. I believe that we have an individual body and soul and that, we are a true microcosm of the macrocosm. In order to be a microcosm of the macrocosm, the macrocosm must be a differentiated individual. I believe that God also has a body and soul, just like we do. I also believe that lesser beings like us exist as parts within his body. I will try to explain why I believe this.

(1) First of all, God is inherently a creator. Creation is integral to the nature of spririt. The material world entails creating systems with bodies. (All creation is 'material' creation. Anything differentiated is material.)

(2) God does not create what is not real. Being does not create non-being. God did not create us not-to-be. He created us to be. We are real, and reality does not reside merely in the substrate. As individuals we are real. Our wholeness (our soul) and our body (our matter) are both real.

(3) Being is Unity. These two are so interdependent that it is not even possible for something to be or exist without being unified. That includes the material universe. (It would not express the nature of spirit as Being-Unity if it were not; and it does express the nature of spirit.)

(4) The unification of the material world has to be different from that of the substrate because the substrate is undifferentiated, and a material world by its very nature is differentiated. The only way that differentiated things can possess both differentiated being and differentiated unity is by being parts of a whole. We use the term 'parts' and 'wholes' frequently and commonly, but do not think about what they really mean. Whenever a group of things are working together coherently, they are parts of a whole. Material wholeness allows differenti-

ation to exist within it by means of its own very special type of unity. The substrate as undifferentiated unity does not permit this. This thing called a 'part-to-whole relationship' is the only type of unity that allows different things to be both differentiated and unified. This type of unity is the foundation of material creation.

(5) It is the soul-to-matter, part-to-whole relationship that produces order. Each law of nature is different. These are the parts of us and of our universe. It is precisley because parts are parts of a whole that order exists. The universe is a whole, and it is lawful and orderly. There would be no order in the universe if it were not a material whole; and wholeness presupposes parts just as parts presuppose wholeness.

(6) What exists but is material and hence differentiated must be *totally* unified to be real. In a dynamic, living world where there are both individuals with parts and individuals who are parts within an external world the processes engendered by part-whole duality lead to hierarchy in nature. By looking at a living body it is possible to see that it is unified through and through, with the highest level in the structure unifying all parts below (precisely because everything below is then a *part* of that whole). This can be seen also by looking at the spatial dimensions. Our cube world unifies surfaces, lines, and points. The highest level unifies all those below as parts within its higher-level wholeness.

This through-and-through type of material unity stabilizes everything. There is an objective reality when everything has its place within the whole. This objectivity is a type of material absoluteness, as was discussed in the chapter on '**Soul**. The over-soul of the universe stabilizes, preserves, protects and unifies everything.

(7) What is true of being is also true of truth and goodness. Whether in ontology, epistemology or ethics, the states of affairs they deal with—being, truth and goodness—cannot exist without being objectified via wholeness. Just as there are universal laws

of nature, there is a universal truth of the matter. There are different perspectives on the truth, but these are objectified within the whole so that they have their place. Truth is not relative; nor is goodness.

(8) The universe at the very top must be a whole rather than an external world composed of parts interacting with other parts, precisely because wholeness is over part-ness, and because wholeness gives being and unity.

(9) The highest level whole is the universe, and as a 'whole' it is an individual, although I unconsciously find myself referring to it as the universe. (I too was born into our current paradigm wherein external worlds are seen as prior to individuals.) The 'universe' I am speaking of here would not be our world, nor the many worlds-within-worlds that exist, but rather the most-high 'world.' In a hierarchically constructed universe (where wholeness is prior to partness), this would be the individual that encompasses all individuals and all worlds, and in that way objectifies and unifies everything materially—allowing it to be differentiated but unified.

(10) The universe is a living being in possession of a body, soul and spirit. We utilize the term universe because it is known that it is a whole or one ('uni-'). We also call it the 'cosmos' to denote its incredibly orderly nature. All order is organic order. That is because the parts of a system are aspects of pure Being. That pure Being is what 'life' is. The universe is no different from any other system, even an atom. It is a living organism.

(11) As for the degree or quality of the being of such an organism, the archetypes show clearly that the higher on the hierarchy of being an individual exists, the more centralized and complex it is and the higher life form it then possesses, including (in higher dimensions) astronomically higher expressions of will, awareness and love. The universe, as the highest organism that exists, would be the 'highest' spatially or hierarchically, and the 'highest' qualitatively.

(12) It is appropriate to speak of this individual as God, not just because this being expresses the highest states of both pure and material forms of spirit, but also because God is infinite. What is infinite lacks nothing. God would be missing something if he did not encompass all individuals in the universe. And he would also be missing something if he was just spirit and did not have a body and soul as well. Furthermore, there is no point in saying he has a spiritual body rather than a material body, any more than there is in saying 'heaven' for us entails having a spiritual body. First of all, all bodies are material. They entail processes. Mind could not think and a person could not move (even in heaven) if they were not material. Secondly, all bodies are spiritual when they are whole. If God is expressed in creation, and most perfectly as the Most-High, he has a material body.

In summary, I believe that we are in God as the substrate. But we are not individuated there. It is within God as a materially actualized being that we are both materially actualized and materially unified. In that sense, not merely God as the substrate, but also God as the Most-High gives us our being. Every system is both a part and a whole. Not only do they have parts, as a whole they are a part within all levels above them and ultimately within the body of God. God is both no-thing (as the substrate) and everything (as the universe). The material world would not express the Being and Unity of God if it were not materially real and materially unified. God as an individual makes this possible. God truly participates in his creation. He has a personal spirit, a personal soul and a personal body.

These are some of my reasons for believing as I do. I always knew in my heart that God was a being, and that he loved me in a very personal way. I know this now intellectually as well.

It is possible for him to be aware of us and to love us in such a personal way, despite our smallness, because he is the highest being, and because the power of pure spirit (the degree of the awareness and love of a being) is increased as complexity, layers, types of bodies, etc., grow in number and degree of expression. We can be assured intellectually as well as via the heart that God is real, that he really does watch over us, and that he really does love us.

We have been told—precisely because God is infinite and absolute—that as finite creatures we cannot really know if God exists. For the same reason, if he does exist, we have also been told that we cannot really know what God is like and why he does what he does.

That may be true in theism, and in a relativistic, materialistic scientific paradigm, but it is simply not true when the cosmology entails a panentheism. One of the 'ways of knowing' then actually becomes the way of science—looking at the world. A panentheism is a natural theology. A 'natural theology' entails finding God in nature, and that includes utilizing everything science has to offer.

Natural theology is also about reasoning to the existence and nature of God, and that is where philosophy comes in. In the panentheism that emerges with the help of the archetypes, all of the concepts cohere with one another, from the centralization that produces organic wholeness, to the idea of part-whole duality, to the concept of hierarchy in nature, to the greater degree of centralization (and all aspects of pure being) within more highly complex systems. Physics and philosophy belong together. Reasoning does not exclude science, but rather utilizes all that is available.

Furthermore, the mental body is itself a body and is a part of nature. Reason (as with any form of order and coherence) is a material expression of spirit as Unity, Being and the Good. It is good to do our best to reason to the existence and nature of God.

It *is* possible to know about the existence and nature of God. But there is more to the knowing than just 'reasoning to.' There is also the experiential route. And here again there is a paradox reconciled. The world is made so as to be simple as well as complex. We cannot begin to understand the complexity of the Most-High expression of God. God is not a person and creation is not for the sole sake of people. There are bodies actualizing potentials far beyond our own current potentials. There are untold dimensions above us. Those bodies have sizes that are astronomically larger than our own. Bodies can also branch off into different vehicles (e.g., physical, mental and emotional), and we only have three at our current level of evolution. And yet, for all that diversity, the process of creation is itself simple—requiring a body and soul to enclose a piece of the substrate, thereby producing a personal spirit and an individual. The overarching soul of even the Most-High is just like our own. It simply has more potentials being actualized. In Chapter 5 on '**Soul**' there was a section discussing evidence that our world (as a whole, in its wholeness) is a single ray of light—singular spatially and qualitatively. This is true of the highest individual and highest world as well. Furthermore, soul and matter are just different ways of looking at the same thing. Due to hierarchy in nature and the way in which complexity is due to layering, creation is simple as well as complex. In Chapter 5, the cosmos, a photon and the human soul were compared to one another. The underlying simplicity is why the microcosm is like the macrocosm and vice versa. In a reality where there is this microcosm-macrocosm aspect to experience, in order to understand our own composition we can examine a photon, realizing of course that a photon is far simpler because it is only one layer. If we want to understand what the cosmos or God is like, we can come to know even the highest system in terms of ourselves, realizing of course that the complexity of God is far beyond us, but that his simplicity is not.

We can know God because we are God, and not just because we are parts in his body. The identity factor goes deeper than just being parts. Because we are wholes, we are truly created in the image of God. When we demonstrate and actualize compassion and love for instance, we can know what God is like. The more perfect our love, the more whole we are (and our world), and the more God-like we become. When we are truly whole within our own dimension of expression—we are a perfect (and complete) expression of God.

We are not just a piece of God. He made us to be independent beings and independent creators. We are literally gods. We can understand what God is like by taking a good look at ourselves—in accordance with our true potentials of course.

The beautiful thing about panentheism is that it is possible to know both that God exists and what he is like. In theism people are told they simply must have faith that God exists. Relative to his nature they are told that they cannot know the ways of God, that again they must just have faith.

I have faith—I know, deep in my heart that God is real. And even in 'faith' there is a paradox reconciled. Faith is needed because we cannot see higher level bodies. Nor can we see pure spirit. We 'see' with the molecules of our eyes and so can only see our own level of matter. But faith is only part of the answer when we possess the divine faculty of reason, and when everything we can see is an image of God, including other people—and ourselves. When we are a microcosm of the macrocosm, and when even pure spirit is within, we are the proof that God exists, and we are the means to know him.

In addition to faith, reason, and the experience of our own being and nature, our awareness is a pipeline directly to God. We need no one to tell us whatever he is saying to us. We just need to listen with the heart, which is a little different from hearing with the molecules of our ears, but still just as real. It is the same with his love. It *is* possible to feel the love of God.

This book is not about God. It is about people. But, because people are divine, and because they are creators, it is not possible to understand their true nature without recourse to a cosmology that includes God. I believe that panentheism can lead to a better world paradigm—a different kind of cosmology that can help people get a sense of what their world is supposed to be like, and empower them to be the very powerful creators (and the good people) that they truly are, so that collectively it becomes possible to create good nations and a good world.

I also believe that a panentheistic paradigm is in process of being born, but that it is having a difficult time in labor. There are several types of panentheistic cosmologies that are available right now. The problem is they do not have the benefit of the archetypes, and hence do not understand the key role of wholeness in producing material bodies that express the infinite, absolute and perfect nature of souls. Looking at the world does not help much given the sad state of conditions.

The biggest philosophical problem panentheists must deal with is the problem of evil; and a close second is the problem of linking that which is infinite, absolute and eternal to that which is finite and relative. (And that includes the question of the 'division' and the 'limitation' that the 'All' is subject to if there are true individuals within the universe as a whole.)

When the physical universe is seen as too problematical to view it as God, panentheism often wanders into panpsychism—the meaning of which is 'all is mind.' Plato did this with his panentheism. He claimed the material world was unreal. It was something to escape from. The soul was entombed within the body. [Panpsychists are often 'idealists' or 'mentalists.' They believe that mind is the only true reality, and hence that God is just mind (or just soul or spirit).]

Teilhard de Chardin was a panentheist. As a scientist, he wanted to combine science with theology, but he too was unable to overcome the above noted philosophical problems relative to physical matter, and so he also resorted to a panpsychism. Teilhard believed that evolution in our beliefs and ideas would lead to a union of all humanity within a higher level mental world. It was there that human beings would also then be linked to God. As a scientist he saw very clearly that in the physical universe there was a 'biosphere' or network of constantly superiorioring (and uniting) systems; but he could not imagine that human beings might be capable of uniting into a higher dimensional physical reality. Furthermore, people were different from other animals. According to Teilhard, humanity would unite in a much more real mental union exclusive to people—the 'noosphere' or sphere of mind. In addition to being a panpsychism, Teilhard's theology was also anthropocentric. As a paleontologist he saw that 'radial energy' was in non-human living systems, centralizing the physical parts and drawing separated systems to one another. He even talked about atoms being moved by radial energy. It caused them to superiorize and evolve. Per Teilhard, all systems, living and non-living alike, manifested a 'law of complexity' or impetus to become more complex.[2] Teilhard apparently even posited the presence of some sort of cognizance in atoms. But he then made a distinction between awareness and self-awareness, claiming that self-awareness was a prerogative of human beings, and that the entire purpose of creation was for the sake of humanity. This departs from the idea that all systems are in God and hence are all divine. [Per the AC, lower level systems may not have a mind yet, but they are aware of the physical universe—and aware of their own internal parts as well. When there is awareness of internal parts, there is *self*-awareness. (Leo rules both awareness of the self and of the external world; and Leo is a universal law of nature.)]

A major problem in a panpsychism is not just that it is an effort to avoid the problem of physical matter. The panpsychisms available today demonstrate a persistent lack of clarity relative to mental states of affairs. There is a habit of not clearly differentiating between things like mind, soul and awareness (or even 'life'). Plato and Teilhard were both guilty of this. The lack of clarity is not new. The term 'psyche' comes from 'breath' and has been utilized historically to describe variously spirit, mind or soul. (Even life is spoken of as the 'breath of life.') Philosophers who do not clarify these concepts in their own mind tend to think that mind has no spatial attributes and hence is more 'spiritual' or 'soul-like' in that sense. It gives them a way to link something infinite (spirit/soul) with the world via the minds of people. It also gives them a way to philosophically link that which is universal (universal mind/soul) with that which is particular (individual minds/souls). It is believed that mind does not have the individual or qualitative distinctions that matter does. In reality, different ideas are quite distinct. In addition to ideas being distinct, so also are the minds of separate individuals. Each person knows the extent of their own mind—and how difficult it is to get into the mind of somebody else. There are boundaries between individuals in mental as well as in physical realms.

Thanks to the archetypes it was possible, in Chapter 3 on '**Spirit**,' to show that awareness and life were the same creative principle, and (through philosophy) why they must be looked upon as pure expressions of absolute being occurring at the center of systems. It was then possible to differentiate between awareness and mind. Mind is clearly a 'body.' Different thoughts are different. Thinking is a *process* of putting different and separated ideas together. A mind is a locus of many different beliefs, some of them coherent, some incoherent. Thanks also to the archetypes and philosophy it was possible to relate soul to wholeness—the wholeness of both the mental and physical bodies. In the AC the qualities and hence the form of a mental body, and the qualities and form of a physical body, both emanate from the same archetypes, i.e., from manifestations emerging from (one) soul.

In terms of the relationship between life and awareness, Teilhard saw that it was life as radial energy that produced superiorization and consequent hierarchy within the physical world. But then he did not stay with the physical universe. It is more difficult to explain hierarchy in nature in a panpsychism. 'Life' is connected primarily to physical bodies. Because panpsychism has 'consciousness' rather than physical 'life' as its foundation principle, this weakens the interrelatedness of concepts like organicism, evolution, hierarchy, microcosm-macrocosm, etc. 'Organic' concepts are related to the physical parts of *living* systems. It is true that the mind can impose order, but the physical body works hard for order too, so that it might live and flourish. Some panpsychisms have even suggested the existence of a mind/soul hierarchical order. For instance, Tomasso Campanella (1568-1639) believed that souls were arranged in a hierarchical manner. It is the same with worlds as with individuals. The so called 'Neo-Platonic' philosophers (Plotinus, Philo Judaeus, Origin, Proclus, Macrobius, Boethius and others) posited a series of hierarchically arranged spheres or worlds. For Plotinus the highest sphere was the One (pure Being). Next was the sphere of Nous, then the World-Soul. But in order to explain why souls and worlds are arranged in a hierarchical manner it is better to have both 'life' and 'awareness' as foundation principles, so that it is possible to utilize the data available in regards to physical bodies and their order and evolution, and hence see centralization and superiorization as due to a truly universal principle. The strongest position philosophically is to recognize both the mental body and the physical body as realities, with both awareness and life as pure forms of spirit giving both Unity and Being to both. [To recognize the reality of the physical world is a strong position scientifically as well. Science deals with the (real) world of physical matter.]

Part of this mind-soul confusion came about because Plato did not know where to place the archetypes. Although Plato was not strictly an idealist, many philosophers who are idealists believe the archetypes are purely in the mind. In the AC, the archetypes are in the soul. They can exist there as absolutes precisely because the soul is a certain set of potentials and is absolute in character.

The most important reason, though, for blurring the distinctions between soul and mind has to do with the hope that in a panpsychism the problem of evil can be better dealt with. Soul has always been linked to that which is spiritual and good. The thought in panpsychism is that, if mind is soul, then mind is better than physical matter. (Mental illness would not exist if this were true.)

When the mind is used to avoid the problem of evil, this invariably leads to a rejection of physical matter. In Neo-Platonism each lower level realm gets farther away from God and perfection, with the material world at the bottom. For Plato and the Neo-Platonists the material world with all of its problems was a shadow world and only reflected the 'true' world—the realm of the soul.

In addition to trying to solve the absolute-relative problem and the problem of evil, in a panpsychism, mind can also be visualized as more creative than physical matter, and that is important in a panentheism, which usually sees people as creators. [I believe that the mind is a body that evolved after the emotional body, which came after the physical body. As the vehicle of a highly evolved organism, the mind has the capacity to reign over the other bodies (and rein them in—especially the emotional body). Because the bodies interpenetrate, the mind can have a dramatic effect over the other two. But the mind is not the only body, and in fact the state of the physical body also affects the mind. They are parallel, interpenetrating bodies. Furthermore, the 'will' is the ultimate source of power, and the will is a form of pure spirit, as was discussed in Chapter 3. The will does have more power in a more highly evolved vehicle like the mind; but a worm has a will too. More to the point, will is not mind.]

Panpsychism leads away from the 'pan' root of panentheism, which means 'all,' and which sees *everything* as spiritual in essence. The purpose of a good, strong panentheism is to redeem matter—not reject it and shy away from it, as in many panpsychisms (and as in theism and Buddhism).

There are actually some new panentheisms arising that do not run away from the physical universe. These are called 'process' theologies. They include the physical universe, just as it is, within the body of God. The British philosopher Alfred North Whitehead (1861-1947) may be looked upon as the father of modern process theologies. Whitehead wanted to link science and spiritual ideas. I agree with him. If we are to understand who and what we are, it is important to utilize modern science. Physics in particular specializes in the physical universe. Where would we be if science had decided the physical world was not real? The basis of Whitehead's process philosophy is in *Process and Reality*.[3] As the title of his book implies, Whitehead believed that the material world of process was the ultimate reality, rather than some infinite, absolute, unchanging, perfect 'soul' realm as in the Platonic view of reality.

Although there was more philosophy than theology in Whitehead's system of thought, he did include God, since the material universe was, not just nature, but the body of God. He spoke of it as the 'consequent nature' of God. Although process was the ultimate reality, God also had a 'primordial nature' within which there were 'eternal objects' (universals). Hence God was transcendent as well as immanent within the world. [Although Whitehead used the term God to describe aspects of his cosmology, it was actually a graduate assistant of his, Charles Hartshorne (1897-2000) who moved the process philosophy of Whitehead more deeply into theology.]

In terms of describing the 'consequent nature of God,' Whitehead was a true relativist. He believed that the world of process was relative rather than absolute, and finite rather than infinite. He asserted the bi-polar nature of God (a primordial and consequent nature), but he did not explain how the finite world of process arose from out of the infinite, for instance how spatial extent arises from out of no-thing. [Whitehead was a mathematician as well as a philosopher. When visualizing spatial dimensions he proceeded from the top down. He saw cubes as converging to an 'ideal' limit (infinitely 'thin' surfaces), surfaces as converging to an 'ideal' limit (infinitely 'thin' lines), and lines as converging to an 'ideal' limit (non-extended, infinitely 'small' points)]

The panentheism that emerges from the archetypes is an especially strong metaphysical system because it reveals that the finite *is* the infinite. In other words, it is possible to, not just assert the bi-polar nature of God as both infinite and finite, but to explain how the finite world of process arises from out of the infinite. In terms of spatial extent, the AC begins with the reality of the infinite (the point). This is a real thing—not just an 'ideal.' This then creates a line via the point being expressed to an infinite degree. The other dimensions then grow upward from the line to the surface to the cube and beyond. In other words, *we* grow, and we grow upwards. That is what evolution is all about. It is about ever higher wholeness concurrent with ever higher dimensions.

There is so much new data arising in both mathematics and physics that speaks of the reality of the infinite. It is not just an 'ideal' that we imagine or utilize to understand the concept of dimensions. Neither is the concept of that which is absolute. Light may be relative as a particle, but it has a wave-form as well. The speed of light is not relative. This is a clue that energy has two sides to it—a relative-absolute, soul-matter, part-whole, wave-particle duality. Similarly, the size of the world may be finite, but the space-time continuum of the world is unending. The physical cosmos has another side to it that is a higher-dimensional soul side. Not merely individual rays of light, but the cosmos as a whole has part-whole duality.

Whitehead spoke of his cosmology as 'the philosophy of

organism.' The astrological cosmology may be spoken of as 'the philosophy of wholeness.' Organic unity is qualitative wholeness, but the much more expressive term 'wholeness' speaks of the part-whole duality of systems. From that base it is possible to reconcile all the paradoxes that arise in philosophy. Matter and soul are the same thing, depending upon the level of experience; and they are integral to one another.

There are many dilemmas and paradoxes that can be overcome once part-whole duality is recognized as an integral aspect of the world experience, including the theological problem of God's perfection. Even if the world possessed nothing in it that was bad, it is a world of process. With process things change. The idea of completeness is linked to perfection. Thus the question arises as to how God can be perfect and complete, and still change. Whitehead accepted the imperfect and incomplete nature of the world without any attempt to reconcile it.

The reconciliation of this paradox was touched upon in Chapter 5 on '**Soul**' when wholeness was correlated to being a world, and in that way to a world being 'complete' (actually to an individual being complete). Today, in modern physics with its concepts of dimensions and many worlds, including worlds within worlds, it is possible to see how creation reconciles this seeming opposition. The soul is a singularity and a higher dimension of light. Higher dimensions beyond our present one do not exist for us. Our material parts reflect their soul. When we are whole, at whatever level of being, we are complete. So is God as the Most-High.

[In addition to natural completeness, any positively directed movement such as superiorization is a material expression of the Good. The archetypes show clearly that everything in experience is positively directed. The qualitative domain, which rules part-whole duality, is prior to and necessary for the existence of the other two. Even ordinary motion-change entails a connection to soul and is a movement into the future and towards betterment. So long as the parts of a whole are not degrading they are superiorizing. In this way everything is constantly expressing the nature of the Good.]

As helpful as it is to see how something can be perfect and complete, and still grow, it is not so much the supposed imperfection of incompleteness, as it is the type of imperfection known as badness and evil that really troubles modern-day panentheists. This is true even of traditional theisms. Theists ask, if God created the world, and God is good, why are there bad things in the world? Panentheists ask, if God is the world, how can there be bad things in the body of God?

For Whitehead, the real world—the world that science studies—has imperfections in it, and the imperfections he saw as a natural part of the world of process. Hartshorne took this idea to its ultimate logical theological conclusion. If God is relative, finite, changing and imperfect, then his body is imperfect. Hartshorne creates a picture of God as unable to alleviate the suffering in the world (*Omnipotence and Other Theological Mistakes*).[4] Per Hartshorne, God loves us but he suffers too. (This stands to reason if he is perceived as having wounds in his body.) According to Hartshorne, our consolation is that he does love us, and that he is a fellow sufferer. He sympathizes with our state because it is his state as well.

The AC differs with this view. Not in terms of sympathy or love, but in terms of the body of God. The archetypes indicate that matter is a set of relationships, and specifically relationships at a particular level of being. (Quantum physics also now visualizes 'matter' as dynamic interrelationships.) Per the AC, the character of each level is up to the systems at that level. In other words, my matter is how I put things together. And yes these 'things' are composed of atoms, but my matter is the relationship between the atoms, not the atoms themselves. There is nothing wrong with atoms when I do something wrong, or with the rest of the universe. (Even relative to us, God as the Most-High utilizes as *his* matter, the *souls* of those beneath Him, just as we use the soul or wholeness of an atom for our matter. There is nothing wrong with our soul and nothing wrong with the body of God.) Each level is independent *as a whole*. Our body is *our* body. Our world is *our* world.

In our wholeness, we are an *independent* creator. This is what the concept of levels of wholeness indicates. This idea of wholeness, and levels of being, tied in with a concept of will, reconciles the paradoxes in panentheism that have existed to date. The archetype called Leo indicates that there is a spark of pure spirit at the center of the body, and that this is the will. It is a creative principle. Each person is the creator of their own body and (collectively) of their external world. [In addition to levels of personal ownership and responsibility, the notion of the soul as a probability field (and as the source of character and circumstances upon reincarnating) really brings home the concept of utter and total personal responsibility.] On a par with awareness, our creative free will is the greatest gift we have. If we did not have it, we would be a machine. A mechanistic paradigm does not allow for the existence of free will. It encourages people to see themselves as victims. An organic, microcosm-macrocosm paradigm on the other hand allows and requires people to be self-creators. But what having a free will means is that we are truly free, even to using the laws of nature in an incoherent manner. Spirit (the will) is prior to any material circumstance. The laws of nature (the types of being) are all types of unity intended to express wholeness. But because the will is stronger than law, law can be violated.

It was not merely the old mechanistic paradigm that created a stumbling block relative to personal creativity. The role of God in creation presented a dilemma in theism as well. It has even presented a problem for panentheists. Teilhard used the concept of self-awareness as a way to explain mankind's emerging *partic-*

ipation in creation, but he did not utilize his own knowledge of centralization, superiorization, and subsequent hierarchy in nature to help him understand about levels of responsibility. As a priest he had to be careful about visualizing people as (absolute) creators, since that was supposed to be a prerogative of God as creator. For people to create would make of them a god. Thus he spoke of co-creation with God.

[The reconciliation of this paradox resides in the fact that both are true. We are creators but it is God within that is the creator. Of course, what is not of Unity is not of God. We lose a part of our divinity when we are not good.]

When trying to decide if people are free, not merely God's role, but also that of the external world must be considered. When trying to determine if people were truly free creators, Whitehead did in fact see every bit of process or 'actual occasion' as a creative act. The problem is, without a notion of part-whole duality, he could not make a clear distinction between being independent (and hence absolute) and being relative (and hence dependent). This makes it questionable as to how free people really are. In regards to the question of people doing evil Whitehead's answer was that people are not 'purely' free and are subject to the rest of the universe. But to say people are both is not to explain how and why they are both, or which is prior—especially when relativity is emphasized as much as it is in Whitehead's process theology.

In the AC, people are both parts and wholes. Furthermore, wholeness is prior to part-ness, and being an independent self-creator is prior to being affected by other things. We do not have to be tossed about by the world. We are responsible for what we do, think, say and feel—and ultimately for what we are, including the external circumstances we find ourselves in.

Although he struggled with the relative nature of freedom, Whitehead was a true relativist in regards to the problem of evil, claiming that what appeared to be evil, might not really be so. He claimed there could be a bigger picture that we are not aware of. This latter idea is tantamount to saying that one person's evil is another person's good, that good and evil are relative (even blending into one another), and ultimately that there is no such thing in the world as 'the' good—real goodness. The archetypes indicate that such a purely relativist view is not the ultimate truth. The world is literally created from the Good. We may have to choose the lesser of two evils in our world right now due to the overall lack of coherence, but the ways in which we construct ourselves and our world can be totally coherent, and will be totally good again someday, simply because to Be is to be Good, and because Being is a power as well as a substance. Ultimate power resides in the Good.

The best that panentheistic philosophies have been able to do to date is to deny the ultimate existence of evil (e.g., via panpsychism) or, even worse, to accept evil (and see it as a relative thing in a relative world). These are not just problems for philosophers. The problem of evil affects us deeply even when the effects come through philosophers trying to figure this out. If the Neo-Platonic philosophers (not to mention Plato himself) had not been mystified as to how pure Goodness (God) could be in our not-so-good world, Christianity might have evolved into a panentheism. There is no way to judge two thousand years later what our world might have been like if we had recognized sooner our own divine nature and been able to see it in one another. I personally think we would have treated one another (and all the parts within our earthly ecosystem) much better, and would have tried much harder to be who we are—to be good. And I think we would have better accepted responsibility for our own lives instead of blaming God or blaming fate or blaming other people.

It is incredibly important to understand the place and origin of bad things in our world. How can something be eliminated if there is a belief that the foe is invincible and cannot be beaten?

This is a why a new paradigm is needed. And the archetypes that were present within the old pantheisms are precisely the means to a philosophy with the strength to arrive at a more satisfactory panentheism.

It is important that all means for human elevation be utilized as we formulate a new world view—a new paradigm—including philosophy, science and religion. I do not know what part organized religions will play in the cosmology that is emerging. In the old pantheistic nature religions there were temples. In theism there are temples, churches and mosques. It is important to share beliefs with one another, and it is uplifting to worship together. Collective upliftment is exponentially stronger than individual self-upliftment.For that reason I believe that unified panentheism, as a shared way to think about and thank God, will become important; but I wonder if it will just be via informal groups getting together, via new churches, via old churches or what. The 'new thought' churches have panentheistic doctrines. For that reason I joined the Unity Church in my town. Unity Churches are small churches filled with love. They emphasize positive thinking and the divine creative power of an individual mind. Because of this they are doing much good. But their doctrine is also a panpsychism. (They have not jelled into a creed yet, so I am free to be a physical panentheist.) I am still technically a Lutheran and a Methodist too I guess. But when on Christmas Eve I go with relatives to a local Lutheran Church it bothers me to recite the apostle's creed. I just cannot say that I believe this is the one true church. And it is not just that I believe all the various Christian denominations have part of the truth. I think the Muslim and Jewish religions also have part of the truth. I believe we love the same God. But that is not so much the problem as the fact that I am a panentheist, and these three are all theist doctrines.

My beliefs are actually closer to Hinduism. Their description of God within their epic of creation is fascinating. Of the four major world religions it is the only one that is a panentheism. Hinduism is fascinating, not just because it is a panentheism, but also because of the correlations between the gods of Hinduism and the gods of Greece and Rome. (Varuna is equivalent to Uranus, Lakshmi to Venus, Indra to Jupiter, Shani to Saturn, Ganesha to mercury, etc.) These gods are expressions of the archetypes behind experience. Given this, it is no wonder that Hinduism is a panentheism. (Because Hinduism is a panentheism, I suspect that many of the old religions were panentheisms rather than pantheisms.)

Because of all the gods in Hinduism, it has been accused of being polytheistic. A hierarchical panentheism (with microcosm-macrocosm attributes) *is* polytheistic. Not only is the Most-High being God, so is every microcosmic replication of that macrocosm. At the same time, a hierarchical polytheism is also monotheistic. That is because all systems are parts in the body of God.

In a hierarchical panentheism there are many levels of gods. In Christian, Jewish and Muslim tradition there are hierarchies of angels (e.g., thrones, dominions, cherubim, seraphim, archangels, angels, etc.). These are 'angels' because it would be blasphemy in theistic religions to call even perfect spiritual beings 'gods.'In the three theistic religions, the hierarchies of angelic beings live in different level worlds or spheres. (Would it not be interesting if, in addition to being gods, we were also fallen angels needing to find our true angelic nature once again?)

In India people do not worry about calling perfect beings gods rather than archangels or 'cherubim.' (In the pantheistic nature-astral religions one name for the gods was the 'Kerubim.') When the Hindu story of creation is examined, the character and manner of creation follows upon that given by the archetypes used in this book to create a cosmology. In Hinduism God transcendent is called Brahman. He is the eternal substance at base of all creation. He is pure Being but he is not a being, not even God as the Most-High being. Brahman is non-individuated and undifferentiated. It is not possible to come up with any words to describe his nature, and yet he is experienced by people. (In comparison to the principles in the AC, Brahman would be the substrate.)

When Brahman takes on qualities and creates a universe he is called Brahma (variously Saguna Brahman, i.e., 'Brahman with qualities'). He is then God as a being. Brahma is creator of the universe. His title, Prajapati, means 'Lord of Creation.' He is the first created being and is himself the universe. Brahma arises from out of Brahman, and he is specifically described as arising from out of a golden seed ('hiranya-garbha'). [The seed of life is often represented as the male principle being planted within the 'womb' of nature. It also correlates to the idea of a 'center' as the tiniest nothing from which all things arise.]

According to the Hindu story of creation, Brahman/Brahma placed the golden seed in the waters of the deep where it germinated becoming Brahma himself. [In many of the old religions the 'deep' or the 'primal ocean' was the name for the substance from which all things arose. In astrology the substrate is represented by Pisces. The planet that correlates to Pisces is Neptune. In mythology, Neptune is the god of the sea.]

As the seed is germinating to create Brahma, it is described as dividing into two parts. Sometimes the division is said to be into body (earth) and soul (heaven). Sometimes it is said to be a division into male and female halves.

The other half of Brahma's soul—the love of his life—is Sarasvati (variously 'Brahmi' or 'Vach'). Her name as 'Vach' refers to speech, and she is called the 'Logos.' This is reminiscent of the belief in other religions that God spoke the world into existence. (In Christianity, the Son of God is the 'Logos.' Christian apologists who incline towards Neo-Platonism have correlated the Logos or Cosmic Christ to the World-Soul which Plato spoke of.)

[The idea of a primal pair has the philosophical capacity to make of even the Most-High individual a part as well as a whole, and hence to answer the question of whether the highest world, the universe, is an external world or an individual. (Another dilemma reconciled.) A ray of light has positive and negative poles. This is a primal type of division that is yet a unity, because it is a polarity. Because every soul is a single ray of light with positive and negative poles, the soul may be looked upon as having male and female attributes—a 'twin-soul' manifestation. God manifesting in the world would likewise be positive and negative, male and female. Precisely because of being the Most-High, this individual's 'external' interactions (and their collective or external world) would be the interactions between the twin-souls—their own world—the world of the material Father and Mother of all, relating to one another via their love to create both their own external world and the means for lower-level worlds and beings as well.]

In the Hindu epic of creation, together Brahma and Brahmi are seen as creating the universe. She is connected most closely to the creative-potential aspect of Brahman (as the substrate). She is the wisdom of the deep (as expressed through Pisces). The actual beginning of creation is in Brahma's hands. It is in the form of hiranya-garbha, the golden seed (after the nature of Leo). This seed is said to be in every living thing. It is Brahman. In the Bhagavad-Gita the Lord says the seed is in us [tat tuam asi ('it is you')], but at the same time paradoxically it is uncatchable.

In addition to the name hiranya-garbha, the portion of Brahman within people is referred to as 'atman.' This is a very important concept in Hinduism and even in Buddhism. When trying to explain what 'atman' is the description is elusive in that it sometimes sounds like a spark of spirit at the center of an individual and sometimes like the soul of an individual. This lack of clarity, however, is no different from the lack of clarification about these two concepts (spirit-versus-soul) in the other world religions as well. The concept of atman brings God, not just into the world, but specifically into people.

Hinduism actually has a triune godhead, just as in Christianity. If there were a comparison between these two religions, you would think it would be that God the Father could be compared to Brahman, God the Son (in his larger life as the Logos) could be comparable to Brahma, and God the Holy Spirit could be comparable to atman. Unfortunately nothing is that simple. In Hinduism the trifold godhead includes Brahma, Vishnu and Siva. Brahma is the creative aspect of God. Siva corresponds to the power of God (comparable to the Scorpio-Pluto archetype, which principle rules 'power' 'change,' and constant 're-creation' or 'regeneration'). Vishnu is very similar to Brahma. He is a being, but much closer to human beings. If there are many worlds, and even worlds within words, Brahma could be described as the God of the universe. He is more distant in character. Vishnu is closer—like the God of the world, and specifically our world. He has great compassion and concern for mankind. When mankind is threatened by an impending deluge or a great catastrophe he is asked to come to earth to help—and he does so. He has been compared to Jesus. He is referred to as the preserver of life.

As true as its depiction of creation may be, Hinduism is really not of much interest to Westerners. Buddhism has actually attracted more attention. Although Hinduism has Brahman as an undifferentiated, indescribable God, it is Buddhism that has best attempted to grapple with the character of God as the substrate. Buddhism is not classified as a religion because it does not personalize God. On the other hand, it does paradoxically have to be classified as a religion because it sees the character of the substrate, not just as infinite, absolute and eternal, but also as pure Awareness as well as pure Being. It is very difficult to connect a substance to 'awareness' and not have a type of religion. Buddhism cannot be classified as a pantheism or panentheism because the material world is not seen as God. Rather the world is seen as something to escape from, and as an illusion. Nevertheless, the 'atman' concept is utilized in Buddhism and is visualized as the spirit at the center of a person. Yoga as a discipline began in Hinduism, but has really blossomed via Buddhism. More and more people in the West are utilizing yoga in order to obtain some sense of their own divinity.

As a practical practice, yoga can be very helpful to people, and I have tried certain forms of it; however, the theoretical base or philosophy behind Buddhism to me deserts matter, just like theism does. As for Hinduism, I have been to the Hindu temple in our town, but I feel out of place. I am a Westerner and a Christian.

Teilhard was a Christian panentheist. As such he has helped bring panentheism into the Christian sphere. And actually, because he was unclear and sometimes inconsistent in his philosophy, there is a certain physical panentheism to Teilhard's cosmology. He was a vitalist as well as an oganicist, and as a biologist and evolutionist he explored the concept of centralization, organic unity, and superiorization within physical bodies. He also was a Catholic priest. Symbolically he connected the 'blood' (life, spirit) of the Eucharist to a vitalist perspective, and the 'bread' (body) of the Eucharist to a physical panentheistic philosophy. Teilhard was a mystic, and in *Hymn on the Universe* he relates a story of a priest who saw the bread of the Eucharist expand to include the entire material universe. It is popularly believed this vision might have been his.[5]

Christianity has always had elements of true physical panentheism in it. There are references in the New Testament to Jesus having a much larger identity. In the book of John (*John* 1:1-3) the Christ is referred to as the 'Logos.' In it John says that the Logos existed before our world, and that everything was created through him. Paul also describes the Christ as a being "within whom we live and move and have our being" (*Acts* 17:28). On the night of his arrest Jesus spoke with his disciples. The depth and intensity of those last words in *John* 14-15 come through to the reader. He told his disciples that he was in the Father and the Father was in him. He told them that he was going to leave them but that, as his disciples, he would send the Truth to them, and that, "At that day ye shall know that I am in my Father, and ye in me, and I in you." (*John* 14:16-20) But he also said on that night, two thousand years ago, that the world in general was not yet ready for that truth.

I would like to see panentheism brought into the Christian sphere of influence. Although Teilhard has helped bring panentheism to Christianity, the inroads have been slow. Even as a twentieth century priest, Teilhard was shipped off to China to work and he was forbidden by the church to publish his books. He fared far better than the Dominican monk, Giordano Bruno (1548-1600). Bruno was a panentheist, and when he refused to recant his convictions in regards to panentheism was excommunicated, jailed for six years, and eventually burned at the stake. The Dominican Tomasso Campanella spent twenty-seven years in prison for being a panentheist.

In an organic, microcosm-macrocosm panentheism, people are not just pieces within the body of God. They are gods themselves. It has been very dangerous to say something like this. When theism conquered pantheism, it became 'blasphemy.' Blasphemy was claiming an attribute that was the prerogative only of God. Blasphemy was seen as an insult to God. It was punishable by death. Jesus pointed out to Pontius Pilate, the Roman prefect who sentenced him to death for saying this, that it was written in the Hebrew law (*John* 10:34, *Psalms* 82:6). Because of the term 'gods' in *Psalms* 82 ("ye are gods"), it is evident that Jesus was not just talking about himself. [We *are* gods, and were created as such by the unbelievable love of a being who kept nothing from us, truly creating us in his image, and even allowing us to be him in our own way.]

In addition to blasphemy, when theism became the dominant

religion politically, 'heresy' also became a serious offense. Heresy is a religious opinion contrary to the codices and doctrines established by the authorities of a church. When establishing Christianity as a religion, groups of men met together (e.g., the council of Nicaea) and decided upon the correct way to image God. This included the all-important question of whether Christians were henceforth to image God as transcendent to or immanent within the world. As this relates to the 'Logos' concept (as the Cosmic Christ) it was the question of whether the Logos was God, whether the Logos was the soul of the world, the body of the world, just the means to the creation of the world, or what—and of course this included the question of how Jesus was related to the Cosmic Christ.

The primary objections that pantheism and panentheism have raised in the past were not just due to the apparent evil nature of the material world, and the idea that God (if his nature included matter) might have something bad in him. Nor were they just due to the fact that God was visualized as infinite, absolute and eternal, while matter was seen as finite, relative and limited. The real problem has always centered around the concept of 'worship'—the idea that we should worship only God, and that it is a 'mortal' sin ('subject to bring death') to worship any god but God. Included in the idea of worship is, not just acts of veneration practiced in a particular church, mosque or synagogue, but more importantly, 'belief'—the idea that it is a mortal sin to believe in and hence mentally perceive God in an incorrect manner.

It certainly has brought death to millions of men, women and children to not believe 'correctly.' Theism (witness the inquisition) has been an especially cruel dominant belief. But given how beliefs cycle, perhaps pantheists were just as bad in the far distant past. Perhaps they destroyed any theists they encountered. And yet I do not see how pantheists could be quite as cruel, since proponents of pantheism would have to know that they were destroying parts of the body of God.

What is nice about panentheism is that it reconciles the extremes present in pantheism versus theism, and the truth generally does just that. But even if this battle between immanence and transcendence continues, it is time to question the 'battle' concept—the idea that God wants us to kill (and torture) people who do not image him correctly. I do believe that God wants us to image him in truth, but this idea of an incorrect image being heresy or blasphemy to God or a danger to us—a mortal sin; we may need to really rethink ideas like that, especially given the horrible things (imposed upon the body of God) that these concepts have led to. We do not worship God by killing off pieces of God.

As for doing this to 'save' people, there are two senses of 'save.' One has to do with the healing of wounds in ourselves and our collectives. The idea, on the other hand, of being in mortal danger due to imaging God incorrectly is a different sense of save that people have used to justify the use of even torture for the sins of blasphemy and heresy.

What the archetypes indicate is that God is literally Love. No God like this would want people killed for imaging him wrong and thus 'insulting' him. Nor would he want cells in his body subjected to everlasting pain in hell for insulting him. We are the ones who came up with all this business about heresy and blasphemy—and the right name and the right way to worship.

I believe in the first (healing of wounds) sense of 'save.' In terms of the other sense of 'save,' I consider myself to be a Christian; but I do not believe anyone has to be a Christian (much less a particular kind of Christian within a particular church) to be saved. Nor do I think that it is necessary to believe that Jesus was the Christ in order to 'go to heaven.' I believe heaven like hell is a state of being—and specifically a state of *our* being. Because of the levels and dimensions, there are many levels of heaven; and (since individuals are prior to external worlds) as Jesus pointed out, the kingdom of heaven truly is within. Our external world mirrors our own state. When we and our loved ones are whole, that is paradise. It is a state of individual and collective being. When we are not whole, that is hell. Hell is a place where there is sorrow and pain. It too is right here. We get ourselves into these un-wholly places through wrong thoughts, feelings and acts. God does not put us in hell (and especially not for imaging him 'incorrectly'). God is Love; and he loves each of us even more than we love our own children. He would never dole out pain, much less everlasting pain.

God is not the source of pain, suffering and hell; but the spirit of God, what the churches have referred to as the Holy Spirit, is the source of healing. Spirit is the source of all power. All unwholesome conditions (any and all hells) are temporary, not everlasting. Only real being is eternal. The darkness involves wounds, and these will be healed. Jesus showed the way to heaven. It can be achieved through the healing of the wounds within ourselves and our collectives. This requires love, goodness, and work, not just correct imaging, although the latter certainly helps with the work.

I do not mind sharing with you my own thoughts in regards to Jesus, and why I call myself a Christian. It is because I tend to think that, although he took on the body of a person, he was not just a person. My primary reason for believing this is because of the things he said and did, the simplicity and depth of the philosophy behind his words, the character that shines through with such majesty and with such pure goodness, and also because everything that he did during his ministry had layers and layers of (larger-than-life) symbolism, as well as being at the very center of everything.

Relative to layers, the bread and wine of the Eucharist is so aptly representative of matter and spirit. The host was and was not a *symbol* of his body. He said it *was* his body. Or consider his role as a healer. It is through the healing of wounds that we get to heaven. Particularly meaningful was his constant healing of the blind. In terms of knowing who and what we are, mental as well as physical blindness is a serious issue. These are just a few examples. The point is that everything he said and did has these layers that, for me, grow as I grow in my own understanding. They are incredibly alive. Truth is Being. What has being is organic and grows, especially when it is an expression of the Good, which is Being and is Life. The words he spoke—they are the words of Life and of Truth.

Relative to centers, I personally suspect that Jerusalem may be the center of our world-of-living-creatures, or at least a group center for humanity. Where the center is, matter becomes more and more compressed, including the state of that matter (and that includes the state of the people and society at that center). When things are not as they should be, even in a group or whole-in-the-making, they become more so as the center is approached. What I have come to believe is that Jesus possessed no wounds (of his own making) and was the one person on earth that did not deserve to be in the center or midst of all that is wrong with us. What I think is that, somehow, to help us, he subjected himself to what is at the heart of our problems. I think that he, not only received wounds that were not of his doing, but also that his life and his (physical, bodily) resurrection to everlasting life was a way-showing, in other words, showing us the way to live, the way to heal our wounds, and the way to get back to a state of (physical, bodily) everlasting life. The symbol of the cross is an ancient symbol for matter. I would be willing to bet that his cross was placed at the center of the earth. In addition to matter being symbolized by a cross, the time of Jesus' death (and more importantly of his resurrection) occurred at three 'crossing' points—the beginning of the natural day (sunrise), the beginning of the natural year (the vernal equinox), and the beginning of the 'great year' of 25,630 years (reference Chapter 9). These things were no accident.

There were layers of meaning in both the life and the death of Jesus, even to the charge of blasphemy that led to his death. At the end, the insults his body went through—being spat upon, ridiculed, being beaten and crucified—and the very fact that he was tortured and killed on a charge of blasphemy, may be because the inquisition, torture and death that charges of blasphemy have led to, entail one of the biggest ongoing forms of blindness and sin the human race has ever drenched itself in.

Given all the fighting and killing that has taken place over which God to worship by which name and with what image, the word 'worship' should perhaps be changed to mean 'revere.' Reverence entails awe mixed in with gratitude—with the awe involving an appreciation of this incredible being that we cannot begin to image, and yet that we can image. There is awe in just getting a tiny glimpse at the majesty of creation, and the inexpressible power at base of all this. There is awe at the fact that love is that power and that basis. What if it had been hatred? But by definition it never could be, because love is Unity—the unifying power that gives the types of being their organic character and their very being.

Reverence infers gratitude as well as awe, and not just gratitude at being created of love, but gratitude at being loved, and in such a personal way. We cannot begin to imagine what the Most-High expression of God in his fullness is like, but we can know that the degree of love and awareness must be so high and powerful that it is capable of being aware of us personally and loving us personally. That we can know, and that is where humility somehow creeps into the reverence—that and the desire, from gratitude, to try to give love back. And not just love to the Most-High or to the transcendent Father—but to every cell in the body of God. Jesus said that the law above all laws was to love God with all of our strength, and to love each other as ourselves. Not only are all systems parts of God, in an organic panentheism (that also speaks of a future hierarchical concrescence) other living beings *are* our selves.

'Worship' is about love as well as appreciation. It is not possible to love and honor God, and simultaneously treat pieces of his being badly—from ourselves to others. It is hypocrisy to say we love God, and not do our utmost to manifest Love, Unity, Goodness and Being.

I have no doubt that God wants us to know what and who he is—but not so much for his sake as for ours. I think we can do better with how we image God—and ourselves. We are divine beings. We need to see ourselves much more clearly so that we can begin to live up to our true potential. In the process of seeing ourselves as composed of this spiritual substance, which *is* God, a different image of God—a different belief—does arise. But each person must decide for themselves how to image God. To share ideas is one thing; to impose ideas upon one another is something else. Down that path lies the way of eventually killing God in the name of God.

This now completes the chapter on God, and it also completes the eight chapters dedicated to the laying down of a new paradigm. Although it has ended with God, the new paradigm is not about God or about imaging God correctly. It is about seeing who we are.

We are gods, and not just we ourselves. Every creature inhabiting the earth is a piece of God. Each has the potential (in its wholeness) of being absolute, infinite and everlasting. Its body was made to reflect these attributes, and to be immortal just like its soul.

Until we respect ourselves, other human beings, and every living creature in accordance with who and what they really are, nothing will change for any of us.

We are brighter, and more evolved than many creatures. This makes it possible for us to be a leader within our particular world-of-living creatures. To be a leader is to serve, not subjugate. We have the intelligence to show the way, and to lead the way. And the Way is nothing less than the path back to everlasting, pain-free life and living. It is our own happiness that is waiting for us. We are powerful spiritual beings. We are creators, and we are good. We are a microcosm of the macrocosm. We love God when we manifest his love and goodness. The kingdom of heaven is within, and it is within our reach. By knowing who we are, and being who we are, together we can transform ourselves and our collective world.

—Sharon Miller, www.newparadigmforlife.com
info@newparadigmforlife.com

All of the ideas in this paradigm are of course just a hypothesis, as even our best scientific cosmologies are. But I believe the model to be a valid one. Finding truth entails a combination of using reason plus any empirical data from observation that can be used to anchor understanding in. The empiricism utilized in the AC has included, not merely thinking about some important new ideas arising in the sciences, but also examining things that we personally experience, like our will.

Concurrent with reason and empiricism, the cosmology that was constructed has utilized the twelve archetypes to be found in astrology to guide the sequence of ideas that have evolved. **Part II** to follow is actually more of an appendix than a part of the cosmology. **Part II** deals strictly with astrology as a practice and science in and of itself. The reason I am including it is because I have studied astrology, and I know that this strange 'pseudo-science' works. Part of me would like to see astrology investigated scientifically. I say 'part' because everything in our gray world has both good and bad possibilities, and the higher the potential good that resides in something (especially something that is powerful), the higher the potential for misuse exists as well. Given our level of evolution, sometimes I think it would be best if astrology continues to be ignored; because it does involve power, and not just through understanding situations and people, and timing events. It can reveal things about the laws of nature, and we have abused this knowledge before in the past. Despite the risks, and precisely because the human race and its future on the earth is at a tipping point, it may be time for people to have this tool and see what they will do with it.

The real reason I would like to see astrology studied though is not to advance science, not even astrology as a science, but for the realization of a more important goal. I believe that if we truly incorporate a new spiritual paradigm into our belief system we cannot help but change for the better, and in that way change our world. Via astrology the principles at base of this new paradigm can be verified. They can be shown to be real and true. This is a way to show that the cosmology itself is true.

I do not know if astrology actually will be researched. And even if it is, it could take decades to validate its veracity (especially in the face of strong traditional opposition in scientific circles). That is why I took the time to list at the beginning of this book all the normal means we use to find truth. Apart from what is or is not done in regards to astrology, these criteria for truth have been continually used and remarked upon in this thesis—especially coherence, simplicity, comprehensiveness and empiricism. It is the presence of these means to proofs that make me feel so strongly about the truth of this cosmology.

In terms of science itself, right now we are making do with a scientific cosmology that is blooming with new ideas and new information, but which has many inadequacies and inconsistencies. When the ideas of spirit and soul are introduced into that cosmology, it is possible to explain so many more things that we currently cannot explain, and in a simple and coherent manner. As was discussed in Chapters 4 and 5, this cosmology has the strength to point the way towards a sound interpretation of quantum-electrodynamics that can reconcile all the paradoxes that currently plague QED. It also reconciles quantum theory with the theory of relativity. It reconciles the problem of the infinite as wholeness and zero, and hence the problem of renormalization encountered when quantum theory deals with the infinities it encounters in its mathematical computations. Aside from mathematical formulas, this is the real problem entailed in real life with the existence of zero points and wholeness. A philosophy of true wholeness explains how non-local causation can occur. It reconciles electromagnetic theory (as real waves) with quantum theory (in terms of a real wave-form). It reconciles the problem of how a particle can be both a particle and a wave (a part and a whole), depending upon the level of observation and experience. It explains why a particle is either visible or invisible, depending upon whether we see it as a part of our world or not. It explains why there is hierarchy in nature. It explains why electromagnetic forces separate into layers. It explains how complex matter can be created in terms of a few laws of nature. It explains how probability fields register information, and on and on. Most importantly of all, it explains how there can be both unity and division in the world, and hence true individuals such as you and me. And of course there are all the things mentioned in the section on pure spirit that science cannot account for including our awareness, our life, our ability to conceptualize, and our conscience.

An unlimited number of things can be explained once spirit, wholeness and hierarchy are inserted into a cosmology, and the explanations then become simple rather than complex, unlike so many of the scientific theories out today, such as 'superstring theory' which attempts to explain how the world is connected without explaining how the continuity (and extent) of even one of its 'strings' can exist in the first place. Why should we put up with a complex cosmology that is inconsistent, and that is not comprehensive when, if we begin with the two unities of the point-instant and the substrate, everything can be explained in a much simpler and better way? It is like preferring darkness to light, when all that is necessary is to turn around and face the other direction.

It is not merely in science that this cosmology opens new doors, but in philosophy as well. Traditionally philosophers have divided their discipline into the fields of ontology (the study of Being), ethics (the study of good-bad, right-wrong), sociopolitical philosophy (the study of groups), epistemology (the study of knowledge and of truth), and religious philosophy (which involves proofs of the existence of God). These can be unified; and each benefits from a spiritual paradigm.

Relative to ontology, the introduction of a concept of 'spirit' as pure Being helps explain things about the physical world—the *types* of being—that cannot be explained without such a concept. One of the most powerful aspects of this new paradigm resides in the way it differentiates between the two expressions of pure Being—the infinitely small and the infinitely large. Philosophy, science and math thus far just use a general term ('the infinite'), visualizing it vaguely as something infinitely large, whereas it is infinitely small as well. The difference is important, since with adequate discrimination it is then possible to use these two together to explain how the world is made.

Another powerful aspect of this new ontology is the way it links Being to Unity. Although that concept had been considered before by both Aristotle and Plato, the astrological cosmology gets much more specific and down to earth. It posits three types of Being or substances and three types of Unity—spirit, soul and matter. Relative to matter, it shows that there are twelve types of being in the world, that these are concurrently twelve types of unity, and that together they produce three types of wholeness—spatial, qualitative and kinematical wholeness. It also shows how these three types of unity as wholeness are each produced through the pure unity of spirit via its two poles. In other words, it shows how spirit as Unity produces soul as wholeness, which in turn presides over matter as types of unity materially actualizing and expressing the wholeness of soul.

It is the same with ethics as with ontology. In our grey world, systems have dark, disconnected spots. They are not whole, and this brings up the subject of good and bad, right and wrong, which is the domain of ethics in philosophy. Because (via the archetypes) the laws of nature can be seen to be types of unity promoting three types of wholeness, that which is 'good' is then natural (having to do with natural law). The 'good' is that which promotes wholeness in a system, and hence the system's being and unity. That which is 'right' is that which promotes wholeness, not only in a particular system, but also in the various higher-level wholes-in-the-making it participates in, such as the 'molecules' we call families, nations and humanity as a whole. Through the archetypes it is possible to formulate an ethical system based upon the laws of nature (and of course on the freedom of living systems to follow natural law or not).

Ethics is the highest and most important area of philosophy. The real impact of a spirit-based cosmology is not to prove that spirit exists. The importance lies in understanding wholeness and what this has to do with happiness, health, and even the furtherance of life. Philosophy is not, or should not be, a theoretical discipline. It should be capable of helping people, families, nations and the world-of-living-creatures.

It is this that makes of the new paradigm an important contribution in the area of political, social and economic philosophy. This is a cosmology that has elements in it of vitalism and organicism. It not only helps explain what we ourselves must do to stay healthy, whole and alive; it explains what is needed for groups to do so as well. In an organic paradigm each part in either a whole or a whole-in-the-making is an intrinsically valuable, indispensable part.

Sociopolitical (and economic) philosophy has always faced the problem of dealing with the seeming opposition between personal self-interest-versus-social concerns. The astrological cosmology reconciles this opposition (as with so many other seeming oppositions). We do need latitude for self-realization, and to develop our personal, *independent*, inner strength as a whole. But we are also parts in external groups. Furthermore, these are true wholes-in-the-making, and we are that future whole. Other systems are future parts of us. We depend upon each and every living system we encounter, not only for the overall health of our external world-of-living creatures, but also for our own eventual evolution into a higher experience of self.

Another area of philosophy is epistemology. Epistemology is the study of knowledge. One of the things epistemology deals with is conceptualization—with trying to understand, for instance, how we know two red things are red, or how the higher level (Form) of pure Being helps us be cognizant of the types of being—that they are types of being and types of unity—as well as how we connect them. Epistemology is incredibly enhanced through the data available via the astrological category system. That system indicates that the things philosophers have variously called archetypes, Forms or universals do exist. The epistemology of the AC shows that these exist in the soul, and in having a concept of soul, shows clearly how these relate to subjects.

Another powerful aspect of a spiritual epistemology is its capacity to distinguish between the awareness and the mind. This is one of those areas that have not yet been adequately clarified.

Epistemology has to do, not just with how we 'know' things, or with 'awareness,' but also with how we know them to be true. A spiritual epistemology emphasizes the absolutist perspective, making it clear, not only that there is a real truth to be found, but also showing how it can be found. The types of being, when true, must correspond to reality (to real Being). Truth must also correspond to Unity. It is possible to see the truth from different perspectives, but each perspective must fit into the whole. The concept of wholeness allows truth to be seen as related to Unity as well as to Being. In terms of order in the world (and in our minds) when concepts are logical they cohere with one another. It is when we see many different concepts in many different areas all falling into place that we know something is true. That is how I know the AC is true—because the laws of nature it depicts make it possible to create a cosmology that sheds light on all the major areas of science and philosophy.

Another area of philosophy deals with religion, which itself has two concerns—the relationship between moral law and science, moral law and God's laws, etc.—as well as dealing with the question of whether or not there are rational proofs for the existence of God. In the AC, moral law is directly related to the laws of physics, not a different domain. Given that the types of being derive from absolute Being, the laws of physics are spiritual laws and (given the connection between God and spirit) they are God's laws.

As for proofs of the existence of God, materialists have told us that there is no way to know the truth about the reality or nature of God. Under a materialistic paradigm that is an appropriate philosophy. However, when the paradigm changes an entirely different situation emerges. Pantheism (and of course panentheism) is the one theology that has the capacity to be empirically verified. We are the proof both *that* God exists, and the means to empirical verification of what he is like—a small but important example of his nature.

It is possible to create a philosophical system of thought that comprehensively and coherently covers all these areas of philosophy—and by means of the archetypes at base of astrology. I did not make up this philosophical system of thought. I discovered it hidden within the signs. They show how the world is created, and that creation is far more beautiful than me or any philosopher from olden or modern times could ever have concocted or imagined.

Because of all that can be learned from the signs, it amazes me that in such an advanced society the scientific community continues to miss out on what astrology has to offer. The prejudice is so great and has been going on for so long. The possibility of benefits is vast. Even given the *possibility* of such great benefit, this type of continuing response on the part of the scientific community entails close-mindedness to say the least, and apparently band wagon practices as well, i.e., going along with what other scientists say cannot possibly be true so as to be 'in' with the establishment perspective and hence professionally safe.

The most important benefit astrology offers is the validation of a new and very different cosmology. It would be a good thing to establish astrology as a science in and of itself. But it would be far better to validate the truth of a spiritual paradigm. It is a good thing to know that there is an attractive force wherever there are centers—a force that produces gravity and thereby physical planets in the heavens. It is an even better thing to begin to understand what this thing called 'love' consists of which originates in the human heart. It is a good to know that we have a probability field. It is an even better thing to understand that we have a soul, that it really is everlasting, and that it is a Book of Life that records, preserves, and helps create our individual being. It is a good thing to know that the solar system, as a probability field, is what astrology is all about, and that it depicts the wholeness of our collective world-of-living-creatures. It is an even better thing to really understand deep within our heart that each living creature is a part within this whole, and that together we will never be a coherent whole, until we *do* understand this. It is a good thing to know that the signs of the zodiac are physical laws of nature—and that these are twelve specific types of unity producing wholeness. It is an even better thing to understand that we are engaged (in a very personal sense) in trying to establish individual and collective wholeness. Understanding wholeness is the means to understanding what unwholesomeness is. We are not whole, yet we are intended (by natural law) to be whole. Lack of organic wholeness in individual bodies is responsible for pain, sickness and death. Lack of group organic wholeness is responsible for the political, social, economic and ecological crises we are currently facing. These too will bring pain, suffering and death.

If there is, via astrology, one more way to establish the truth of a spiritual paradigm, we need to get to it. There is nothing more important than seeing ourselves and our fellow living creatures as tremendously important pieces in an incredibly beautiful and utterly spiritual material universe.

Part II

The Practice of Astrology

In the history of mankind, people have looked to the skies, and have been fascinated by the 'wandering sheep'—the stars that move and are not fixed. These of course are our planets as opposed to the relatively motionless stars that are the suns in other solar systems. In looking up at the heavens people noticed that the positions of the planets corresponded to certain states of affairs on earth. In this way the science of astrology was apparently born, or at least that is how the practice was preserved.

The second part of this book is devoted to the actual practice of astrology. Only one chart is looked at, and it has a very short interpretation. It is not possible to go deeply into this subject in this particular book precisely because *The New Paradigm* is concerned with the philosophical foundation for a metaphysical system.

In **Part II** of this book, there are two chapters. Chapter 9, **'The Language of Astrology'** picks up where **'The Language of Light'** leaves off in Chapter 2. This chapter explains things like signs, planets, houses and aspects, relating these to phase-frequency differentials within light. In that sense Chapter 9 is still somewhat theoretical in character. In Chapter 10 the chart of a well-known celebrity, Marilyn Monroe, is 'read.' In that chapter some of the human-level personality characteristics related to the signs are given so that the reader can assess the degree to which these correspond to the archetypes as were given in Chapter 2. Hopefully, also, the reader will get an introductory glance into astrology itself.

Chapter 9

The Language of Astrology

Phases and Frequencies

The language of astrology is a very ancient one. It involves the manner in which the character of a group probability field can be read and understood. A probability field is a ray of light in wave-form with 'coloration' potentials. To understand how such a field unites and produces different types of being for the world we live in is not essentially different from understanding how the probability field of a photon produces different kinds of colors. Each color has a different frequency. In astrology, the qualitative component or 'meaning' in each of the twelve basic types of being is expressed spatially through different frequencies and phase interactions.

In a way phases are more basic than frequencies because they involve the potential divisions (and hence potential parts) *within* a single electromagnetic wave. On the other hand, a particular wave of a specific frequency intersects with another wave of a different frequency in terms of a specific phase point within the second. Thus the allowable phases are defined by the allowable frequencies or wave-lengths.

The potential frequencies in light in wave-form entail a continuum of possibilities. It is the same with phases. As with differing frequencies, it is possible to divide a cyclic electromagnetic interaction internally anywhere along its path. In a three dimensional world, there are only twelve basic actualized potentials despite the fact that there is a continuum of ultimate possibilities. The key divisions occur precisely because the matter in our world is three-dimensional. The manner in which waves take up space in a centered body is limited by the nature of space. For instance, as was touched upon in Chapter 4 relative to the atomic level of being, it is electrons that give atoms their primary defining qualitative characteristics. The electrons reside out from the center in orbits called shells. They cannot reside just anywhere. They are standing waves and these can only fit around the atom in certain orbits because of the nature of three dimensional space. The different placements involve jumps in frequencies.

These jumps not only entail different frequencies of standing waves, they also produce different energy amounts concurrently. The fact that there are jumps make scientists look upon these as bundles of energy. This is where the term 'quanta' arises from.

Although quanta are viewed as basically quantitative in character, these particles are looked upon as ultimate qualitative explanatory units as well. It is believed that somehow their character entails the laws of nature and hence produces the qualities of matter. Since light is a wave as well as a particle, frequency and phase differentials within a wave may be looked upon as ultimate qualitative determinates as well. The jumps in definite wave intersection points occur for the same reason as jumps occur in quanta. These two ways of interpreting the basic types of being do not oppose one another. Quanta have geometrical correspondences. (Planck's constant is based upon the relationship between quanta of energy and vibration frequencies.) Because astrology entails an examination of a probability field—light in wave-form—it is more appropriate in this science to look at the fundamental types of being in terms of frequencies and phases.

The Sun and Its Planets

A probability field is a specific set of material *potentials*. The sun and planets above us have to do with a group probability field, one that we are 'parts' within. This group is a set of systems that collectively have a history with one another, and a certain degree of coherence. The systems in our group are evolving towards a higher sense of self, as has been discussed in previous chapters.

That group probability field is composed of three-dimensional, living systems. It is an external field but, even though it is merely a whole-in-the-making, it is manifesting the same number of actualized dimensions as the individuals making it up—three dimensions. Three dimensions utilize twelve basic types of being to exist and be what they are. Within this group probability field there are twelve centers or points where the actualized types of being connect to the archetypes in their probability field. These centers are sources of centrifugal and centripetal forces. The centripetal forces are gravitational in nature. They have attracted atomic-level material, and that has produced our physical sun and planets. This makes it possible to see the centers. The primary center has very strong attractive effects, and has produced the huge mass of material that is our physical sun. The secondary centers have produced the planets.

Astrology and astronomy are integrally related. The difference is that astronomy deals with the planets and sun as physical objects composed of molecules whereas astrology deals with centers in a probability field. Thus whether a physical planet is composed of water (ice) or normal planetary material does not matter. It does not even matter if conditions do not allow for the formation of a planet, nor if a physical planet gets destroyed through a comet or other space debris. The center is still there.

The sun is at the center of the centers, and there are eleven other centers. These produce twelve types of waves. The sun-center produces the over-wave or fundamental, which is an immediate mode of absolute being and unity. This sun center expresses the nature of one of the twelve laws (the Leo law of centralization). The other eleven are as waves within a wave. They have different frequencies in terms of the type of being they represent.

The probability field in the heavens has a causal influence

upon affairs on earth, just like an atomic level probability field does. With any cause there are concurrently first-final, formal and mechanical causal influences. Relative to first-final causation, the centers in the heavens not only unify affairs below, they produce the things that occur. In other words, they entail first as well as final causation. Spirit creates matter as well as unifying it. The centers in a probability field are transposition places where light in wave-form converts into energy. An actual material interaction or relationship is produced in this way. As 'quanta' these are higher level photons. [In addition to different types of photons (different colors), there are different levels of photons as well. Scientists use the generic term 'boson' to refer to a family of photon-like particles. There are mesons in the nuclear field of nuclear interactions, and photons in the atomic-level field of atomic interactions. Both are bosons. A boson or photon is simply a temporary electromagnetic or electromagnetic-like relationship. When a higher level whole acts as a causal agent, there are higher level relationships generated. These are photons.]

The energy for our interactions with other living creatures comes, not only from our center, but also from the center of our group probability field. Actually the energy comes from a particular *type* of center in us and a corresponding center of that *type* in the solar system. We affect the world through the group field of our external world, and it affects us. (Medieval astrologers claimed there were 'cosmic rays' emitted from the planets that affected affairs on earth.)

The difference between levels is of course tremendous. The relative sizes of the electromagnetic vibrations at the various levels can be visualized by looking at the relative sizes of atoms, ourselves, and the solar system. The atoms that we use to construct our parts are 1,750,000,000,000,000 times smaller than we are. As for how we compare to our solar system, if we squeeze all of our parts together, we are only about three *feet* in diameter (depending upon how rotund we are), while the solar system is 7,350,000,000 *miles* in diameter. Although the differences between levels of being are vast, the principles of differentiation taking place in any probability field are the same, as are the interactions between levels of fields as was discussed in Chapter 6. The world is created by light, and in a basically unchanged form. In a microcosm-macrocosm cosmology, every single ray of light in wave-form is basically the same. A group probability field is light in wave-form, and it emits photons to provide energy-matter for the relationships generated within the parts in the world it encompasses.

In addition to first-final causation via centers, the astrological phenomenon also entails formal causation. 'Form' involves the actualization of a set of relationship potentials. No single type of being exists alone, and the character of any state of affairs is determined by all the basic types of being manifesting in a particular type of world.

Because of their differing frequencies the twelve basic types of being emerging from our group field interrelate with one another in different ways at different times. The changing interrelationships between the complete set of frequencies produces a unique quality-of-the-times at any one moment in time. The present has a 'form' in that it involves a plexus or confluence of distinct yet ever converging rivers, each with their specific wave-motions interacting with the other merging tributaries, all combining to produce the spirit-of-the-times—the mighty Mississippi of the temporal present. This is what astrology studies. We can read the spirit-of-the-times because, as noted, visible planets have formed at the involved centers.

The outcome of any current act will be in terms of the field as a whole. It is possible to tell which types of actions will work out well at particular times by studying the set of relationship potentials (the planetary foci) as they are relating to one another (and to us) at any one moment in time. How they are currently interrelating qualifies whatever it is that we do. It is possible to draw the energy to do something at an advantageous time. The creation of any particular relationship is best done when all the parts will facilitate the nature of the involved type of act or state of affairs.

That form also needs to fit well with our form. Despite the fact that we are embedded in the probability field of our external world, we are also individuals. Sometimes the form of what we create externally fits well with our own form; sometimes not. That higher level probability field holds us in its grip as surely as our own individual probability fields do—those that create our bodies. Those two fields constantly interact. That is why our interrelationships with one another so dramatically affect our mental, emotional and physical bodies.

As its name implies, a probability field does not determine what affairs will be like on earth. It merely sets the stage. Even formal causation is dependent upon final causation. It is the will which determines how probabilties will manifest. The will is the means throughwhich we create ourselves and (collectively) our world.

Aspects

Electromagnetic phase interrelationships are the basis of the language of astrology. A term used in astrology for phase interactions between two different frequencies is 'aspect.' When an electromagnetic wave interacts with three-dimensional matter, it will do so in terms of twelve basic segments. The cycling of the wave in three-dimensional experience becomes a circling phenomenon. A circle has 360° to it. The degrees in a circle or cycle are 'parts' so to speak, and because there are only twelve basic parts within our world, the qualitatively meaningful segments are 30° each. In looking at the relationship between two different things (e.g., between two planets in the heavens), these are expressed quantitatively in terms of spatial degree differences called aspects. The relationships or aspects have specific names. For instance, when astrologers say that Venus is in a 'trine' aspect to Mars, they mean that the two are relating in terms of a 120° difference. [The term 'aspect' is used, not only to depict how the phases in the twelve basic wave-forms are relating to one another, but also how they are relating to us at any one moment in time. Thus the idea of phase interrelationships is at base of the concept of 'houses' and 'signs' as well, and we will talk about that in a minute.] The names for the aspects in astrology are the 'conjunction,' which is the fundamental or crossing point which occurs when two things are experienced and seen (by somebody somewhere) as 0° apart. A 'semi-sextile' occurs when two things are 30° apart. This is actually called a '2nd house' semi-sextile, and the reason for that will be evident in the section on '**Houses and Signs**.' The next relationship potential is the (3rd house) 'sextile' which occurs at 60°. Then there is the (4th house) 'square' which occurs at 90°, the (5th house) 'trine' which occurs at 120°, the (6th house) 'quincunx' which occurs at 150°, and the (7th house) 'opposition' which occurs when two things are 180° apart. These (0°, 30°, 60°, 90°, 120°, 150° and 180°) are the seven basic relationship potentials in astrology. The other relationship potentials then repeat the earlier ones (in reverse order) as the circle turns back to its beginning point. When two phase relationships are 210° apart they are quincunx again (called in astrology an 8th house quincunx to differentiate it from a 6th house quincunx). When the phase or angle is at 240° relative to its beginning point it is producing a 9th house trine (to differentiate it from a 5th house trine). When a phase has progressed to 270° it is in 10th house square (as opposed to a 4th house square). When it is at 300° it is in 11th house sextile, when at 330° in 12th house semi-sextile, and when at 360° it is again conjunct the beginning point or 0°.

Although an 'aspect' occurs every 30°, each is different qualitatively, even the two that are the same degree from the starting

(0°) point (e.g., the 90° and 270° aspects). That is why they are called a 4th house square versus a 10th house square. Qualitatively, the former aspect has to do with 4th house matters while the latter has to do with 10th house matters. As can be seen by the use of the term 5th house trine versus 9th house trine, this type of division is the basis of the 'houses' as well. In a chart, the houses progress around the circle in the same way. At the equator, the 1st house is from 0° to 30°, the 2nd house is from 30° to 60°, etc.

Houses and Signs

We are parts within a group. By looking at the solar system from a perspective, we can tell something about what kind of part we are. Astrology looks at the whole from a particular spatial perspective, and then converts that into its qualitative correspondent. This is its particular brand of the 'Fourier' language of wave-forms.

Astrology is not concerned with how the planets relate to the sun (as in astronomy) but with how all the centers in this probability field (including that at the center of the sun) relate to us. We exist in a particular place on the earth. Thus an astrological chart is earth-centered. Even more specifically, a chart is drawn for an exact place on the earth that corresponds to where we are on the earth. Because of this, the view of the solar system used in charts must be even more earth-centered than in pre-Copernican astronomy. All movements and placements are centered, not just upon the earth, but upon a point on the earth.

We are not only a part of the whole; the whole in a way is in us, precisely because we *are* a perspective on the whole. By looking at it from a perspective, we can tell much about ourselves. (This was discussed in the metaphor of the hologram in Chapter 4.)

How we fit into the probability field of our external world depends upon (1) where we are on earth and (2) where the earth is relative to the solar system. We see and experience the relationship between us and the probability field of our external world through a 'frame'—actually two frames of reference. Where the earth is relative to the sun produces the 'signs,' and where we are on the earth relative to the sun produces the 'houses.'

Relative to the former, the earth revolves around the sun and this creates the frame of reference that astrologers call the signs. This frame of reference depicts how the earth as a whole relates to the sun. This set of perspectives (temporally speaking) gives us our experience of a 'year.' The twelve signs are twelve natural divisions of a year.

The signs are relative, and they are determined by a relationship. Although they are divisions of the year they are related to the means for the division of the day. The equator is a line that expresses the base frequency of the rotation of the earth around its axis, producing the 'day.' This daily rotation produces a lower-level frame of reference that we will talk about in a minute—that of houses. Relative to the signs, the beginning of this set (0° Aries) is defined by where the rotation of the earth around the sun intersects with the rotation of the earth on its axis. The rest of the signs (the months of the natural year) then follow upon Aries in 30° (approximately 30 day) increments. These twelve signs are referred to as the tropical zodiac.

When an electromagnetic wave is traveling through space-time, the wave has no 'beginning' or 'end' until it interrelates with (interferes with) something else, and it does so in terms of the point of connection. The point of interaction between different levels or sets of waves defines where the fundamental relates to its twelve-fold parts, and this crossing point produces a 'beginning' and 'end.' Beginnings and ends are relationships. Relationships are what matter is all about. This can be thought of as the creation of time as well. The center of a wave is no-thing and is eternal temporally (without beginning or end). The wave (as a whole) is everlasting (also without beginning or end). Time periods (with beginnings and endings to them) involve relationships. The spin of the earth occurs at the poles, and the 'belt' of the earth (at the equator) is the fundamental of that wave. There is also a 'belt' produced by the earth as it rotates around the sun each year. The point where these two belts intersect defines (and produces) the beginning of the signs.

Matter entails relationships, relationships are relative, and the frame of reference called the signs is hence a purely relativistic one. All living creatures born above the equator at the time of the spring equinox will have their sun depicted in their birth chart as occurring at 0° Aries. All living creatures born below the equator on the same day will have their sun in 0° Libra, the sign opposite Aries. In other words, from their own perspective, what those in northern hemispheres call 0° Libra is 0° Aries for those below the equator. (Spring and fall occur at opposite times, depending upon whether you live above or below the equator.)

[Unlike the *tropical* zodiac, which depicts the relationship of our solar system to the earth, the *sidereal* zodiac of star groupings in the sky called Aries, Taurus, etc., have to do with the relationship of our solar system to the galaxy. The sun-galaxy relationship is a connection between our solar system and a still higher level probability field—that of our galaxy. This interrelationship produces a 'great year' of approximately 25,630 years. The 'months' of the great year are 2,135 years long. These are called 'ages.']

The second frame of reference used in astrology is the 'houses.' The earth is a wave within the over-wave of the sun. As such it is a part of the solar system. But it is a whole in its own right as well. It has phases of its own. These are the parts of the photon in wave-form that is the earth's individual wholeness. The earth manifests two sets of interaction potentials relative to the sun—a yearly circle around the sun (signs) and a daily spin on its axis (houses). In terms of the latter, one complete rotation on its axis is a 'day.' In a three dimensional world these interaction potentials are divided into twelve natural spatiotemporal divisions. Temporally they are twelve 'hours.' (The twenty-four hour division of the day that we currently use is not natural for a three-dimensional world.) In astrology the twelve spatiotemporal phases are called 'houses.'

We live on the earth. We are a part of it. Astrology looks at where we are so as to be able to translate this spatial *where* into a qualitative *what*. The houses in a chart for us depict very specifically where we are on the surface of the earth, both relative to longitude and latitude.

Relative to longitude, an astrological chart draws the first point of the first house (our sunrise) where *our* longitude, which is *our* eastern horizon, exists. Our houses begin where our day begins, at sunrise. At the point where the center of the sun (when rising) exactly bisects our residential center on the earth, this is the point where the first degree of our first house begins. Our other houses follow upon our first house in order.

Our point on the surface of the earth is specific to latitude as well as to longitude. If we lived on the equator each house would entail an equal 30° increment. However, we live at a place that involves a point of latitude north or south of the equator. The houses (called in astrology the Placidus House System) take this into account and vary in the number of degrees in each house in accordance with where on the earth we live.

An astrological chart is relative to a particular point in space and time on the earth, and so each and every person, place or thing on the earth at any one particular moment in time will have a different astrological chart from any and every other person, place or thing. Everything is a different perspective on the whole. In astrology, the notations of 'signs' and 'houses' as frames are ways of taking this into consideration. In a chart, a particular planet will thus be said to be 'in' a particular sign and 'in' a particular house.

The perspective called 'houses' is closer to us than the signs, including via the fact that houses have smaller space-time phases. The houses involve situations very close to our day-to-day affairs. That is why it is so important, when doing astrology, to have the time of the day that a person was born, and not just the day of the month and the month of the year.

Although they are different, houses and signs interrelate, and each has an impact on the whole; thus even the signs are 'in' different houses at different times of the day and in different places on the earth. Both signs and houses must be taken into consideration when looking at how we relate to our group probability field. Even in a physical sense, the spin of the earth on its own axis affects the rotation of the earth around the sun. [This is true of electron spin and rotation as well. Initially, when Schrödinger calculated the amount of energy in electron waves to arrive at his theory of 'wave mechanics,' the energies did not match the waves that atoms demonstrate in their spectral analyses or colors (which in turn have to do with their electron orbits). His calculations were not wrong, but he did not take into account the spin of electrons as well as their orbits. Later, Dirac did so and found that the calculations then exactly matched the spectral lines.]

The Twelve Basic Relationship Potentials (RPs)

All of the aforementioned terms (signs, houses, planets, and aspects) express twelve basic relationship potentials ('RPs') in one way or another. Because there are only twelve basic archetypes, a relationship of a particular type is immediately related to others of the same type. Thus the third house is said to be ruled by the third sign (Gemini), while both are related to the 3rd house sextile (60°) aspect, and all these in turn are related to the center of the planet called Mercury.

When we get into Chapter 10 and actually begin reading a chart, it will be evident that any one of the above ways of expressing relationship potentials is capable of expressing qualitative meaning. It is important to remember this when reading a chart. For instance, in a human being, anger involves an interaction that entails one person rubbing another the wrong way. The action of the interaction is not whole. There is lack of smoothness and some mental-emotional 'sticking' (irritation) in the serial nature of the dynamic relationship between the two. Irritableness and anger are ruled by Aries, which has to do with the holistic, serial nature of actions. In order to understand anger issues in a person, one looks to the Aries RP. In addition to Aries, it is important to also look to Mars and the first house as well. Even the aspect known as the 'conjunction' speaks of this archetype because it is a first house aspect. Signs, planets, houses and aspects are different ways to express the same relationship potential.

Because there are only twelve basic types of being in our three dimensional world, people are in many ways the same, but they are different too—and specifically in terms of the emphases in their particular soul. When a particular trait is emphasized in a personal chart it will be shown in several ways.

[Because a particular chart looks to emphasis, which is a form of 'more-ness,' I find it helpful to look to the RP that rules quantities as well, since it will show if there is a 'lot' of something present. Sagittarius rules quantities. Quantitative 'much-ness' is ruled collectively by the Sagittarius-Jupiter-9^{th}house RP. Thus if a particular person has anger issues, they may have Mars in Sagittarius, Jupiter in Aries, Mars in the 9th house, etc. And if anger is a major problem in their lives, they may have, for instance, both Jupiter in Aries *and* Mars in the 9th house. It is very common for charts to show duplications and triplications of these RPs when they are emphasized in a particular personality.]

Just as astrology expresses the twelve fundamental laws of nature variously as houses, signs, planets and aspects, traditional scientists look for the laws of nature via simple regularities that can manifest in different ways. The fundamental laws of nature are to be found in certain particle regularities, be the 'particles' atoms, electrons or photons. Regularities are a signal that law is involved. The periodic table of the atomic elements, for instance, entails mathematical regularities that speak of the presence of law. These element regularities are directly linked to the qualitative nature of our world. They reflect, among other things, the number of electrons atoms possess, which in turn have much to do with the nature of each atomic element. Electron orbitals or shells entail frequencies.

It is the same with spin as with orbitals. (An electron has a spin as well as an orbital or 'shell' frequency.) Scientists believe that the laws of physics (including presumably Newton's laws of motion) may be found one day in the spin of an electron or photon. The spin of a particle is a phase differential (which in turn is a differing interaction potential). Frequencies and phases are integrally related, as was mentioned previously. The spin of a particle is restricted to values involving small whole numbers—mesons to 0, 1, 2, etc., baryons to 1/2, 3/2, 5/2, etc. Simple ratios (1/1, 2/1, 1/2, 3/2, 5/2, etc.) express simple interrelationships. Math corresponds to geometry. These simple ratios can be looked upon as frequency relationships as well as in terms of spin.

If these ratios express laws of nature, then these simple spatial relationships would appear to have a qualitative component as well as a quantitative-spatial component. Astrology entails the same presumption.

Astrology correlates the domains of kinematics, space and quality so that the qualitative nature of some state of affairs can be determined by its kinematical-spatial characteristics. This chapter has depicted the manner in which the things called signs, houses and planets can be perceived as derived from the spatial attributes of huge electromagnetic wave interactions. The next chapter links the signs, houses and planets to the qualities that human beings possess, and to the types of human-level conditions that they encounter in their day-to-day lives. There are any numbers of questions attendant upon accepting the possibility of this. The question could be asked, for instance, what cycles (or phases), which are spatial and energy realities, have to do with any quality that humans possess or experience. For that matter, in what way do cycles (or phases) have to do with any of the laws of nature that produce the qualities in our atomic-level world, be it the law of inertia, the third law of motion, or the second law of thermodynamics? Scientist must ask themselves, when they state that they hope to find the fundamental laws of nature in, for instance the spin of a particle, what a particular spatial orientation or spatial phase like spin has to do with the quality of experience expressed as, "For every action there is an equal and opposite reaction?" If physicists are going to see the laws of nature in terms of spatial orientation or spin, they should not object when astrologers claim that planetary spin and spatial orbits in the skies have qualitative meaning and a qualitative causal influence.

What makes me so certain that these frequency and phase interactions in the heavens express fundamental laws of physics is because each can be correlated to a fundamental law of nature. What makes me even more certain that these frequency and phase interactions in the skies express fundamental principles is that they are so simple, comprehensive and coherent in their nature. The laws of physics make the world what it is. The signs not only account for the qualities in our human level world, but also for those in the atomic level world. And in doing so, they make it possible to construct a logically powerful cosmology. There have been numerous philosophers who have invented category systems that they believed reflected and expressed the fundamental qualities of the world. None of them hold a candle to the astrological category system.

What makes me even more certain that these frequency and

phase interactions in the skies express fundamental laws of nature is due to the fact that the twelve signs not only justify the new paradigm being constructed, they help explain their own existence. The signs and their character attest to the existence of wholeness and of hierarchy in nature, which in turn attests to the nature of astrology as entailing a group probability field for a whole-in-the-making.

What makes me even more certain that these frequency and phase interactions in the skies express fundamental laws of nature is when I see them in act—when for instance I see the transiting Pluto (which rules 'endings' and has a key word of 'leaving behind-ness') oppose somebody's natal Venus ('pairing'), and I see their marriage break up. After 35 years of looking at charts, I have a hard time saying, "It may be that … ," "It could be that … ," "It appears that." I know the science of astrology is a true one, which means it just waits for a proper scientific explanation and study.

If you have never actually studied and used astrology, you may not be personally convinced of its validity. And if so, as you read the next chapter, you may fall back into a measure of skepticism. That is because you will now be confronted with the language of astrology as it is traditionally encountered in books written about people—their sun sign characteristics and so forth. It is a real jump, philosophically speaking, to go from a discussion of Libra as the third law of motion ("for every action there is an equal and opposite reaction") to speaking of human beings with Libra strong in their birth chart, and describing these people as being very reactive emotionally. The laws that are studied in physics are atomic laws; but the twelve types of being utilized in astrology really are the same laws, and they do control and produce very basic kinds of traits and behaviors in living systems as well.

As the switch is now made to traditional astrology, be skeptical if you will, but remain open as well. It is so important that intelligent people get past the gridlock that several hundred years of an atomistic, mechanistic, materialistic paradigm has produced. Scientists have been wrong about astrology. Astrology works. And when it is recognized as real and true, it will validate the astrological cosmology.

Chapter 10

Reading a Chart

In order to read a chart, it is necessary to first know the symbols that are used for the signs and planets. (The houses are shown by the numbers within the inner circle of the chart.) It is also important to know which RP goes with which, since a particular personality trait can be expressed by any one of four ways of expressing an RP. For instance, the Aries principle corresponds to Mars, which corresponds to the first house and to the conjunction aspect.

SIGNS	HOUSES	PLANETS	ASPECTS	
♈ Aries →	The First House →	♂ Mars →	Conjunction	0°
♉ Taurus →	The Second House →	⊗ The Earth →	Semi-Sextile	30°
♊ Gemini →	The Third House →	☿ Mercury →	Sextile	60°
♋ Cancer →	The Fourth House →	☽ The Moon →	Square	90°
♌ Leo →	The Fifth House →	☉ The Sun →	Trine	120°
♍ Virgo →	The Sixth House →	Vulcan →	Quincunx	150°
♎ Libra →	The Seventh House →	♀ Venus →	Opposition	180°
♏ Scorpio →	The Eighth House →	♇ Pluto →	Quincunx	210°
♐ Sagittarius →	The Ninth House →	♃ Jupiter →	Trine	240°
♑ Capricorn →	The Tenth House →	♄ Saturn →	Square	270°
♒ Aquarius →	The Eleventh House →	♅ Uranus →	Sextile	300°
♓ Pisces →	The Twelfth House →	♆ Neptune →	Semi-Sextile	330°

You will notice that I have named one center (planet) Vulcan. According to what is called the Titius-Bode Law, which predicts the spacing between planets, there should be a planet between Mars and Jupiter. Instead there is an asteroid belt with the largest asteroid named Ceres. When I first learned astrology, my teacher advised me that a planet called Vulcan ruled Virgo. Because of the mythology surrounding both Ceres and Vulcan (as this mythology relates to Virgo), it is possible that a planet was once in this place and was destroyed by a comet or something of that nature. For now I prefer not to use the placement of Ceres, as the case is uncertain. Unlike a planet, it is difficult to know whether or not an asteroid is encasing a center.

My teacher also claimed that the earth was the planet connected to the sign Taurus. This I do accept because the earth is a planet and, where any planet exists, there is a center for one of the twelve principles. . The earth provides for the matter we use for our bodies. The primary attribute of matter is mass (inertia). Taurus rules inertia, so I accept the rulership as well. My teacher said that the correct placement for the earth in a chart, since we live on the earth, was opposite the sun. I have found that placement to be valid through experience.

Below is the birth chart of actress Marilyn Monroe. At her exact moment of birth she became a 'part' within a group—the world-of-living-creatures to which we all belong. By drawing a map of that precise time and place, it is possible to see how, as a part, she related to that larger field. Marilyn's chart is especially helpful as an example of how astrology works because everyone knows what she was like, and the reader can compare the details of her personality with the symbols for the archetypes in her chart—especially those that played a primary role in forming her character. If the reader is not familiar with Marilyn, there are many good biographies available. Those I used are in the bibliography.[1]

Norma Jean Baker (Marilyn Monroe)
(Tropical Zodiac, Placidus House System)
Los Angeles, California
June 1, 1926
9:30 a.m.

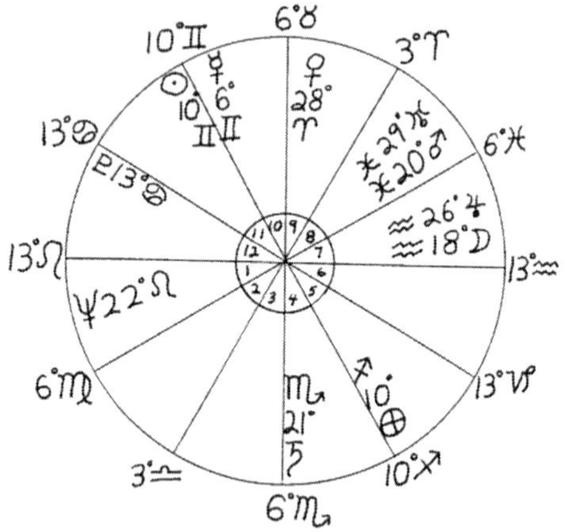

In order to see how the symbols for the RPs express Marilyn's character and life circumstances, it is helpful to have an overview of her life. Marilyn was born June 1, 1926. She was given the name Norma Jean Baker by her mother Gladys Baker. Marilyn did not know who her father was, but believed that he was C. Stanley Gifford, a man who had worked with her mother. The lack of a father did not impact her life as dramatically as the lack of a mother. Right after birth, Marilyn's mother left her in a town called Hawthorne (California) with a babysitter by the name of Ida Bolender. Gladys then went back to work in Hollywood. It is not known how often Gladys came to see Marilyn, but visits apparently depended upon whether Gladys had male attachments and week-end dates in Hollywood at the time or not. Because of this lack of mothering, Marilyn may have suffered from an attachment problem. She liked having people around her, was good at making contacts, and had numerous superficial friendships; but she had trouble with strong, deep attachments.

In addition to an apparent lack of interest in mothering, Gladys also had mental health problems. She finally brought Marilyn to live with her when she was five, but three months later had a psychotic breakdown and was admitted to a mental health institution. Except for brief, intermittent periods, Gladys remained in one mental hospital or another for the remainder of her life. (Della, Marilyn's maternal grandmother also spent most of her life in mental institutions, and died in one at the age of 51.)

After Gladys had a mental break-down, someone in the boarding house where Gladys had been living kept Marilyn for close to a year. Then she lived in foster homes for a while—per her own report about ten different ones. When she was seven, Marilyn went to live in the Los Angeles Orphans Home for two years, after which she was in another foster home for a year. After that, the court appointed a guardian by the name of Grace McKee. Grace and Gladys had been friends. After Grace's husband reportedly hugged Marilyn, and demonstrated evidence of inappropriate behaviors, Grace sent her to live with an unmarried aunt—an Anna Lower. A neighborhood boy by the name of James Dougherty became interested in Marilyn. She married him in 1942 when she was sixteen. He left in 1944 to go overseas when war broke out. Marilyn divorced him in 1947 after she signed a contract with Fox.

Marilyn had been working in a factory during the war. A photographer came to the factory to take pictures of the women doing war-work. He was impressed by Marilyn's good looks, and offered her a job freelance modeling. Later she got a job with a modeling agency in Hollywood. This led to a contract as a stock player with Twentieth Century Fox Studios in 1946. They gave her the name Marilyn Monroe.

It was as a model and stock player that she met a number of influential men who helped her with her career including Johnny Hyde who was vice president of the William Morris Agency.

Marilyn was still involved with Hyde in 1950 when supposedly he helped her get plastic surgery on her nose to make it slightly smaller and more pugged.[2] During her career as an actress she married and divorced baseball player Joe DiMaggio and playwright Arthur Miller.

In 1947 Fox did not renew her contract, but she picked up another one with Columbia Pictures in 1948. After that, as everyone knows, she became a highly successful film actress.

Marilyn herself had mental health problems and suffered from depression. She became addicted to barbiturates and alcohol. She attempted suicide several times; although these were apparently just calls for help, since she notified other people right after taking the substances. She died of an overdose on August 4, 1962. Then too her phone records showed that she attempted to call a number of people but did not make contact and apparently became sleepy and passed out.

When looking at a chart, the two most important RPs are the sign the sun is in and the sign on the ascendant (the cusp of the first house). The sun is important because it shows the ruling archetype in a person's life—the characteristics (and life circumstances) around which everything else centers. (The sun entails the Leo principle of centralization.)

The sun is a clue to the true nature. The ascendant is important because (as the house of the Aries archetype) it rules things that are to-the-fore in a person's life. The ascendant rules the 'face' of a person. It shows how they greet the world. This is the 'persona.' If there is a major difference between the characteristics of the sun sign and the sign on the ascendant (or between the signs and symbols that actually *are* in the first house), then the persona may be a 'front' rather than connected in an integral manner to the true self.

Marilyn had Leo, the sign of the sun, on her ascendant. Leo rules gravity. In people it produces charisma. Leo is usually a very prominent sign in the chart of actors and actresses. Marilyn came across to people as very charismatic.

Marilyn's sun is in Gemini. It is also conjunct Mercury the planet that goes with Gemini. Thus there is a duplication of RPs. The Gemini-Mercury principles rules connections, as was discussed in Chapter 2. Marilyn was good at making connections. Gemini's generally like to be around people. They are very amiable and accommodating. They do tend to be very changeable. (This is known as the sign of the chameleon.) The reason that they are changeable is because they need to adapt in order to make the connections that are important to them. Gemini's will often take on the characteristics of whomever they are around. Being able to adjust herself to the conditions within the (multiple) foster homes she found herself in probably enhanced this characteristic within her nature.

Marilyn's sun is right on the cusp of her 11th house. The 11th house is the house of the Aquarian principle. It thus governs part-whole relationships. It is important to look at this house to see what roles people play as parts within the various wholes they belong to, from their family to their nation. Because Marilyn's sun is there (in Gemini, the sign of adaptation), the playing of roles was primary in her life. Mercury is there as well. As the ruler of her sun, there is as was noted a duplication of RPs. Mercury rules the early childhood years. Marilyn probably began playing roles as a child in order to adapt to the people within whose homes she lived. This would have helped her become a good actress.

Marilyn's sun is trine the 7th house cusp. The 7th house has to do with Libra matters. As was noted, the Leo-sun principle governs charisma. The 7th house governs attractiveness—especially in regards to the opposite sex. Libra governs polar attraction. (The seventh house also rules audiences in general—the 'other' in the way of other people.) Marilyn's 7th house has Aquarius on the cusp, and her sun is on the cusp of the 11th house (which is the natural house of Aquarius). This shows the constant playing of roles with her audiences. Part of Marilyn's attraction was due to being different, and standing out from the crowd. Aquarius rules coherence but also incoherence and incongruence, since it has to do with how all the parts fit within a whole. Marilyn used her natural attractiveness, but she emphasized it in ways (role-playing ways) that were a fascinating contradiction, including combining childishness (Gemini-Mercury) with voluptuous sexuality. Thus she made her nose even more child-like by shortening it, and affected a child-like whisper which was concurrently sexual in character.

In astrology, the sun-Leo principle, the Venus-Libra principle, and the Sagittarian-Jupiter principle are said to depict beneficial things. The Saturn-Capricorn, Neptune-Pisces, and Pluto-Scorpio principles, on the other hand, are often connected to negative things. In actuality, none of these are good or bad. That would be like saying one law of nature is good—another bad. At the same time, in a gray world, some of the signs are more closely connected to unfortunate circumstances. The second law of thermodynamics (Pisces) can drag a person down and be their *self*-undoing so to speak. (Leo, the polar opposite to Pisces, rules the self; Pisces rules loss of individual distinctions, including loss of a sense of self.) As with Pisces, the Pluto-Scorpio principle merely rules change as a leaving-behind; but in a gray world it can be involved with leaving things behind that people want and need. It shows where there is loss and rejection. In a similar manner, the Saturn-Capricorn principle merely rules boundaries, and these do not entail areas of nothing-ness; but in a gray world there can be real lack. This is why Neptune, Pluto and Saturn are sometimes spoken of as 'malefics.'

Marilyn has a series of these so-called malefics involved with her Cancer-Moon-4th house principle. This archetype traditionally has to do with mother and mothering. Cancer rules early nurturing, the home, and belonging. These are human-level expressions of the principle of within-ness ruled by Cancer.

Emotional and physical 'feeding' provide the within that makes the without possible.

In looking at the Cancer-moon-4th house RP, Marilyn has her moon in 4th house square to Saturn. She also has her moon in 8th house quincunx (a Scorpio aspect) to Pluto in Cancer. These are duplications of RPs having to do with rejection, lack and loss in areas of early nurturing. In addition to these things, she also has the sign Scorpio (rejection, loss) right on her 4th house cusp (the house of the moon & mother). In addition, Saturn (lack) is in Scorpio (rejection), and is in the 4th house (mother, nurturing). In addition to all that, her Pluto (rejection) is in Cancer (mother), and is on the cusp of the 12th house (the house of Pisces). [The 12th house is also the house that rules institutions. That is due to the fact that, in institutions, people lose their personal identity.] There are many, many duplications of RPs in Marilyn's chart showing the degree to which her lack of early nurturing and belonging contributed to the person that she became.

Marilyn has Leo on her first house cusp. As was noted, this is the house that rules the persona. If the signs within it are very different, it indicates the existence of a complex personality, and one where the persona may be a 'front.' Because Leo rules charisma, as was noted, many actors are Leos. If they have a very strong personality they may take on strong character roles, precisely because those types of roles match their self-confident temperament. Marilyn, however, has Neptune in the first house (the sign opposite to Leo). Leo's have a strong sense of self. Leo governs self-awareness and the 'I am' or ego. The Neptune-Pisces principle on the other hand has to do with loss of a sense of self, and lack of a true sense of individual value and identity. The fact that Leo and Neptune are both in Marilyn's 1st house would tend to indicate that the Leo aspect of her character was utilized as a persona in the sense of a defense mechanism or front. Marilyn evidently lost her sense of self due to all the roles she played, and all the adaptation she had to manifest as a very young child. This no doubt helped her with later role-playing as an actress. She could forget herself and immerse herself in various roles. But I suspect that it also left her wondering who on earth she really was—that plus the sun in Gemini on the house of roles. Marilyn acted even when she was not acting. She played typical Marilyn roles with reporters, but then she hated it when they treated her like the 'dumb blonde' character she often was type-cast as—and she especially disliked the fact that the studio stereotyped her in this way as well.

Marilyn's 10th house is very interesting. This is said to be the house of career. (That is because it is Capricorn's house, and rules limits per se including 'uppermost limits.' It is usually within a career that people are able to 'scale the heights' so to speak. Marilyn certainly did.) Marilyn has Taurus on her 10th house cusp. Taurus governs mass, and for that reason rules the body per se. Marilyn's body was closely connected to her career. The ruler of her 10th—the earth—is on the 5th house cusp in Marilyn's chart. This is the house of the sun, and rules acting. Her career was acting. The sign Sagittarius is on the 5th house cusp. The keynote of Sagittarius is projection outward. The business of an actor or actress is to project themselves to the audience. (Note also Jupiter in the house of 'audience.') Sagittarius is also the sign of salesmanship. That is because it rules extent or 'much-ness.' Salespeople make their products look larger than life, and hence valuable. Marilyn was a good saleswoman. She sold herself to people. She was called 'The Body' in the trade, and her body was the primary thing that she sold. That is why the earth is there in the 5th house and in the sign Sagittarius. (Marilyn told a friend that she prostituted for a time prior to being a movie star).[3]

Before finishing up with this look into Marilyn's Book-of-Life, there is another duplication of RPs that is very interesting. People have always wondered if Marilyn really killed herself or was killed. The chart indicates that she was her own self-undoing. Mars rules motion as energy. Wherever the Mars-Aries-1st house principle is to be found, there will be a strong impulse to act. The Neptune-Pisces-12th house principle has to do with suicide because it governs self-undoing. The combination of the two principles Pisces and Aries is often to be found in the chart of suicides or people who are very impulsive about acts of self-undoing. Instead of thinking, they just act. Marilyn not only has Neptune in the 1st house, she also has Mars in Pisces in the 8th house, which rules death. She also has Pluto (death) in the 12th house (self-undoing). Marilyn's death was a suicide.

There are many, many more things that can be 'read' in this chart, but this will give the reader an idea of how astrology works. Everyone has all the laws of nature in their character, but the emphases really tell the story.

If you found this necessarily brief introduction to astrology interesting, you might want to study it more in depth, if nothing more than to learn about the archetypes. Astrology will tell you much about yourself as a particular living creature. The archetypes will tell you even more about your overall nature as a spiritual being. The archetypes are truly fascinating. The extant myths utilizing these archetypes go back 5,000 years or more. This is the oldest philosophical category system in existence, and it is still with us in the form of astrology.

Appendix – Zeno's Paradoxes

Zeno of Elea (490-430 BC) was a Greek philosopher. He was a disciple of Parmenides (515-450 BC) and supported his master's belief that 'real' reality was a changeless Absolute. Zeno and Parmenides were true absolutists. I speak of myself as an 'absolutist' philosopher, but at least I believe in the reality of that which is relative. Relationships are relative, and they are real. They just ultimately entail internal relationships involving what is absolute relating to itself.

To not believe in that which is relative is to not believe in the reality of the material universe. The reason Zeno and Parmenides did not believe in the existence of the material world was because of the paradoxes that arise when trying to account for spatial extent, motion, and any zero point reached by material things. Zeno 'solved' the paradox by claiming the material world did not exist. Quantum theory 'solves' the paradox by going the other direction.

The term 'paradox' is actually supposed to refer to something that only seems to entail an opposition. The supposition is that there is a way for the two opposing sides to be reconciled. That is what a philosophy of wholeness (and of spirit) is able to do.

I will give an abbreviated version of two of Zeno's paradoxes. The more complete stories and arguments behind them can be found in Aristotle's *Physics* (226a-b, 238b, 239b, 263a).

Achilles and the Tortoise: Achilles runs much faster than a tortoise. Achilles is behind the tortoise. It makes sense to think that Achilles will pass the tortoise. But to do so he must traverse half the distance between them, then half of that distance, then half of that distance. An infinite number of divisions are required to reach the zero point where he passes the tortoise. Since an infinite number of things cannot be accomplished in a finite time, he can never catch up. (The conclusion of Zeno was that extent did not exist.)

This paradox is applicable to both the spaces between systems and the spatial extent of systems. Relative to the spatial extent between systems, Zeno decided that since the intervening distance must be divided to an infinite degree, then zero point can never be reached by finite systems. He is correct unless no-thing (zero point) exists, and the world is both material and spiritual in nature.

When considering the extent of a single system rather than the space between systems, Zeno reasoned that a finite extent would be a real material thing only if the extent of the parts disappeared in a finite number of subdivisions. Since they do not, they could be lacking in extent. But, if an extent is produced by a finite number of parts lacking magnitude, Zeno concluded that the extent itself could not exist.

[Aristotle says in *Physics* 239b20 "everything continuous is divided into an infinite number of parts." What it was not possible for these philosophers to see without the archetypes (and the mathematics of the continuum) is that matter is the relationship between two expressions of spirit, and that the ultimate 'parts' *are* non-extended. Time too has an aspect that is non-extended. It is what we experience as 'now' and it is zero point in time. No-thing is real. Achilles can 'reach' this infinitely small thing (zero point in space and time) precisely because he is a spiritual being as well as a material being, and even the matter producing the space between systems involves expressions of the absolute.]

The Moving Arrow: This was another of Zeno's paradoxes. Either an arrow is moving or it is not. To be *what* it is (to have self-identity) and to be *where* it is, it must occupy a specific place. If it exists, it is where it is. But if it is moving, it is in a place where it is not, and that does not make sense.

This is a denial of individual 'place' and of individual self-identity as well as of motion. I enjoy Zeno's thoughts in regards to this paradox because it points out the need to include a concept of soul as well as of spirit in order to understand individual motion—i.e., the fact that it is whole, three-dimensional individuals that move. It is easy to see how spirit is involved with motion per se, just as with extent (particularly relative to one simple dimension or line). With both motion and extent it takes an infinite number of point instants to make them serial and continuous. In this paradox, Zeno also brings in the concept of the more common, three-dimensional individual unity and self-identity to be found within the day-to-day world. Per Zeno, anything in motion has to be something and hence has to have, as a minimum, the unity of self-identity (e.g., as an 'arrow'). With that self-identity it will, as a whole, also be in a particular place. If it was moving in the place where it was, it would not be there (or be what it was). But if it is moving, then it cannot occupy a place and have self-identity.

[In the AC, it is possible via a concept of part-whole duality to have both place (self-identity) and also move. We are a whole, but we move as a 'part' of higher level realities. This paradox is present today in the problem of 'complementariety' that quantum physics cannot explain, i.e., the fact that it is not possible to measure both place and movement at the same time. As was mentioned in Chapter 4, part-whole duality reconciles this paradox.]

The Negation of Plurality: Per Zeno, if there were multiple things, then they would have to possess, at the very minimum, unity and self-identity. But whatever exists has parts, and whatever has parts is not a unity. For these reasons, both Parmenides and Zeno believed that only pure absolute Being and Unity was real.

To deny plurality is, not only to deny motion, extent, and all the qualities in the material world, it is to deny the real existence of you and I as well. This type of concept led to the philosophy behind Buddhism. On the other hand, if a cosmology recognizes the existence of the material world, but does not concurrently recognize the real existence of zero point in space and time (and also the part-whole duality of systems) then it too gets into many difficulties. And this is the problem that quantum field theory encounters.

The answer resides in a cosmology that recognizes the existence of spirit with its two poles or expressions of the infinite, and of the soul-matter relationship with its two poles of part-whole duality. And of course the ultimate answer resides in the identity between all these things. Matter and spirit are one. Like the horn of the unicorn, what presents as dual when seen from below is actually one at the tip.

A cosmology that recognizes the existence of spiritual things within the physical universe is a panentheism. A panentheism in turn is like the unicorn. According to medieval tradition, the unicorn symbolized that the Christ was one with God.[1] In a sound panentheism, it also symbolizes that the material universe is one with spirit.

Notes

Preface

1. Lovelock, *Gaia: A New Look at Life on Earth.*
2. Ho, *The Rainbow and the Worm: The Physics of Organisms.*
3. Bohm, *Wholeness and the Implicate Order.*
4. Zukav, *The Dancing Wu-Li Masters.*
5. Capra, *The Tao of Physics.*
6. Goswami, *The Self-Aware Universe.*
7. Davies, *God and the New Physics.*
8. Dalai Lama, *The Universe in a Single Atom: The Convergence of Science and Spirituality.*
9. McFague, *Models of God: Theology for an Ecological, Nuclear Age.*
10. Ruether, *Gaia and God: An Ecofeminist Theology of Earth Healing.*
11. Lipton, *The Biology of Belief.*
12. Borge, *Meeting Jesus Again for the First Time.*
13. Borge, *The God We Never Knew.*
14. Peacocke, *Theology for a Scientific Age: Being and Becoming—Natural, Divine and Human.*
15. Clayton and Peacocke, *In Whom We Live and Move and Have Our Being: Panentheistic Reflections on God's Presence in a Scientific World.*
16. Bracken, *The One in the Many: A Contemporary Reconstruction of the God-World Relationship.*
17. Gregersen, *Encyclopedia of Science and Religion.*
18. Page, *God and the Web of Creation.*
19. Hermann, *Expanding Humanity's Vision of God: New Thoughts on Science and Religion.*
20. Beckwith, *Spiritual Liberation: Fulfilling Your Soul's Potential.*
21. Laszlo and Dennis, *The New Science and Spirituality Reader.*
22. Braden, *The Divine Matrix: Bridging Time, Space, Miracles and Belief.*

23. Chopra, *War of the Worldviews: Science vs. Spirituality.*
24. Dennis and Laszlo, *New Consciousness for a New World.*
25. Russell, *From Science to God: A Physicist's Journey into the Mystery of Consciousness.*
26. Hubbard, *Birth 2012 and Beyond: Humanity's Great Shift to the Age of Conscious Evolution.*
27. Wilson, *Consilience: The Unity of Knowledge.*
28. Whitehead, *Process and Reality.*
29. Hartshorne, *A Natural Theology for Our Time.*
30. Hartshorne, *Beyond Humanism.*
31. Hartshorne, *Omnipotence and Other Theological Mistakes.*
32. Wilson, *Consilience: The Unity of Knowledge*, p. 294.

Chapter 1: A New Paradigm

1. *Webster's New Collegiate Dictionary*, 'astrology.'
2. Ibid, 'soul.'
3. Ibid, 'spirit.'

Chapter 2: The Astrological Category System

1. Davidson, "*New Light on Signs of the Zodiac.*"
2. Teilhard de Chardin, *The Phenomenon of Man.*

Chapter 3: Spirit

1. Barrow, *The Book of Nothing: Vacuums, Voids, and the Latest Ideas about the Origins of the Universe.*
2. Bragdon, *A Primer of Higher Space.*
3. Cantor, *Contributions to the Founding of the Theory of Transfinite Numbers.*
4. Boyer, *A History of Mathematics*, p. 91.
5. Davidson, "New Light on Signs of the Zodiac."
6. See Teilhard de Chardin; see McElwain.
7. *Webster's New Collegiate Dictionary*, 'spirit.'
8. Pribram, "The Neurophysiology of Remembering."
9. Ho, *The Rainbow and the Worm: The Physics of Organisms.*

10. McTaggart, *The Field: The Quest for the Secret Force of the Universe.*

Chapter 4: The Role of Science in the AC

1. McKeon , ed., *The Basic Works of Aristotle*, "On the Soul."
2. Waldman, *Introduction to Light: The Physics of Light, Vision and Colors*, pp. 27-28.
3. Dirac, *The Principles of Quantum Mechanics.*
4. Heisenberg, *Physics and Philosophy*, p. 107.
5. Bohm, *Wholeness and the Implicate Order*, p. 170.
6. Ibid, p. 161.
7. Einstein, Podolsky & Rosen, *Can Quantum-Mechanical Description of Physical Reality be Considered Complete?"*
8. Bohm, *Wholeness and the Implicate Order*, pp. 15-18, 156-57, 170-177.
9. Hiley & Peat, "The Development of David Bohm's Ideas from the Plasma to the Implicate Order."
10. Bohm, *Wholeness and the Implicate Order*, p. 16.
11. Ibid, pp. 237-238.
12. What the archetypes would seem to indicate is that, in a singularity at the center of a centralized system, the curvature of space becomes infinite, ceasing to be a 'curvature' or field (soul) and becoming at that place a point-instant (spirit).
13. Bohm, *Wholeness and the Implicate Order*, pp. 237-238.
14. Bragdon, *A Primer of Higher Space.*

Chapter 5: Soul

1. Davidson, "New Light on Signs of the Zodiac."
2. Teilhard, *Hymn of the Universe.*
3. Sheldrake, *A New Science of Life: The Hypothesis of Morphic Resonance.*

Chapter 6: Power and Causation

1. Ho, *The Rainbow and the Worm: The Physics of Organisms.*
2. Leadbeater, *The Chakras.*
3. Brennan, *Hands of Light: A Guide to Healing through the Human Energy Field.*
4. Davidson, "New Light on Signs of the Zodiac."
5. Leibniz, *The Monadology: An Edition for Students.*
6. Cantor, *Contributions to the Founding of the Theory of Transfinite Numbers.*
7. Bragdon, *A Primer of Higher Space.*
8. Kaufman, *The Portable Nietzsche.*
9. Thompson, *On Growth and Form.*
10. Needham, *Biochemistry and Morphogenesis.*
11. Weis, *Principles of Development.*
12. Goodwin, "A Phase-Shift Model for the Spatial and Temporal Organization of Developing Systems."
13. Sheldrake, *Presence of the Past: Morphic Resonance and the Habits of Nature.*
14. Watson, *Lifetide.*
15. Dewey & Mandino, *Cycles: The Mysterious Forces that Trigger Events.*
16. Luce, *Biological Rhythms in Human and Animal Physiology.*
17. Brown, "Living Clocks."
18. Maddox, "No Patterns yet for Snowflakes."
19. Marshall, "ESP and Memory: A Physical Theory."
20. Einstein, Podolsky & Rosen, "Can Quantum-Mechanical Description of Physical Reality be Considered Complete?"
21. Bohm, *Wholeness and the Implicate Order.*
22. Pibram, *Languages of the Brain.*
23. Wilbur, *The Holographic Paradigm.*
24. Ho, *The Rainbow and the Worm: The Physics of Organisms*, pp. 163-166.
25. Omrani, B., & Leeming, M., *Afghanistan: A Companion and Guide.*

26. Jung, *Synchronicity: An Acausal Connecting Principle*, pp. 111-113.
27. Jeans, *Science and Music*, pp. 25,166.

Chapter 7: Matter

1. Ho, *The Rainbow and the Worm: The Physics of Organisms.*

Chapter 8: God

1. Krause, *The Ideal of Humanity and Universal Federation.*
2. Teilhard, *The Phenomenon of Man.*
3. Whitehead, *Process and Reality.*
4. Hartshorne, *Omnipotence and Other Theological Mistakes.*
5. Teilhard, *Hymn of the Universe.*

Chapter 10: Doing Astrology

1. See Mailer, Mellen, or Summers.
2. Mellen, *Marilyn Monroe*, p. 26.
3. Summers, *Goddess: The Secret Lives of Marilyn Monroe*, p. 18.

Appendix: Zeno's Paradoxes

1. Brewer, *The Dictionary of Phrase and Fable*, p. 1261.

Bibliography & Further Reading

Aharonov, Y; Bohm, D, "Significance of Electromagnetic Potentials in Quantum Theory". *Physical Review*, 1959, **115**: 485–491.

Barbour, Ian G., *Religion and Science*, HarperCollins, New York, 1997.

Barrow, John D., *The Book of Nothing: Vacuums, Voids, and the Latest Ideas about the Origins of the Universe*, Pantheon Books, New York 2000.

Beckwith, Michael, *Spiritual Liberation: Fulfilling Your Soul's Potential*, Beyond Words, Hillsboro, Oregon, 2008.

Bergson, Henri, tr. Arthur Mitchell, *Creative Evolution*, Henry Holt, New York, 1911.

Bible (KJV), commonly known as the Authorized (King James) Version.

Bohn, David, *Quantum Theory*, Dover Publications, New York, 1989.

Bohm, David, *Wholeness and the Implicate Order*, Routledge & Kegan Paul, New York, 1980.

Borg, Marcus J., *Meeting Jesus Again for the First Time*, HarperCollins, New York, 1995.

Borg, Marcus J., *The God We Never Knew*, HarperCollins, New York, 1998.

Boyer, Carl B., *A History of Mathematics*, 2nd ed., John Wiley & Sons, Inc., New York, 1991.

Bracken, Joseph, *The One in the Many: A Contemporary Reconstruction of the God-World Relationship*, Erdmans Publishing Co., Cambridge, 2001.

Braden, Gregg, *The Divine Matrix: Bridging Time, Space, Miracles and Belief*, Hay House, New York, 2007.

Bragdon, Claude, *A Primer of Higher Space (The Fourth dimension)*, Omen Press, Tucson, 1972.

Bragdon, Claude, *Explorations into the Fourth Dimension*, CSA

Press, Lakemont, GA, 1972.

Brennan, Barbara A., *Hands of Light: A Guide to Healing Through the Human Energy Field*, Bantam Books, New York, 1987.

Brewer, E. Cobham, *The Dictionary of Phrase and Fable*, Henry Altemus Co., Philadelphia, 1898.

Brown, Frank A., "Living Clocks," *Science*, vol. 130, no. 338, pp. 1535-1544., 1959.

Cantor, Georg, tr. P.E.B. Jourdain, *Contributions to the Founding of the Theory of Transfinite Numbers*, Open Court Publishing Co., LaSalle, IL, 1952.

Capra, Fritjof, *The Tao of Physics*, Shambhala Publications, Inc., Boston, 2000.

Capra, Fritjof, *The Web of Life*, Anchor Books, New York, 1996.

Chopra, Deepak, *The Third Jesus: The Christ We Cannot Ignore*, Three Rivers Press, New York, 2008.

Chopra, Deepak, and Leonard Mlodinow, *War of the Worldviews: Science vs. Spirituality*, Harmony Books, New York, 2011.

Clayton, Philip, and Arthur Peacock, eds., *In Whom We Live and Move and Have Our Being: Panentheistic Reflections on God's Presence in a Scientific World*, Eerdmans Publishing Co., Grand Rapids, MI, 2004.

Cooper, John W., *Panentheism: The Other God of the Philosophers*, Baker Academics, Grand Rapids, MI, 2006.

Dalai Lama, *The Universe in a Single Atom: The Convergence of Science and Spirituality*, Broadway Books, New York, 2005.

Davidson, William M., "New Light on the Signs of the Zodiac," *Yearbook of the American Federation of Astrologers*, 1950.

Davies, Paul, *God and The New Physics*, Touchstone, New York, 1984.

Davies, Paul, *The Mind of God*, Touchstone, New York, 1993.

Dennis, Kingsley, and Ervin Laszlo, *New Consciousness for a New World*, Inner Traditions, Rochester, VT, 2011.

Dewey, E. R. and Mandino, O. *Cycles: The Mysterious Forces that Trigger Events*, Manor Books, New York, 1973.

Dirac, Paul, *The Principles of Quantum Mechanics*, Clarendon Press, New York, 1981.

Einstein, Albert, tr. R. W. Lawson, *Relativity: The Special and the General Theory*, Bonanza Books, New York, 1961.

Einstein, Albert, Boris Podolsky, and Nathan Rosen, "Can Quantum-Mechanical Description of Physical Reality Be Considered Complete?" *Physical Review* 47, 1935.

Fechner, Gustav Theodor, tr. Hugo Wernekke, *On Life After Death*, Sampson, Low, Marston, Searle & Rivington, London, 1882.

Feynman, Richard, *QED: The Strange Theory of Light and Matter*, Princeton University Press, New York, 1985.

Goodwin, Brian, and M.H. Cohen, "A Phase-Shift Model for the Spatial and Temporal Organization of Developing systems," in C. H. Waddington, ed., *Towards a Theoretical Biology*, vol. 3, 231-232., 1970.

Goswami, Amit, *The Self Aware Universe: How Consciousness Creates the Material World*, Penguin Putnam, New York, 1993.

Gregersen, Niels Henrik, Wentzel Van Huyssteen, Nancy R. Howell and Wesley J. Wildman, eds., *Encyclopedia of Science and Religion*, MacMillan Reference USA, 2003.

Gribbon, John, *Q is for Quantum: An Encyclopedia of Particle Physics*, The Free Press, New York, 1998.

Hamilton, Edith & Huntington Cairns, *The Collected Dialogues of Plato*, Princeton University Press, 1961.

Hartshorne, Charles, *A Natural Theology for Our Time*, Open Court, LaSalle, IL, 1973.

Hartshorne, Charles, *Beyond Humanism: Essays in the Philosophy of Nature*, University of Nebraska Press, Lincoln, 1969.

Hartshorne, Charles, *Omnipotence and Other Theological Mistakes*, State University of New York Press, Albany, 1984.

Hartshorne, Charles, and William L. Reese, eds., *Philosophers Speak of God*, University of Chicago Press, 1953.

Heisenberg, Warner, *Physics and Philosophy*, Harper Torchbooks,

New York, 1958.

Hermann, Robert, *Expanding Humanity's Vision of God: New Thoughts on Science and Religion*, Templeton Foundation, West Conshohocken, PA, 2001

Hiley, Basil J., and F. David Peat, "The Development of David Bohm's Ideas from the Plasma to the Implicate Order," in *Quantum Implications*, ed. Basil J. Hiley and F. David Peat, Routledge & Kegan Paul, London, 1987.

Ho, Mae-Wan, *The Rainbow and the Worm: The Physics of Organisms*, World Scientific Publishing Co., New Jersey, 1999.

Hubbard, Barbara Marx, *Birth 2012 and Beyond: Humanity's Great Shift to the Age of Conscious Evolution*, Shift Books, 2012.

Jeans, James, *Science & Music*, Dover Publications, New York, 1968.

Jung, Karl, *Synchronicity: An Acausal Connecting Principle*, Bollinger Foundation, Bollinger, Switzerland, 1973.

Kaufman, Walter, tr., *The Portable Nietzsche*, Viking Penguin, 1968.

Krause, Karl Christian Friedrich, tr. W. Hastie, D.D., *The Ideal of Humanity and Universal Federation*, Forgotten Books, 2012.

Laszlo, Ervin, and Kingsley Dennis, eds., *The New Science and Spirituality Reader*, Inner Traditions, Rochester, VT, 2012.

Leadbeater, Charles W., *The Chakras*, Theosophical Publishing House, Wheaton, IL, 1994.

Leibniz, Gottfried, tr. Nicholas Rescher, *The Monadology: An Edition for Students*, University of Pittsburg Press, 1991.

Lipton, Bruce, *The Biology of Belief*, Hay House, New York, 2005.

Lovelock, James, *Gaia: A New Look at Life on Earth*, Oxford U. Press, New York, 2000.

Luce, Gay Gaer, *Biological Rhythms in Human and Animal Physiology*, Dover Publications, New York, 1971.

Maddox, J., "No Pattern Yet for Snowflakes," *Nature*, vol. 313, p. 93, 1985.

Mailer, Norman, *Marilyn: A Biography*, Grosset & Dunlap, Alskog, Inc., & Norman Mailer, New York, 1973.

Marshall, Ninian, "ESP and Memory: A Physical Theory," *The British Journal for the Philosophy of Science*, Vol. 10, No. 40, pp. 265-286, 1960.

McElwain, Hugh, *Introduction to Teilhard de Chardin*, Argus Communications Co., Chicago, 1967.

McFague, Sallie, *The Body of God: An Ecological Theology*, Fortress Press, Minneapolis, 1993.

McFague, Sallie, *Models of God: Theology for an Ecological, Nuclear Age*, Fortress Press, Philadelphia, 1989.

McKeon, Richard, ed., *The Basic Works of Aristotle*, Random House, New York, 1941.

McTaggart, Lynn, *The Field: The Quest for the Secret Force of the Universe*, HarperCollins, New York, 2002.

Miller, Sharon A., *"The Astrological Cosmology: A Unified Field Theory,"* Master's Thesis, Department of Philosophy, Southern Illinois University, Edwardsville, 1981.

Needham, Rodney, *Right and Left: Essays on Dual Symbolic Classification*, University of Chicago Press, 1973.

Omrani, B., & Leeming, M., *Afghanistan: A Companion and Guide*, W. W. Norton & Co., Inc., New York, 2005.

Page, Ruth, *God and the Web of Creation*, S.C.M. Press, 2009.

Peacocke, Arthur, *Theology for a Scientific Age: Being and Becoming—Natural, Divine, and Human*, First Fortress Press, Minneapolis, 1993.

Pribram, Karl, "The Neurophysiology of Remembering," *Scientific American*, Vol. 220, January 1969.

Pribram, Karl, *Languages of the Brain*, Wadsworth Publishing, Monterey, CA, 1977.

Ruether, Rosemary Radford, *Gaia and God: An Ecofeminist Theology of Earth Healing*, HarperCollins, New York, 1992.

Russell, Peter, *From Science to God: A Physicist's Journey Into the Mystery of Consciousness*, New World Library, Novato, CA, 2004.

Schroeder, Gerald L., *The Science of God*, Broadway Books, New

York, 1998.

Sheldrake, Rupert, *A New Science of Life: The Hypothesis of Morphic Resonance*, Park Street Press, Vermont, 1981.

Sheldrake, Rupert, *The Presence of the Past: Morphic Resonance and the Habits of Nature*. Park Street Press, Vermont, 1988.

Summers, Anthony, *Goddess: The Secret Lives of Marilyn Monroe*, Macmillan, New York, 1985.

Talbot, Michael, *The Holographic Universe*, Harper Perennial, New York, 1992

Teilhard de Chardin, Pierre, *Hymn of the Universe*, Harper & Row, New York, 1969.

Teilhard de Chardin, Pierre, *The Phenomenon of Man*, Harper & Row, New York, 1965.

Thompson, D'Arcy, *On Growth and Form*, Cambridge University Press, 1961.

Waldman, Gary, *Introduction to Light: The Physics of Light, Vision and Colors*, Prentice-Hall, New York, 1983.

Watson, Lyall, *Lifetide*, Bantam Books, New York, 1980.

Webster's New Collegiate Dictionary, G & C Merrian Co., Riverside Press, Cambridge, 1956.

Weiss, Paul, *Principles of Development*, Henry Holt & Co., New York, 1939.

Whitehead, Alfred North, *Process and Reality: An Essay in Cosmology*, Free Press, New York 1978.

Whyte, L. L., Wilson, A., and Wilson, E., *Hierarchical Structures*, American Elsevier Publishing Co., New York, 1969.

Wilber, Ken, *The Holographic Paradigm and other Paradoxes*, Shambhala Publications, Boulder, CO, 1982.

Wilson, Edward O., *Consilience: The Unity of Knowledge*, Vintage Books, New York, 1998.

Yates, Frances A., *Giordano Bruno and the Hermetic Tradition*, University of Chicago Press, 1964.

Zukav, Gary, *The Dancing Wu Li Masters: An Overview of the New Physics*, HarperCollins, New York, 2001.

BOOKS

Iff Books is interested in ideas and reasoning. It publishes material on science, philosophy and law. Iff Books aims to work with authors and titles that augment our understanding of the human condition, society and civilisation, and the world or universe in which we live.